Level 3

CARPENTRY & JOINERY

NVQ/SVQ and Diploma

carillion

www.pearsonschoolsandfe.co.uk

✓ Free online support
✓ Useful weblinks
✓ 24 hour online ordering

0845 630 44 44

Part of Pearson

Heinemann is an imprint of Pearson Education Limited, a company incorporated in England and Wales, having its registered office at Edinburgh Gate, Harlow, Essex, CM20 2JE. Registered company number: 872828

www.pearsonschoolsandfecolleges.co.uk

Heinemann is a registered trademark of Pearson Education Limited

First published 2011

14 13 12 11
10 9 8 7 6 5 4 3 2 1

British Library Cataloguing in Publication Data
A catalogue record for this book is available from the British Library

ISBN 978 0 435047 19 1

Typeset by Phoenix Photosetting, Chatham, Kent ME4 4TZ
Original illustrations © Pearson Education Ltd
Illustrated by Oxford Designers and Illustrators
Cover design by Wooden Ark
Printed in Spain by Grafos

Acknowledgements
Every effort has been made to contact copyright holders of material reproduced in this book. Any omissions will be rectified in subsequent printings if notice is given to the publishers.

Websites
The websites used in this book were correct and up-to-date at the time of publication. It is essential for tutors to preview each website before using it in class so as to ensure that the URL is still accurate, relevant and appropriate. We suggest that tutors bookmark useful websites and consider enabling students to access them through the school/college intranet.

The information and activities in this book have been prepared according to the standards reasonably to be expected of a competent trainer in the relevant subject matter. However, you should be aware that errors and omissions can be made and that different employers may adopt different standards and practices over time. Before doing any practical activity, you should always carry out your own risk assessment and make your own enquiries and investigations into appropriate standards and practices to be observed.

Acknowledgements

Carillion would like to thank Kevin Jarvis for his hard work and dedication in preparing the content of this book. Carillion would also like to thank John McLaughlin Harvie for preparing the functional skills content for this book.

Pearson would like to thank Kevin Diett, Bernard Deale and Nigel Edwards of Sussex Downs College and John Spalding of Fareham College for their comprehensive review work. Pearson would also like to thank the staff and students at Carillion's Hull Training Centre for their co-operation in helping to set up some of the photos used in this book.

Photo acknowledgements
The author and publisher would like to thank the following individuals and organizations for their kind permission to reproduce photographs:

(Key: b-bottom; c-centre; l-left; r-right; t-top)

Alamy Images: Andy Marshall 240tr, AT Willett 124, David Barton 87br, David J. Green 12, David Lawrence 96tl, Eric Nathan 64, Image Farm Inc. 297t, Image Source 127, Images of Birmingham Premium 1, Justin Kase 96bl, 126, 131, Kris Mercer 118, Libby Welch 128t, Pintail Pictures 65, Roger Bamber 217bl, Steve Welsh 27, We Shoot 164c; **Construction Photography:** Adrian Sherratt 94tl, 96cl, Buildpix 95, 123, 217tl, Damian Gillie 164t, David Burrows 87c, David Potter 62, 237, David Stewart-Smith 94br, 240bl, DIY Photolibrary 129, 319, Grant Smith 85t, Jean-Francois Cardella 15, 240c, Simon Turner 240br; **Corbis:** Martin Meyer 141; **Creatas:** 58; **CSCS:** 4; **Getty Images:** PhotoDisc 125; **iStockphoto:** Ales Veluscek 154/6, Bill Noll 154/5, 155/1, 155/4, 155/5, 155/6, 156/1, 156/2, 156/3, David White 154/1, Dmitriy Pochitalin 155/2, Ivan Vasilev 154/2, John Weise 240tl, Nancy Nehring 154/4, Tibor Nagy 155/3; **Masterfile UK Ltd:** Boden / Ledingham 173; **Pearson Education Ltd:** Gareth Boden 9, 76, 79, 80/1, 80/2, 80/3, 85b, 86b, 87tl, 128b, 144, 145, 161br, 162, 163, 167, 168, 169, 176, 177, 182, 184, 186, 187tl, 187tr, 187bl, 188tl, 188tr, 188bl, 189, 190, 194, 195, 196, 198, 199, 200, 201, 210, 278, 310, 324c, 324b, 325cr, 325bl, 326, 327c, 328, 329, Jules Selmes 165, 180, 181, 183, 258tl, 258cl, 258cr, 259, 274, Naki Photography 80/4, 80/5, 160, 161tl, 161tr, 161bl, 187br, 188br, 260; **Robert Clare:** 239c; **Science Photo Library Ltd:** Alex Bartel 30, Carlos Dominguez 40; **Shutterstock.com:** Alfgar 296c, Andrr 317, auremar 269, Brandon Bourdages 205, Cynthia Farmer 239t, David Hughes 135, David Lee 114, digitalreflections 281, Edd Westmacott 325br, Fekete Tibor 96cr, Frances A. Miller 10, Gilles Lougassi 57, Joachim Wendler 324t, Martina Orlich 154/3, Naturaldigital 297c, R-Studio 296b, Risto Viita 291, Stavklem 133, stocksnap 327t, Yuri Arcurs 4 (Inset); **Will Burwell:** 325tr

Cover images: *Front:* **Alamy Images:** Image Source (Drillbit); **Shutterstock.com:** Diego Cervo (Man)

All other images © Pearson Education Ltd

Picture research by: Chrissie Martin

Content

Introduction

Welcome to NVQ/SVQ CAA Diploma Level 3 Carpentry & Joinery!

Carpentry & Joinery combines many different practical and visual skills with knowledge of specialised materials and techniques. This book will introduce you to the construction trade and in particular the knowledge and skills needed for interpreting site documents, basic woodworking joints, and safe storage and use of a selection of hand and power tools.

About this book

This book has been produced to help you build a sound knowledge and understanding of all aspects of the Diploma and NVQ requirements associated with Carpentry & Joinery.

The information in this book covers the information you will need to attain your Level 3 qualification in Carpentry & Joinery. Each chapter of the book relates to a particular unit of the CAA Diploma and provides the information needed to form the required knowledge and understanding of that area. The book is also designed to support those undertaking the NVQ at Level 3.

This book has been written based on a concept used by Carillion Training Centres for many years. The concept is about providing learners with the necessary information they need to support their studies and ensuring it is presented in a style which is both manageable and relevant.

This book will also be a useful reference tool for you in your professional life once you have gained your qualifications and are a practising carpenter or joiner.

Qualifications for the construction industry

There are many ways of entering the construction industry, but the most common method is as an apprentice.

Apprenticeships

You can become an apprentice by being employed:

- directly by a construction company who will send you to college
- by a training provider, such as Carillion, which combines construction training with practical work experience.

Construction Skills is the national training organisation for construction in the UK and is responsible for setting training standards.

The framework of an apprenticeship is based around an NVQ (or SVQ in Scotland). These qualifications are developed and approved by industry experts and will measure your practical skills and job knowledge on-site.

You will also need to achieve:

- a technical certificate
- the Construction Skills health and safety test
- the appropriate level of Functional skills assessment
- an Employees Rights and Responsibilities briefing.

You will also need the right qualifications to get on a construction site, including qualifying for the CSCS card scheme.

CAA Diploma

The Construction Awards Alliance (CAA) Diploma was launched on 1 August 2008 to replace Construction Awards. These Diplomas aim to make you:

- more skilled and knowledgeable
- more confident with moving across projects, contracts and employers.

The CAA Diploma is a common testing strategy with knowledge tests for each unit, a practical assignment and the GOLA (Global Online Assessment) test.

The CAA Diploma meets the requirements of the new Qualifications and Credit Framework (QCF) which bases a qualification on the number of credits achieved (with ten learning hours gaining one credit):

- Award (1 to 12 credits)
- Certificate (13 to 36 credits)
- Diploma (37+ credits).

As part of the CAA Diploma you will gain the skills needed for the NVQ as well as the functional skills knowledge you will need to complete your qualification.

National Vocational Qualifications (NVQs)

NVQs are available to anyone, with no restrictions on age, length or type of training, although learners below a certain age can only perform certain tasks. There are different levels of NVQ (for example 1, 2, 3), which in turn are broken down into units of competence. NVQs are not like traditional examinations in which someone sits an exam paper. An NVQ is a 'doing' qualification, which means it lets the industry know that you have the knowledge, skills and ability to actually 'do' something.

NVQs are made up of both mandatory and optional units and the number of units that you need to complete for an NVQ depends on the level and the occupation.

NVQs are assessed in the workplace, and several types of evidence are used:

- Witness testimony can be provided by individuals who have first-hand knowledge of your work and performance relating to the NVQ.
- Your performance can be observed a number of times in the workplace.
- You can use evidence from past achievements or experience, if it is directly related to the NVQ. This is known as historical evidence.
- Assignments or projects can be used to assess your knowledge and understanding.
- Photographic evidence showing you performing various tasks in the workplace can be used, as long as it is authenticated by your supervisor.

Features of this book

This book has been fully illustrated with artworks and photographs. These will provide additional information about certain concepts or procedures, as well helping you to follow a step-by-step procedure or identify a particular tool or material.

This book also contains a number of different features to help your learning and development.

Functional skills

This feature is designed to support you with your functional skills, by identifying opportunities in your work where you will be able to practise your functional skills.

Key term

These are new or difficult words. They are picked out in **bold** in the text and then defined in the margin.

Working Life

This feature gives you a chance to read about and debate a real life work scenario or problem. Why has the situation occurred? What would you do?

Safety tip

This feature gives you guidance for working safely on the tasks in this book.

Remember

This highlights key facts or concepts, sometimes from earlier in the text, to remind you of important things you will need to think about.

Did you know?

This feature gives you interesting facts about the building trade.

Find out

These are short activities and research opportunities, designed to help you gain further information about, and understanding of, a topic area.

FAQ

These are frequently asked questions appearing at the end of each unit to answer your questions with informative answers from the experts.

Check it out

A series of questions at the end of each unit to check your understanding. Some of these questions may support the collection of evidence for the NVQ.

Getting ready for assessment

This feature provides guidance for preparing for the practical assessment. It will give you advice on using the theory you have learnt about in a practical way.

CHECK YOUR KNOWLEDGE

This is a series of multiple choice questions at the end of each unit, in the style of the GOLA end of unit tests.

Unit 1001

Safe working practices in construction

Health and safety is a vital part of all construction work. All work should be completed in a way that is safe not only for the individual worker, but also for the other workers on the site, people near by and the final users of the building. Health and safety is not optional in your career, but an essential part of working in the industry. Every year there are over 100 fatalities in the construction industry. Don't become one of them.

This unit supports TAP Unit 1 Erect and dismantle working platforms, and delivery of the five generic units.

This unit contains material that supports the following NVQ units:

- VR 01 Conform to general workplace safety
- VR 03 Move and handle general resources

This unit will cover the following learning outcomes:

- Health and safety regulations
- Accident, first aid and emergency procedures and reporting
- Hazards on construction sites
- Health and hygiene
- Safe handling of materials and equipment
- Basic working platforms
- Working with electricity
- Use of appropriate personal protective equipment (PPE)
- Fire and emergency procedures
- Safety signs and notices

K1. Health and safety regulations

Health and safety **legislation** is there not just to protect you – it also states what you must and must not do to ensure that no workers are placed in a situation **hazardous** to themselves or others. You will also use codes of practice and guidance notes (produced by the Health and Safety Executive, or HSE, and by companies themselves).

Key terms

Legislation – a law or set of laws passed by Parliament, often called an Act.

Hazardous – something or a situation that is dangerous or unsafe.

Subcontractors – workers who have been hired by the main contractor to carry out work, usually specialist work.

Suppliers – companies that supply goods, materials or services.

Omission – something that has not been done or has been missed out.

The Health and Safety at Work Act 1974 (HASAWA)

HASAWA applies to all types and places of work and to employers, employees, self-employed people, **subcontractors** and even **suppliers**. The Act protects people at work and the general public, and is designed to ensure their health, safety and welfare. It also controls the use, handling, storage and transportation of explosives and highly flammable substances, and the release of noxious/offensive substances into the atmosphere.

Table 1.01 Duties of employers, employees and suppliers under HASAWA

Employer's duties	Provide a safe place to work with safe plant and machinery. Information, training and supervision supplied to all employees. A written safety policy supplied and risk assessments carried out. Personal protective equipment (PPE) provided to all employees free of charge. Health and safety assured when handling, storing and transporting materials and substances.
Employee's duties	Must take reasonable care for his/her own health and safety, and the health and safety of anyone who may be affected by his/her acts or **omissions**. Co-operate with employer and other persons to meet the law, use materials provided for their safety, and report hazards or accidents.
Supplier's duties	Duty to make sure products are designed and constructed safely, fully tested, and safe to use, handle, transport and store. Information should be provided to the user on all these aspects of the product.

Functional skills

While reading and understanding the text in this unit, you will be practising **FE 2.2.1–2.2.3**, which relate to selecting and using texts, summarising information and ideas, and identifying the purpose of texts.

If there are any words or phrases you do not understand, use a dictionary, look them up on the Internet, or discuss with your tutor.

Health and Safety Executive (HSE)

The HSE is the government body responsible for the encouragement, regulation and enforcement of health, safety and welfare in the workplace in the UK and of HASAWA and other laws, through inspectors who can prosecute people or companies that break the law.

HSE inspectors have the authority to enter and examine any premises at any time, taking samples and possession of any dangerous article/substance. They can issue improvement notices,

Did you know?

The HSE must be contacted if an accident occurs that results in death or major injury. The report must be followed up by a written report in ten days (form F2508).

ordering a company to solve a problem in a certain time, or issue a prohibition notice stopping all work until the site is safe.

Construction (Design and Management) Regulations 2007 (CDM 2007)

The Construction (Design and Management) Regulations 2007 are designed to help improve safety. Employers must plan, manage and monitor work, ensuring employees are competent and provided with training and information. They must also provide adequate welfare facilities for workers. There are also specific requirements relating to lighting, excavations and traffic, etc.

Employees must check their own competence and co-operate to co-ordinate work safely, reporting any obvious risks.

Did you know?

On large projects, a person is appointed as the CDM co-ordinator. This person has overall responsibility for compliance with CDM. There is a general expectation by the HSE that all parties involved in a project will co-operate and co-ordinate with each other.

Provision and Use of Work Equipment Regulations 1998 (PUWER)

These regulations cover all new or existing **work equipment**. PUWER covers starting, stopping, regular use, transport, repair, modification, servicing and cleaning of equipment.

Key term

Work equipment – any machinery, appliance, apparatus or tool, and any assembly of components, that are used in non-domestic premises.

The general duties of the regulations require equipment to be used and maintained in suitable and safe conditions by a trained person. Equipment should be fitted with appropriate warnings and be able to be isolated from sources of energy.

In addition, the regulations also require access to dangerous parts of machinery to be prevented or controlled. Suitable controls for stopping and starting of work equipment, in particular emergency stopping and braking systems, should be installed. Sufficient lighting must be in place for operating equipment.

Other pieces of legislation

Table 1.02 Health and safety legislation

Legislation	Content
Reporting of Injuries, Diseases and Dangerous Occurrences Regulations 1995 (RIDDOR)	Employers have a duty to report accidents, diseases or dangerous occurrences. HSE uses this to identify where and how risks arise and to investigate serious accidents.
Control of Substances Hazardous to Health Regulations 2002 (COSHH)	States how employees and employers should work with, handle, store, transport and dispose of potentially hazardous substances. This includes substances used and generated during work (e.g. paints or dust), naturally occurring substances (e.g. sand) and biological elements (e.g. bacteria).
The Control of Noise at Work Regulations 2005	Employers must assess the risks to the employee and make sure legal limits are not exceeded, noise exposure is reduced, and hearing protection is provided, along with information, instruction and training.

Table 1.02 Health and safety legislation (cont.)

Legislation	Content
The Electricity at Work Regulations 1989	Covers work involving electricity. Employers must keep electrical systems safe and regularly maintained, and reduce the risk of employees coming into contact with live electrical currents.
The Manual Handling Operations Regulations 1992	Covers all work activities involving a person lifting. Manual handling should be avoided wherever possible and a risk assessment must be carried out.
The Personal Protective Equipment at Work Regulations 1992 (PPER)	PPE must be checked by a trained and competent person and must be provided by the employer free of charge with a secure storage place. Employees must know how to use PPE, the risks it will help to protect against, its purpose, how to maintain it and its limitations.
The Work at Height Regulations 2005	Employers must avoid working at height and use equipment that prevents or minimises the danger of falls. Employees must follow training, report hazards and use safety equipment.

Find out

There are several sources for health and safety information. Use the Internet to find out more about each of the following:

- Construction skills
- Royal Society for the Prevention of Accidents (RoSPA)
- Royal Society for the Promotion of Health (RPH).

Remember

As a trainee, once you pass the health and safety test you will qualify for a trainee card and once you have achieved a Level 2 qualification you can then upgrade your card to an experienced worker card. Achieving a Level 3 qualification allows you to apply for a gold card.

Site inductions

Site inductions are the process that individuals undergo in order to accelerate their awareness of the potential health and safety hazards and risks they may face in their working environment; but excludes job-related skills training. Different site inductions will always cover operations on site, health and safety, welfare and emergency arrangements, reporting structure and the process for reporting 'near misses' (see page 5). Records must be kept to ensure all workers have received an induction.

Construction Skills Certification Scheme (CSCS)

The Construction Skills Certification Scheme requires all workers to obtain a CSCS card before working on a building site. There are various levels of cards which indicate competence and skill background. This ensures that only skilled and safe tradespeople can work on site. To get a CSCS card all applicants must pass a health and safety test.

Figure 1.01 CSCS card

K2. Accident, first aid and emergency procedures and reporting

Major emergencies that could occur on site include not only accidents but also fires, security alerts and bomb alerts. Your site induction should make it clear to you what to do in the event of an emergency.

Reporting accidents

All accidents need to be reported and recorded in the accident book, and the injured person must report to a trained first aider. An accident may result in an injury which may be minor (e.g. a cut or a bruise) or major (e.g. loss of a limb). Accidents can also be fatal. When an accident happens, the first thing to do is to be sure the victim is in no further danger, without putting yourself at risk.

As well as reporting accidents, 'near misses' must also be reported. (A 'near miss' is when an accident nearly happened but did not actually occur.) These might identify a problem and prevent accidents from happening in the future.

The accident book is completed by the person who had the accident or someone representing the injured person. You will need to enter some basic details, including:

- who was involved, what happened and where
- the date and time of the accident and any witnesses
- the address of the injured person
- what PPE was being worn and what first-aid treatment was given.

Remember

Companies that have a lot of accidents will have a poor company image for health and safety and will find it increasingly difficult to gain future contracts. Unsafe companies with lots of accidents will also see injured people claiming against their insurance, which will see their premiums rise. This will eventually make them uninsurable, meaning they will not get any work.

Remember

An accident that falls under RIDDOR should be reported by the safety officer or site manager and can be reported to the HSE by telephone (0845 300 9923) or via the RIDDOR website (www.riddor.gov.uk)

K3. Hazards on construction sites

A major part of health and safety at work is being able to identify hazards and to help prevent them in the first place, therefore avoiding the risk of injury. Hazards include falls (see page 10), electricity (see page 12) and fires (see pages 13–14).

Key term

Housekeeping – cleaning up after you and ensuring your work area is clear and tidy. Good housekeeping is vital on a construction site as an unclean work area is dangerous.

Table 1.03 Hazards on construction sites

Hazard	What to do
Tripping	Caused by poor **housekeeping**. Keep workplaces tidy and free from debris.
Chemical spills	Most are small with minimal risk and can be easily cleaned. If the spill is hazardous, take the correct action promptly.
Burns (from fires or chemical materials)	You must be aware of the dangers and take the correct precautions.

Step 1 – Assess the risks to health from hazardous substances.

Step 2 – Decide what precautions are needed.

Step 3 – Prevent employees from being exposed to any hazardous substances. If prevention is impossible, the risk must be adequately controlled.

Step 4 – Ensure control methods are used and maintained properly.

Step 5 – Monitor the exposure of employees to hazardous substances.

Step 6 – Carry out health surveillance to ascertain if any health problems are occurring.

Step 7 – Prepare plans and procedures to deal with accidents such as spillages.

Step 8 – Ensure all employees are properly informed, trained and supervised.

Figure 1.02 Process for dealing with hazardous substances

Remember

Waste on site will need to be correctly identified and disposed of. Hazardous materials will need to be disposed of specially.

Risk assessments and method statements

You must know how to carry out a risk assessment. You may be given direct responsibility for this, and the care and attention you take over it may have a direct impact on the safety of others. You must be aware of the dangers or hazards of any task, and know what can be done to prevent or reduce the risk.

There are five steps in a risk assessment:

Step 1 Identify the hazards.

Step 2 Identify who is at risk.

Step 3 Calculate the risk from the hazard against the likelihood of it taking place.

Step 4 Introduce measures to reduce the risk.

Step 5 Monitor the risk.

A method statement takes information about significant risks from risk assessments and combines them with the job specification to produce a practical and safe working method for the workers to follow for tasks on site. The hazard book can also be used to identify tasks and produce risk assessments.

K4. Health and hygiene

One of the easiest ways to stay healthy is to wash your hands on a regular basis to prevent hazardous substances from entering your body. You should always clean any cuts you may get to prevent infection. Welfare facilities should be provided for employees. These include toilets, washing facilities, drinking water, and storage and lunch areas.

Health effects of noise

Damage to hearing can be caused by one of two things:

- **Intensity** – you can be hurt in an instant from an explosive or very loud noise which can burst your ear drum.
- **Duration** – noise doesn't have to be deafening to harm you; it can be a quieter noise over a longer period, e.g. a 12-hour shift.

Hazardous substances

Hazardous substances are a major health and safety risk on a construction site. They must be handled, stored, transported and disposed of in very specific ways. Ask the supplier or manufacturer for a COSHH (Control of Substances Hazardous to Health) data sheet, outlining the risks involved with a substance. Most substance containers carry a warning sign stating whether the contents are

Figure 1.03 Common safety signs for corrosive, toxic and explosive materials

corrosive, harmful, toxic or bad for the environment. Exposure to chemicals can cause skin problems, such as **leptospirosis** and **dermatitis**.

Waste

You need to identify all the types of waste you create and the best way of disposing of them. The Controlled Waste Regulations 1992 state that only those authorised to dispose of waste may do so, and that they must keep full records.

Several different types of waste are defined by these regulations:

- **household waste** – normal household rubbish
- **commercial waste** – for example, from shops or offices
- **industrial waste** – from factories and industrial sites.

K5. Safe handling of materials and equipment

Manual handling is the lifting and moving of a piece of equipment or material from one place to another without using machinery. This is one of the most common causes of injury at work and can cause injuries such as muscle strain, pulled ligaments and hernias. Spinal injury is the most common injury and is very serious because, very often, there is little doctors can do to correct it.

When lifting a load the correct posture is as follows:

- Feet shoulder width apart, with one foot slightly in front of the other.
- Knees bent with the back straight and arms as close to the body as possible.
- Grip must be firm, using the whole hand and not just the finger tips.

These are the correct techniques to use when lifting:

1. Approach the load squarely, facing the direction of travel.
2. Place hands under the load and pull it close to your body, lifting using your legs, not your back.
3. When lowering bend at the knees, not the back.

Safety tip

Not all substances are labelled, and sometimes the label may not match the contents. If you are in any doubt, do not use or touch the substance.

Key terms

Leptospirosis – an infectious disease that affects humans and animals. The human form is commonly called Weil's disease. The disease can cause fever, muscle pain and jaundice. In severe cases it can affect the liver and kidneys. Leptospirosis is a germ that is spread by the urine of the infected person. It can often be caught from contaminated soil or water that has been urinated in.

Dermatitis – a skin condition where the affected area is red, itchy and sore.

Did you know?

In 2004/2005 there were over 50 000 injuries while handling, lifting or carrying in the UK. (Source: HSE)

Safe storage and handling of tools and equipment

Safety tip

- Hand tools with sharp edges should be covered to prevent cuts.
- Power tools should be carried by the handle.
- Power tools that have gas-powered cartridges must be stored in an area that is safe and away from sources of ignition to prevent explosion. Used cartridges must be disposed of safely.

Tools

All tools need to be stored safely and securely in suitable bags or boxes to protect them from weather and rust. When not in use they should be safely locked away.

Bricks, blocks and paving slabs

Table 1.04 Storage and handling of bricks, blocks and paving slabs

Type	Storage and handling issues
Bricks and blocks	Largely pre-packed in shrink-wrapped plastic and banded using either plastic or metal bands with edges protected by plastic strips. Store on level ground close to where they are required and stack on edge in rows no more than two packs high. Take from a minimum of three packs and mix to prevent changes in colour in final brickwork.
Paving slabs	Normally delivered in wooden crates, covered in shrink-wrapped plastic, or banded and covered on pallets. Do not stack higher than two packs. Store outside and stack on edge to prevent lower slabs being damaged by the weight of the stack. Store on firm, level ground with timber bearers below to prevent damage to edges.

Aggregates, cement and plaster

Aggregates are delivered in tipper lorries or one-tonne bags. They should be stored on a concrete base, with a fall to allow for water to drain away. Cover aggregates with tarpaulin or plastic sheets.

Both cement and plaster are usually available in 25 kg bags. Bags are made from multi-wall layers of paper with a polythene liner. Do not puncture the bags before use. Store in a ventilated, waterproof shed on a dry floor, and no higher than five bags.

Wood and sheet materials

Table 1.05 Storage and handling of wood and sheet materials

Type	Storage and handling issues
Carcassing timber	Store outside under a covered framework, on timber bearers clear of the ground, vegetation-free to reduce ground moisture absorption. Use piling sticks between each layer of timber to provide support and allow air circulation.
Joinery grade and hardwoods	Store under full cover with ventilation to prevent build-up of moisture. Store on bearers on a well-prepared base.
Plywood and sheet materials	Store in a dry, well-ventilated environment. Stack flat on timber cross-bearers, spaced close together to prevent sagging. Do not lean them against walls as this makes the wood bow. For faces, place these against each other to minimise risk of damage. Keep different sizes, grades and qualities of sheet materials separate.

Table 1.05 Storage and handling of wood and sheet materials (cont.)

Type	Storage and handling issues
Joinery components	Doors, frames, etc. should be stored flat on timber bearers under cover to protect them from the weather. In limited space they can be stored upright in a rack, but do not lean them against a wall. Wall and floor units must be stacked on a flat surface no more than two units high. Store inside and use protective sheeting to prevent damage and staining.
Plasterboard	Store in a flat, waterproof area and do not lean plasterboard against a wall.

Adhesives

All adhesives should be stored and used in line with the manufacturer's instructions. They are usually stored on shelves, with labels facing outwards, in a safe, secure area (preferably a lockable store room). It is important to keep the labels facing outwards so that the correct adhesive can be selected.

Paint and decorating equipment

Table 1.06 Storage of paint and powdered materials

Type	Storage issues
Oil- and water-based paint	Store at a constant temperature in date order (new stock at the back) on clearly marked shelves with the labels turned to the front. Regularly **invert** to prevent settlement or separation of ingredients, and keep tightly sealed to prevent **skinning**. Water-based paint should be protected from frost to prevent freezing.
Powdered materials	Heavy bags should be stored at ground level. Smaller items should be stored on shelves with loose materials in sealed containers. Protect from frost, moisture and high humidity.

Figure 1.04 Correct storage of paints

Key terms

Invert – tip and turn upside down.

Skinning – the formation of a skin which occurs when the top layer of paint dries out.

Remember

Decorating materials hazardous to health include spirits, turps, paint thinners and varnish removers. These should be stored on shelves in line with any COSHH requirements. Temperatures must be kept below 15°C to prevent storage containers expanding or blowing up.

K6. Basic working platforms

Remember

There is also a danger of objects falling from height and striking workers and people below. Barriers should be in place to prevent this.

Fall protection

With any task involving working at height, the main danger is falling. There are certain tasks where edge protection or scaffolding simply cannot be used. In these instances some form of fall protection must be used.

Table 1.07 Types of fall protection

Type of fall protection	Description
Harnesses and lanyards	A harness is attached to the worker and a lanyard to a secure beam/eyebolt. If the worker slips, they will fall only the length of the lanyard.
Safety netting	Used on the top floor where there is no point for a lanyard. Nets are attached to the joists to catch any falling workers.
Air bags	Made from interlinked modular air mattresses that expand together to form a soft fall surface. Ideal for short-fall jobs.

Figure 1.05 A harness and lanyard can prevent a worker from falling to the ground

Stepladders and ladders

Ladders should only be set up on ground that is firm and level. All components should be checked fully before use. Do not use ladders to gain extra height on a working platform. There are some common safety checks for the different types of ladder.

Table 1.08 Types of ladder

Type of ladder	Safety issues
Wood	Check for loose screws, nuts, bolts and hinges. Check tie ropes are in good condition. **Never** paint as this will hide defects.
Aluminium	Avoid working near live electricity supplies.
Fibreglass	Once damaged, this type of ladder **cannot be** repaired and must be replaced.

Using a stepladder

Stepladders should only be used for work that will take a few minutes to complete. When work is likely to take longer, use a sturdier alternative. Always open the steps fully and check for the Kitemark (Figure 1.06), which shows the ladder has been tested independently and audited to ensure it meets appropriate standards.

Figure 1.06 British Standards Institution Kitemark

Using ladders

Ladders are not designed for work of a long duration and should be secured in place. One hand should always be free to hold the ladder and you should not have to stretch while using it.

You should also observe the following points when erecting a ladder:

- Ensure that there is at least a four-rung overlap on each extension section.
- Never rest on plastic guttering as it may break, causing the ladder to slip.
- If the base of the ladder is exposed, ensure it is guarded so it is not knocked.
- Secure the ladder at top and bottom. The bottom can be secured by a second person, but they must not leave while the ladder is in use.
- The angle of the ladder should be a ratio of 1:4 (or 75°) – see Figure 1.07.
- The top of the ladder must extend at least 1 m, or five rungs, above its landing point.

4 m
1 m

Figure 1.07 Correct angle for a ladder

> **Did you know?**
>
> On average in the UK, every year 14 people die at work falling from ladders, and nearly 1200 people suffer major injuries. (Source: HSE)

Scaffolding

Tubular scaffold is the most commonly used type of scaffolding within the construction industry. There are two types of tubular scaffold:

- **Independent scaffold** – free-standing and does not rely on the building to support it.
- **Dependent scaffold** – attached to the building via poles (putlogs) into holes left in the brickwork. The poles stay in position until work is complete and give the scaffold extra support.

> **Safety tip**
>
> No one other than a qualified carded scaffolder is allowed to erect or alter scaffolding. However, you must always be sure scaffolding is safe before you work on it.

Mobile tower scaffolds

Mobile tower scaffolds can be moved without being dismantled. They have lockable wheels and are used extensively by many different trades. They are made from either traditional steel tubes and fittings or aluminium, which is lightweight and easy to move. The aluminium type of tower is normally specially designed and is referred to as a 'proprietary tower'. A 'low tower' scaffold is designed for use by one person at 2.5 m height.

Tower scaffolds must have a firm and level base. The stability of the tower depends on the height in relation to the size of the base:

- For use inside a building, the height should be no more than three and a half times the smallest base length.

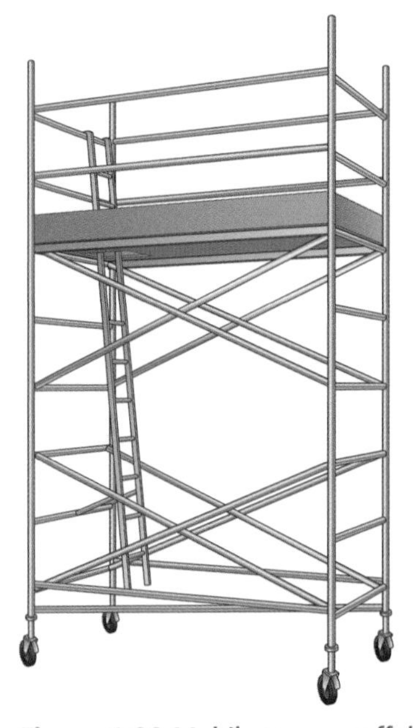

Figure 1.08 Mobile tower scaffold

Safety tip

When using a power tool, always check plugs and connections and be careful how you handle it. Do not pass it by the lead or allow it to get wet. Check the power is off when connecting leads, and keep cables away from sharp edges to avoid damage.

- For outside use, the height should be no more than three times the smallest base length.

Working platforms at any height should be fitted with guardrails and toe boards on all four sides of the platform. Platforms over 9 m must be secured to the structure. Towers should not exceed 12 m unless specially designed to do so. Working platforms should be fully boarded and be at least 600 mm wide.

K7. Working with electricity

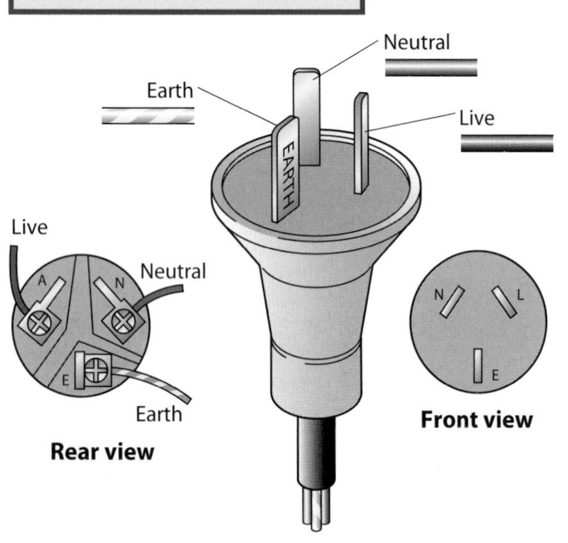

Figure 1.09 Colour-coding of the wires in a 230 V plug

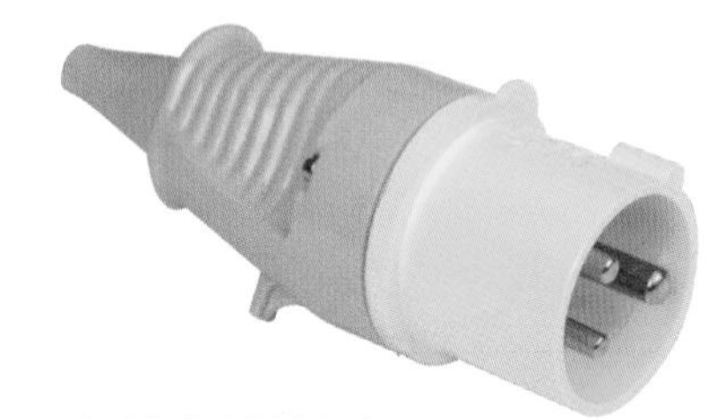

Figure 1.10 A 110 V plug

One of the main problems with electricity is that it is invisible. You do not even have to be working with an electric tool to be electrocuted. Working close to electrical supplies can put you at risk. There are two main types of voltage in the UK. These are 230 V and 110 V. The standard UK power supply is 230 V and this is what all the sockets in your house are.

Contained within the wiring there should be three wires: the live and neutral, which carry the alternating current, and the earth wire, which acts as a safety device. The three wires are colour-coded as follows to make them easy to recognise:

- **live** – brown
- **neutral** – blue
- **earth** – yellow and green.

230 V has been deemed as unsafe on construction sites, so 110 V must be used. This is identified by a yellow cable and different style plug. A transformer converts the 230 V to 110 V. In domestic situations a portable transformer should be used.

Dealing with electric shocks

Always disconnect the power supply if it is safe to do and will not take long to find. Touching the power source may put you in danger. If the victim is in contact with something portable (e.g. a drill), move it away using a non-conductive object such as a wooden broom. Do not attempt to touch the person until he or she is clear of the supply. Be especially careful in wet areas.

People 'hung up' in a live current flow may be unable to make a sound. Their muscles may also contract, preventing them from moving. Use a wooden object to swiftly and strongly knock the person free.

Safety tip

Do not attempt to go near a victim of an electric shock without wearing rubber or some form of insulated sole shoes; bare or socked feet will allow the current to flow through your body.

K8. Using appropriate personal protective equipment (PPE)

Personal protective equipment (PPE) forms a defence against accidents or injury. PPE should be used with all the other methods of staying safe in the workplace. It must be regularly maintained, otherwise its effectiveness may be compromised. This means that PPE needs to be cleaned and examined on a regular basis and, where necessary, replaced or repaired. The cost of maintaining PPE is the responsibility of the employer. Storage facilities must be provided to protect PPE from contamination, loss, damage, damp or sunlight. PPE should be 'CE' marked to show it complies with the PPE Regulations 2002.

Find out

What are the most common forms of PPE on site? Find examples of each type.

K9. Fire and emergency procedures

Fires can start almost anywhere and at any time, but a fire needs all the ingredients of 'the triangle of fire' to burn. Remove one side of the triangle, and the fire will be extinguished. Fire moves by consuming all these ingredients and burns fuel as it moves.

Figure 1.11 The triangle of fire

Fires can be classified according to the type of material that is involved:

- **Class A** – wood, paper, textiles, etc.
- **Class B** – flammable liquids, petrol, oil, etc.
- **Class C** – flammable gases, liquefied petroleum gas (LPG), propane, etc.
- **Class D** – metal, metal powder, etc.
- **Class E** – electrical equipment.

There are several different types of fire extinguisher and it is important that you learn which type should be used on each class of fires.

Table 1.09 Types of fire extinguisher

Fire extinguisher	Colour band	Main use	Details
Water	Red	Class A fires	Never use for an electrical or burning fat/oil fire. Water will conduct electricity and 'explode' oil and fat fires.
Foam	Cream	Class A fires	Can also be used on Class B fires if no liquid is flowing and on Class C fires if gas is in liquid form.
CO_2 (carbon dioxide)	Black	Class E fires	Can also be used on Class A, B and C fires.
Dry powder	Blue	All classes	Commonly used on electrical and liquid fires. Powder puts out the fire by smothering the flames.

Safety tip

For small fires, fire blankets can be used. These are made from fireproof material and work by smothering the fire and stopping any more oxygen from getting to it. A fire blanket can also be used if a person is on fire.

What to do in the event of a fire

During your induction you will be made aware of the fire procedure and the location of fire assembly points. These should be clearly indicated by signs, and a map of their location displayed in the building. On hearing the alarm make your way calmly to the nearest muster point. This is so that everyone can be accounted for and prevents someone searching for you.

K10. Safety signs and notices

There are many different safety signs, but each will usually fit into one of four categories:

- **Prohibition signs** – these tell you that something MUST NOT be done.
- **Mandatory signs** – these tell you something MUST be done.
- **Warning signs** – these signs are there to alert you to a specific hazard.
- **Safe condition signs** (often called information signs) – these give you useful information like the location of things (e.g. a first aid point).

The colour and shape is always the same for each category of sign. Figures 1.12–1.15 show an example of each.

Figure 1.12 A prohibition sign

Figure 1.13 A mandatory sign

Figure 1.14 A warning sign

Figure 1.15 A safe condition (or information) sign

Unit 3002

Knowledge of information, quantities and communicating with others 3

In order to work well in the construction industry, it is important that you are comfortable dealing with a range of information sources. Information should be used effectively to make practical working decisions, both during planning and when working on buildings.

Drawings are a key source of information in the construction industry, both those found in specifications and those used for more detailed work. The information from drawings can be used to put together a more detailed list of the quantities of materials required for work. It will then be an important part of your duties to communicate this information to the people you are working with.

This unit contains material that supports the following NVQ units:

- VR 209 Confirm work activities and resources for the work
- VR 210 Develop and maintain good working relationships
- VR 211 Confirm the occupational method of work

This unit will cover the following learning outcomes:

- Know how to produce drawn information
- Know how to estimate quantities and price work
- Know how to ensure good working relationships

Functional skills

In this unit, you will be practising **FE 2.2.2–2.2.5**, which relate to comparing, selecting, reading and understanding texts and using them to gather information, ideas, arguments and opinions. If there are any words or phrases you do not understand, use a dictionary, look them up on the Internet, or discuss with your tutor.

FM 2.2.1 and **2.2.2** relate to applying a range of mathematics skills to find solutions, using appropriate checking procedures and evaluating their effectiveness at each stage.

Types of drawing

Before looking at producing types of drawing, it is worth revising the different types of drawing that can be used, as well as the process followed to create drawings.

Plans and drawings are vital to any building work as a way of expressing the client's wishes. Drawings are the best way of communicating a lot of detailed information without the need for pages and pages of text. Drawings form part of the contract documents and go through several stages before they are given to tradespeople to use.

Stage 1 The client sits down with an architect and explains his/her requirements.

Stage 2 The architect produces drawings of the work and checks with the client to see if the drawings match what the client wants.

Stage 3 If required, the drawings go to planning to see if they can be allowed, and are also scrutinised by the Building Regulations Authority. It is at this stage that the drawings may need to be altered to meet Planning or Building Regulations.

Stage 4 Once passed, the drawings are given to contractors along with the other contract documents, so that they can prepare their tenders for the contract.

Stage 5 The winning contractor uses the drawings to carry out the job. At this point the drawings will be given to you to work from.

There are three main types of working drawings: location drawings, component drawings and assembly drawings. We will look at each of these in turn.

Location drawings

Location drawings include:

- **block plans**, which identify the proposed site in relation to the surrounding area (see Figure 2.01). These are usually drawn at a scale of 1:2500 or 1:1250.
- **site plans**, which give the position of the proposed building and the general layout of things such as services and drainage (see Figure 2.02). These are usually drawn at a scale of 1:500 or 1:200.
- **general location drawings**, which show different elevations and sections of the building (see Figures 2.03, 2.04 and 2.05). These are usually drawn at a scale of 1:200, 1:100 or 1:50.

Find out

Use information about your local area to complete a block plan for your building and its surrounding buildings, roads, etc.

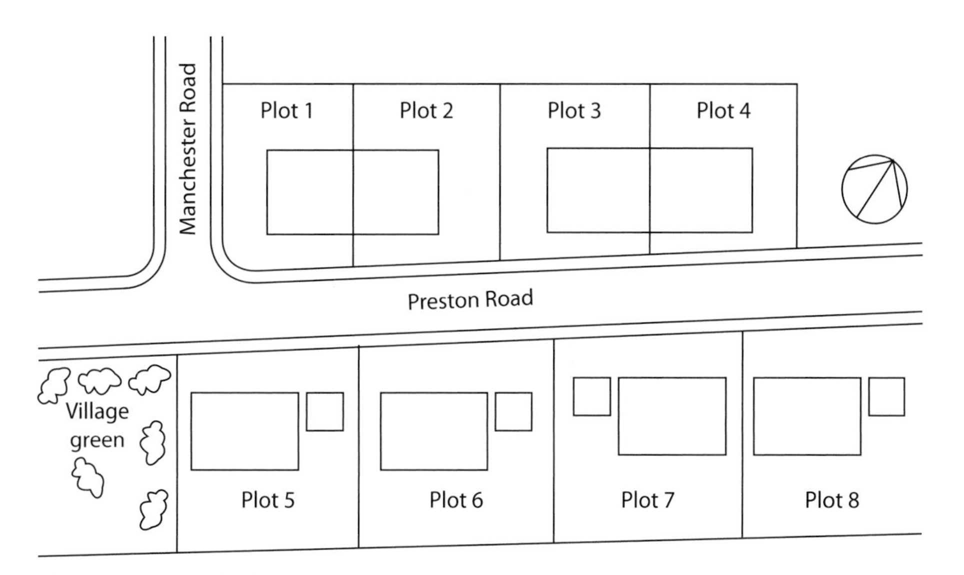

Figure 2.01 Block plan

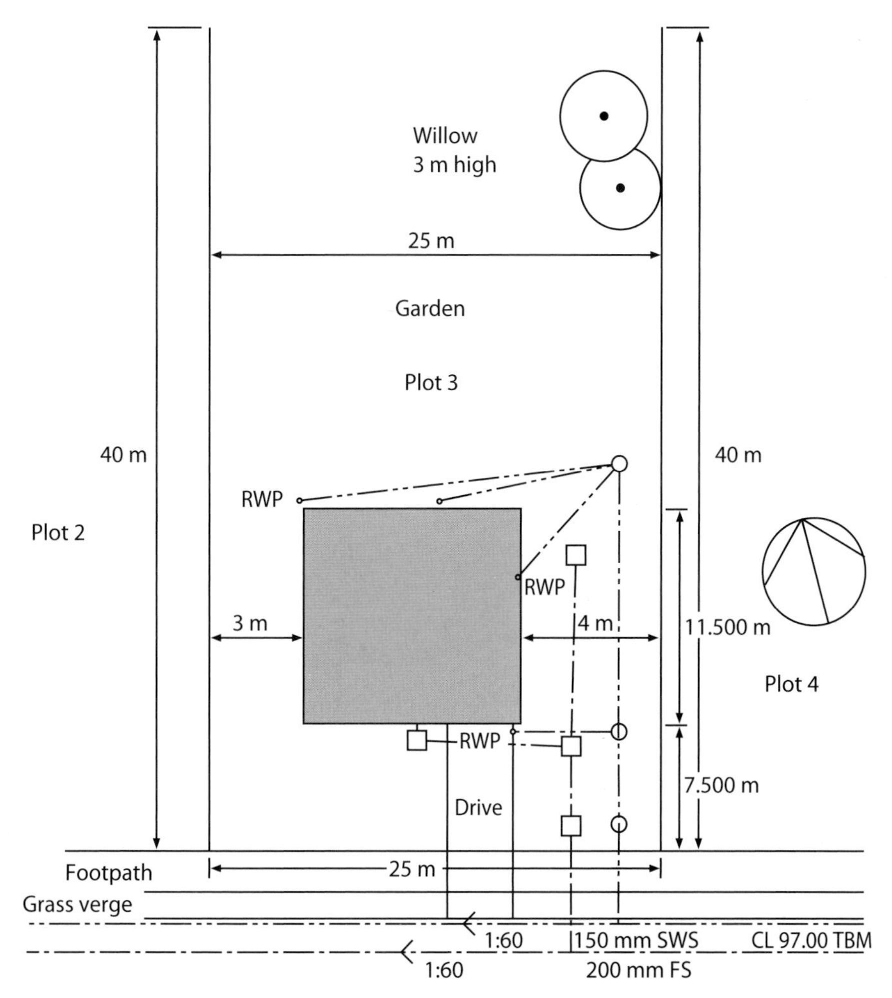

Figure 2.02 Site plan (not drawn to scale)

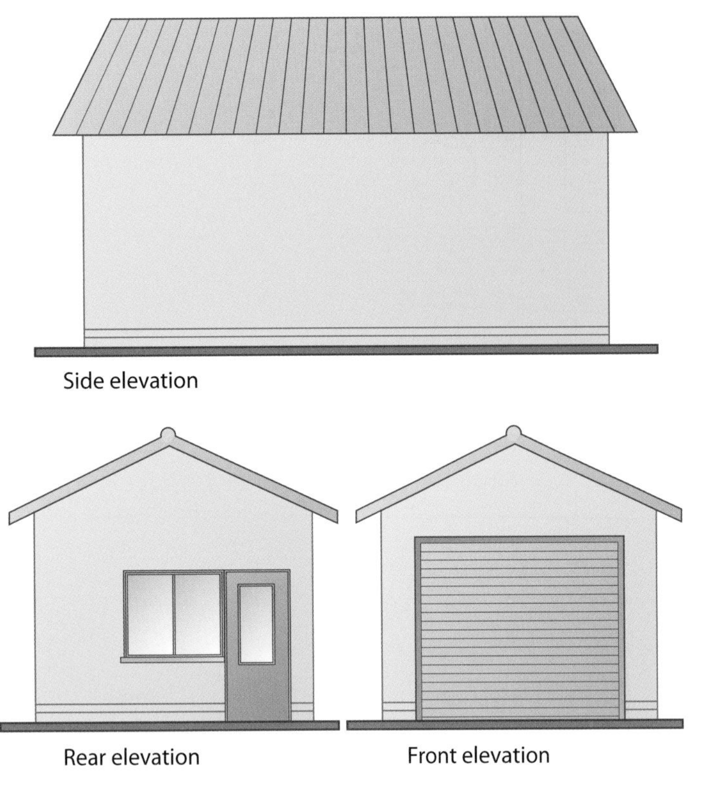

Figure 2.03 General location drawing

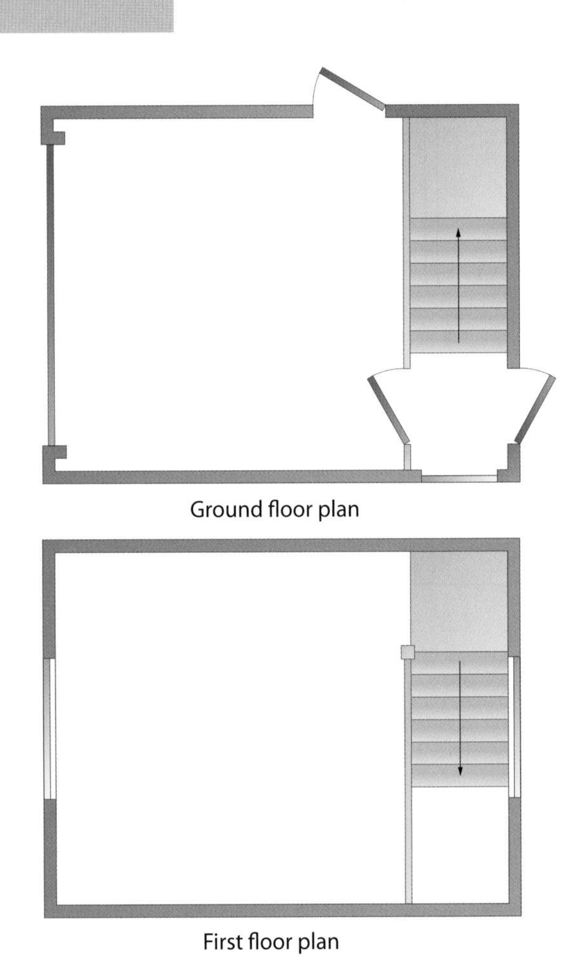

Figure 2.04 Floor plan

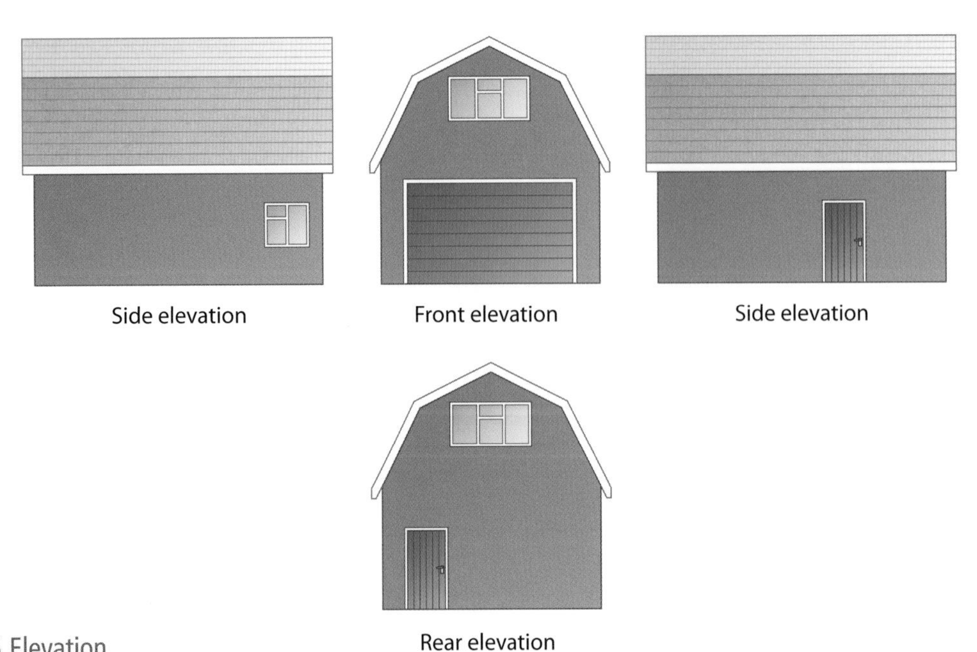

Figure 2.05 Elevation

Component drawings

Component drawings include:

- **range drawings**, which show the different sizes and shapes of a particular range of components (see Figure 2.06). These are usually drawn at a scale of 1:100, 1:50 or 1:20.
- **detail drawings**, which show all the information needed to complete or manufacture a component (see Figure 2.07). These are usually drawn at a scale of 1:10, 1:5 or 1:1.

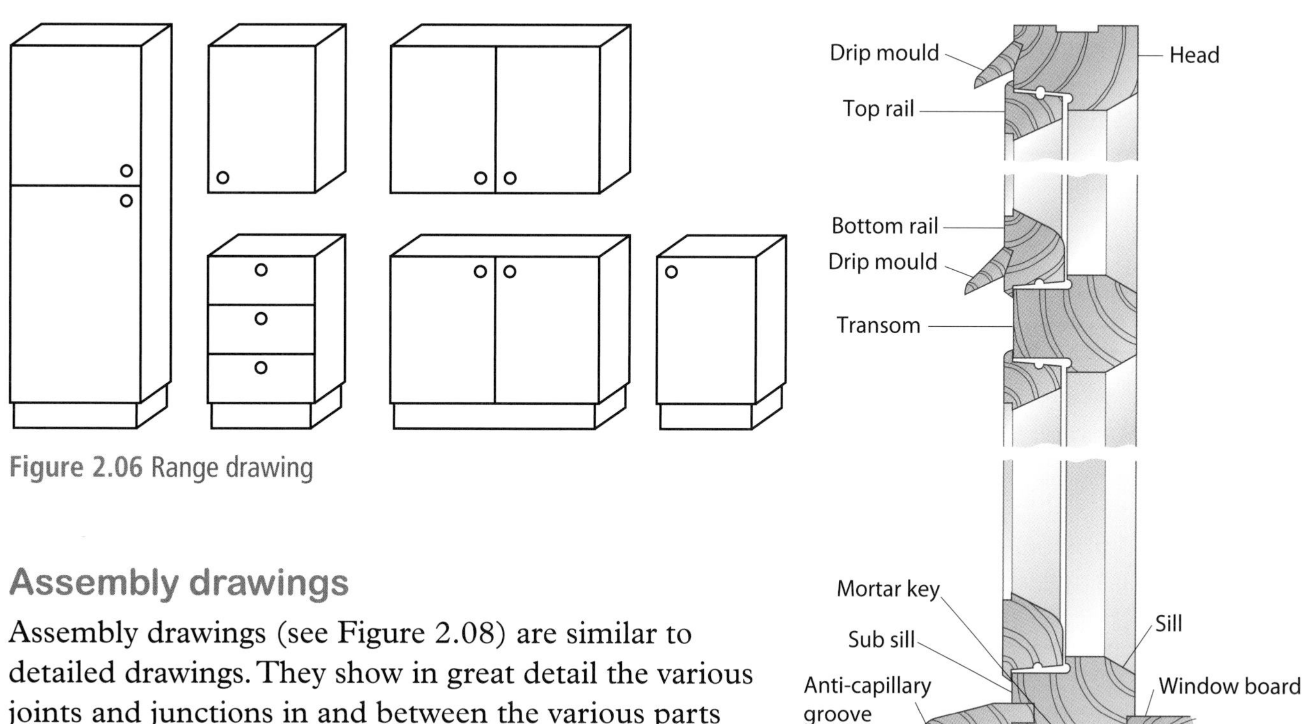

Figure 2.06 Range drawing

Figure 2.07 Detail drawing

Assembly drawings

Assembly drawings (see Figure 2.08) are similar to detailed drawings. They show in great detail the various joints and junctions in and between the various parts and components of a building. Assembly drawings are usually drawn at a scale of 1:20, 1:10 or 1:5.

Joist
Damp walls affecting timber
Wall plate

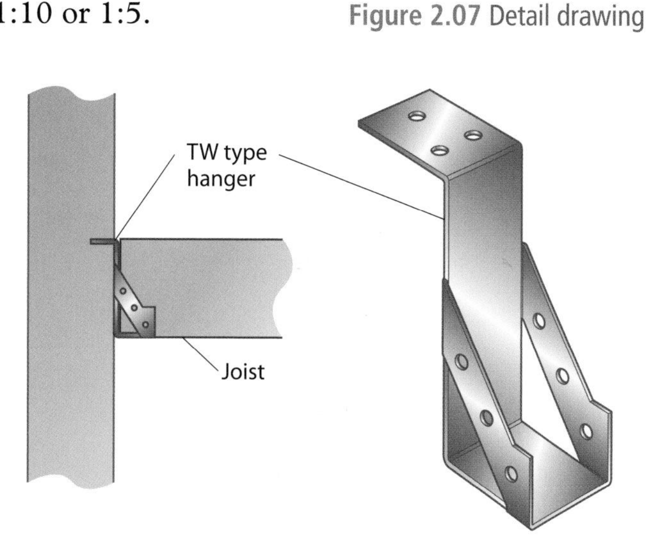

Figure 2.08 Assembly drawing

K1. Know how to produce drawn information

When drawings are mentioned in the construction industry, people generally tend to think of the architect's drawings and plans that form part of the contract documents. These types of drawings are vital in the construction industry as they form part of the legal contract between client and contractor.

Remember

Mistakes, either in design or in interpretation of the design, can be costly.

However, there are other forms of drawings that are just as important. Setting-out drawings are used to mark out for complex procedures such as constructing cut roofing, staircases or brick arches; and with advances in technology, CAD (computer-aided design) is being used more often.

Advantages of CAD

Find out

What different CAD programs are available for use online? Use the Internet to locate some good examples of these programs and use them to produce some simple 2-D diagrams.

CAD is a system in which a draftsperson uses computer technology to help design a part, product or whole building. It is both a visual and symbol-based method of communication, with conventions particular to a specific technical field.

CAD is used particularly at the drafting stage. Drafting can be done in two dimensions (2-D) and three dimensions (3-D).

CAD is one of the many tools used by engineers and designers, and is used in many ways, depending on the profession of the user and the type of software in question.

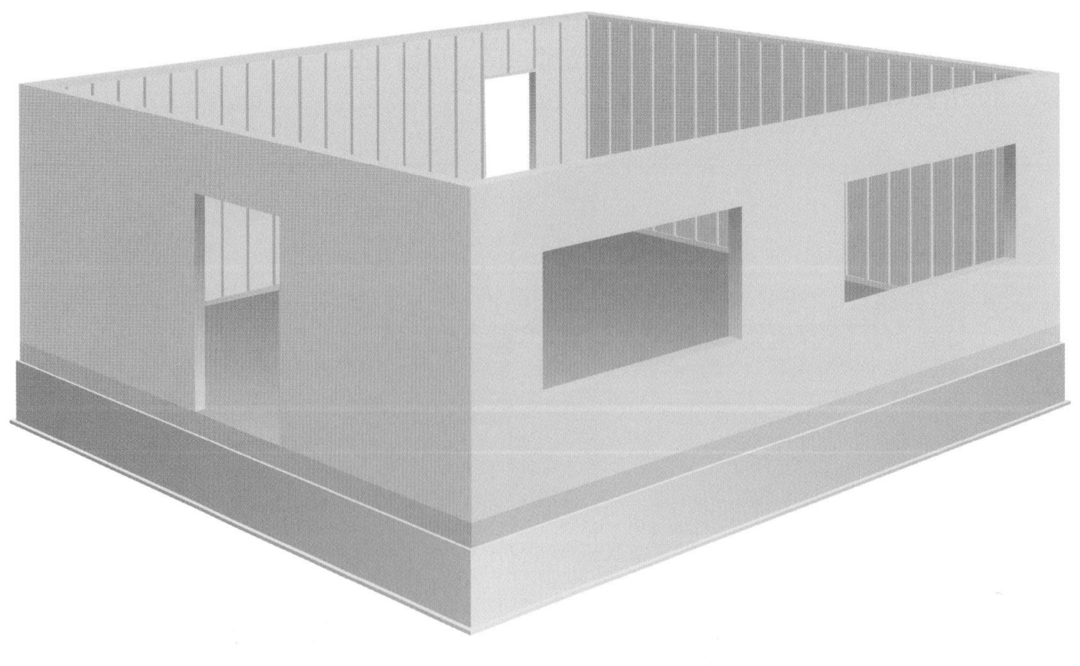

Figure 2.09 CAD drawing

There are several types of CAD. Each requires the operator to think differently about how he or she will use it, and he or she must design the virtual components in a different manner for each.

Many companies produce lower-end 2-D systems, and a number of free and open source programs are available. These make the drawing process easier, because there are no concerns about the scale and placement on the drawing sheet that accompanies hand drafting – these can simply be adjusted as required during the creation of the final draft.

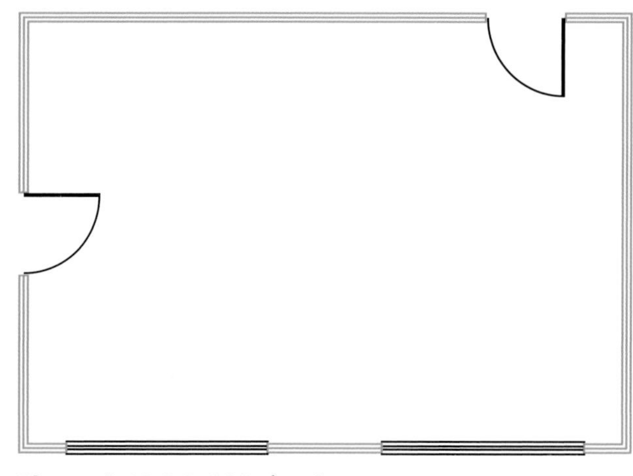

Figure 2.10 2-D CAD drawing

3-D wireframe

3-D wireframe is in essence an extension of 2-D drafting. Each line has to be manually inserted into the drawing. The final product has no mass properties associated with it, and cannot have features directly added to it, such as holes. The operator approaches these in a similar fashion to the 2-D systems, although many 3-D systems allow you to use the wireframe model to make the final engineering drawing views.

Did you know?

Google SketchUp is a simple, free 2-D CAD program, and IKEA and B&Q, among others, operate simple 2-D CAD programs for designing kitchens, etc.

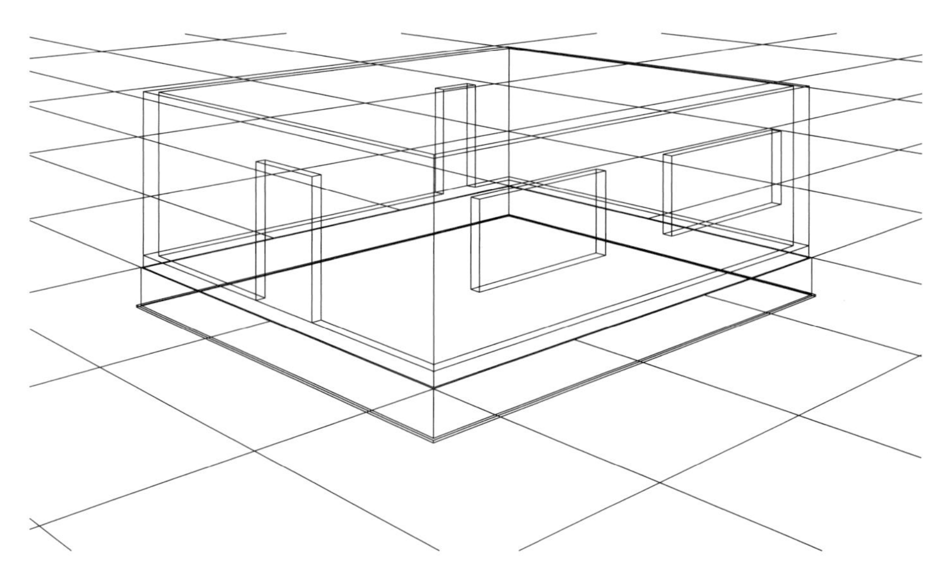

Figure 2.11 3-D wireframe produced using CAD

3-D dumb solids

3-D dumb solids are created in a way corresponding to manipulations of real-world objects. Basic 3-D geometric forms (prisms, cylinders, spheres, etc.) have solid volumes added to or subtracted from them, as if assembling or cutting real-world objects. 2-D projected views can easily be generated from the models. The sorts of basic 3-D solids that are created do not usually include tools to easily allow motion of components, set limits to their motion, or identify interference between components.

Figure 2.12 3-D view of a house produced using CAD

Top-end systems

Top-end systems offer the capabilities to incorporate more organic, aesthetic and ergonomic features into designs. Freeform surface modelling is often combined with solids to allow the designer to create products that fit the human form and visual requirements, as well as the interface with the machine.

Uses of CAD

CAD has become an especially important technology within the scope of computer-aided technologies, with benefits such as lower product development costs and a greatly shortened design cycle. CAD enables designers to lay out and develop work on screen, print it out and save it for future editing, saving time on their drawings.

Details required for floor plans

To complete floor plans you will need to use a range of different information sources. Some of the key pieces of information you will need to know about are covered on the following pages.

Sections

Sectional drawings are useful as they can show details of how certain aspects of a structure are constructed. They show a cross section of the build, using symbols to indicate what materials are used.

These drawings are particularly useful for showing how floors are constructed and also for types of walls, such as sleeper walls.

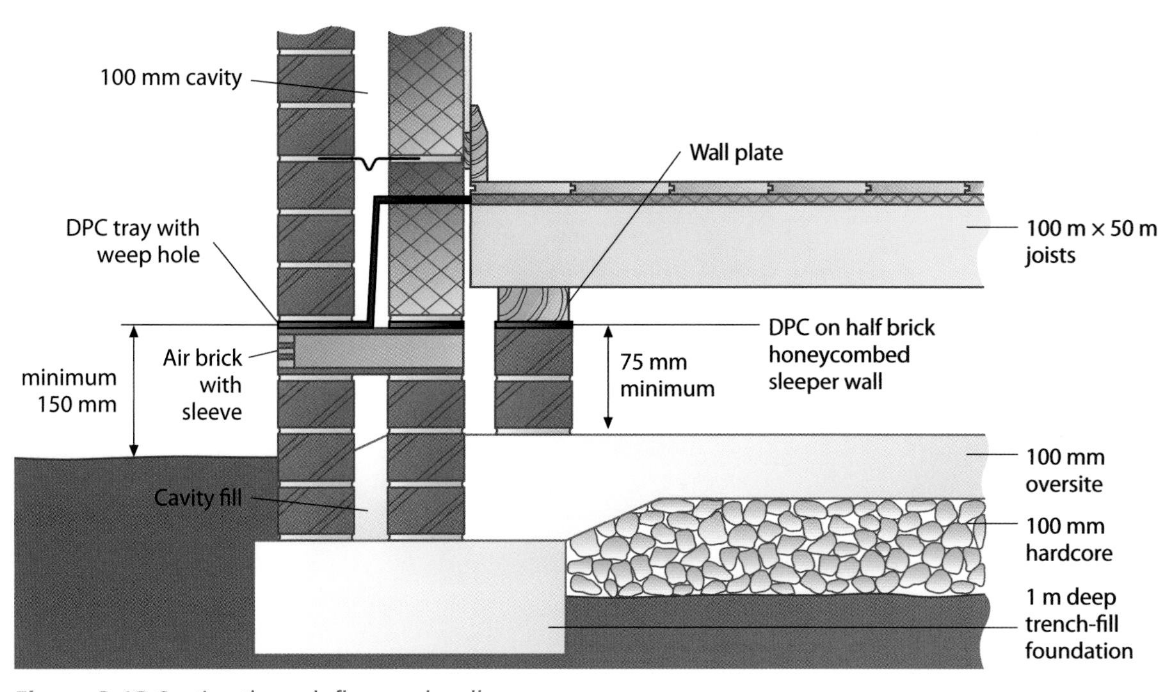

Figure 2.13 Section through floor and wall

Datum points

The need to apply levels is required at the beginning of the construction process and continues right up to the completion of the building. The whole country is mapped in detail and the Ordnance Survey place datum points (bench marks) at suitable locations from which all other levels can be taken.

Ordnance bench mark (OBM)

OBMs are found cut into locations such as walls of churches or public buildings. The height of the OBM can be found on the relevant Ordnance Survey map or by contacting the local authority planning office. Figure 2.14 shows the normal symbol used, although it can appear as shown in Figure 2.15 (on the next page).

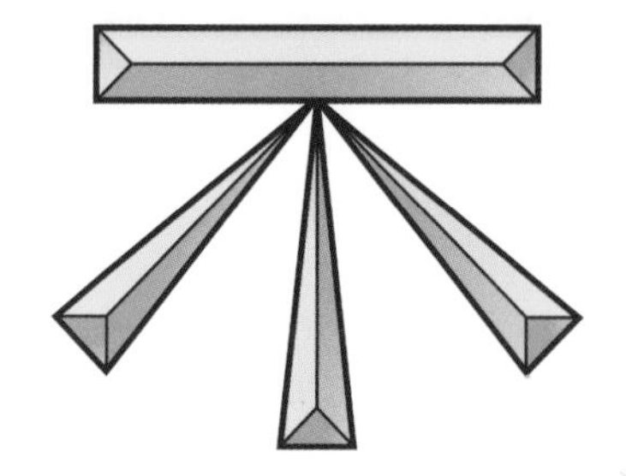

Figure 2.14 Ordnance bench mark

Site datum

It is necessary to have a reference point on site to which all levels can be related. This is known as the site datum. The site datum is usually positioned at a convenient height, such as finished floor level (FFL).

The site datum itself must be set in relation to some known point, preferably an OBM, and must be positioned where it cannot be moved.

Figure 2.15 shows a site datum and OBM, illustrating the height relationship between them.

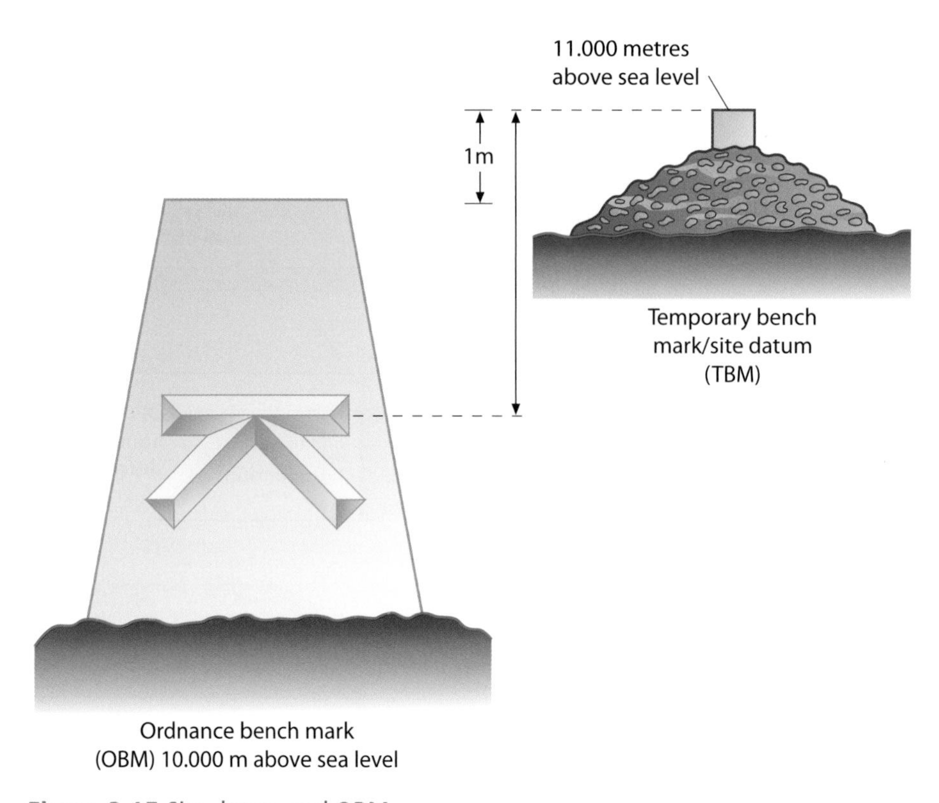

Figure 2.15 Site datum and OBM

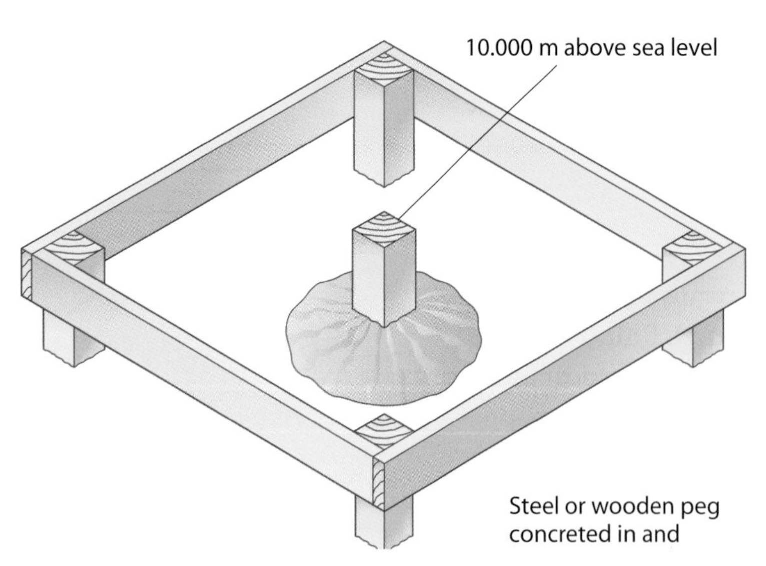

Figure 2.16 A datum peg, suitably protected

A datum peg may be used if no suitable position can be found, its accurate height transferred by surveyors from an OBM as with the site datum. It is normally a piece of timber or steel rod positioned accurately to the required level and then set in concrete. However, it must be adequately protected and is generally surrounded by a small fence for protection, as shown in Figure 2.16.

Temporary bench mark (TBM)

When an OBM cannot be conveniently found near a site, it is usual for a temporary bench mark (TBM) to be set up at a height suitable for the site. Its accurate height is transferred by surveyors from the nearest convenient OBM.

All other site datum points can now be set up from this TBM using datum points, which are shown on the site drawings. Figure 2.17 shows datum points on a drawing.

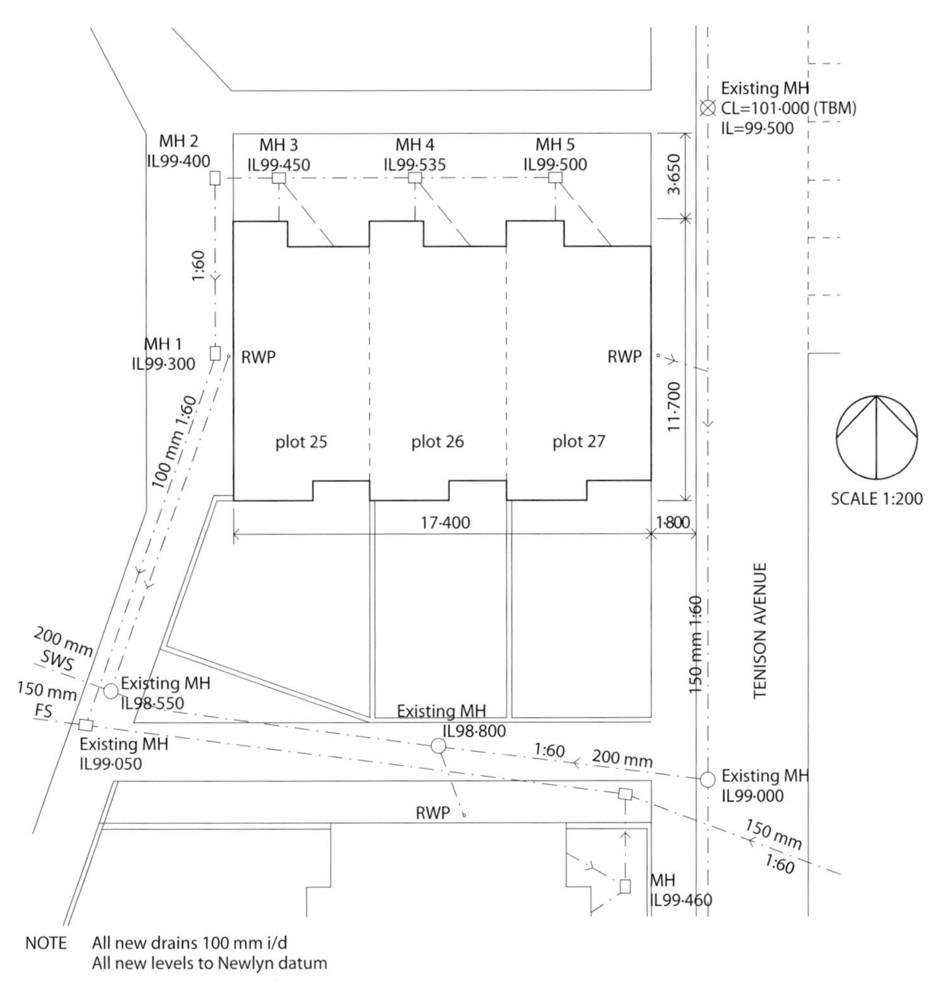

Figure 2.17 Datum points shown on a drawing

Wall constructions

The positioning of internal walling is important when planning out a floor, as internal walling (particularly solid block walling or load-bearing walling) will place strain on the floor. This strain will have to be supported. Internal walling can be built directly onto the foundations. This will break the area of flooring into smaller sections and is usually done in larger buildings. The alternative is that the whole area can be floored and the area which the internal

Key term

Sleeper wall – sometimes known as a dwarf wall, this is a smaller wall built to support the ground floor joists.

walls are to be built on can be reinforced with either **sleeper walls** or reinforced concrete or steel beams.

The position of the internal walling is also used by other trades, as they need to know where the services need to be run.

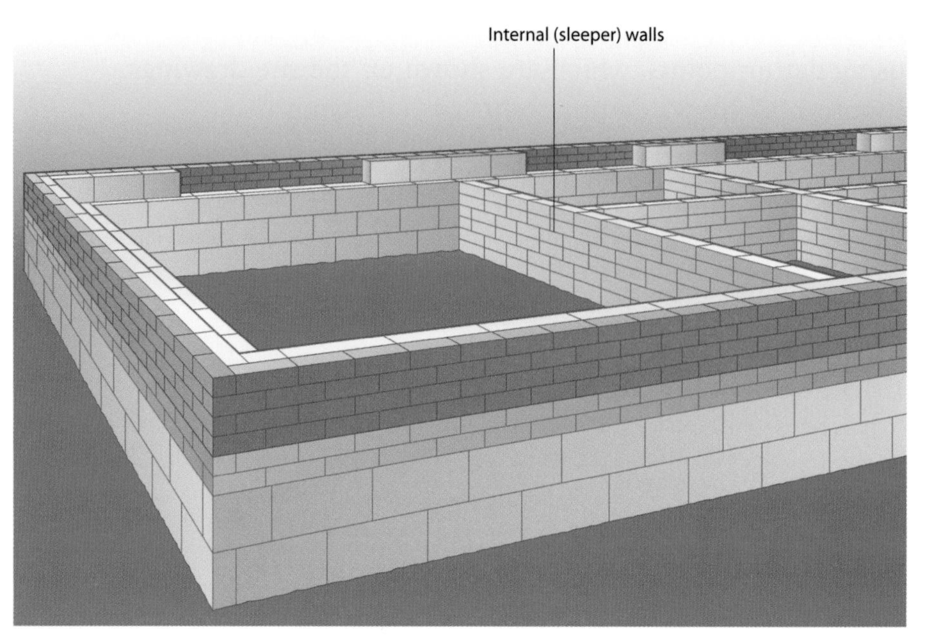

Figure 2.18 Internal (sleeper) walls being built onto foundations

Material codes

Almost all the materials that are used in construction must adhere to British Standards. Each type of material must pass stringent tests so that it can be classified by the BSI and be used. With the move into European markets, some of the codes are now prefixed ISO (International Organization for Standardization).

The following is a list of BS ISO standards that relate to flooring.

Remember

Diagrams and drawings include a great deal of information about the type and quality of materials to be used in the construction process.

Timber

- BS EN 383:2007 – Timber structures. Tested by determining embedment strength and foundation values for dowel type fasteners.
- BS 5268-2:2002 – Structural use of timber with a code of practice for permissible stress design, materials and workmanship.

Concrete, aggregates and masonry

- BS EN 12350-1:2009 – Testing fresh concrete. A sample of concrete is tested.
- BS EN 1097-8:2009 – Tests for mechanical and physical properties of aggregates.

- BS EN 12390-3:2009 – Testing hardened concrete. Specifies a method for testing the compressive strength of test specimens of hardened concrete.
- BS EN 12390-5:2009 – Testing hardened concrete. Specifies a method for testing the flexural strength of specimens of hardened concrete.
- BS EN 12390-7:2009 – Testing hardened concrete. The density of hardened concrete is tested.

Materials that have been stamped with the relevant BS number will have been tested to meet the required standards, or a sample of the materials will have been tested.

Depth dimensions and heights

The depth dimensions and heights of materials used in flooring construction will be identified by the architect. He or she will then use the relevant BSI specification to decide which type and size of materials will be suitable for each job.

Specifications and schedules

Specifications

The specification or 'spec' is a document produced alongside the plans and drawings and is used to show information that cannot be shown on the drawings. Specifications are almost always used, except in the case of very small contracts. A specification should contain details about:

- **the site description** – a brief description of the site, including the address
- **restrictions** – what restrictions apply, such as working hours or limited access
- **services** – what services are available, what services need to be connected and what type of connection should be used
- **materials description** – including type, sizes, quality, moisture content, etc.
- **workmanship** – including methods of fixing, quality of work and finish.

The specification may also name sub-contractors or suppliers, or give details such as how the site should be cleared, and so on.

Figure 2.19 A good spec helps to avoid confusion

Functional skills

This exercise will allow you to practise **FM 2.3.1** Interpret and communicate solutions to multistage practical problems.

Working life

You are working on a job and have received the site plans, which show the layout of the services. You start to dig out for the services and, when you reach the site where the mains gas should be, you find it is not where the drawing shows.

What could have caused this problem? You will need to look at the sources of information you have used to make your decisions, and check to see what could have caused the mistake. What other problems could arise from using faulty information?

What effect could this have financially? You will also need to think about the impact not only on money but also the time that might be needed to carry out the project. What impact could this have on your company's reputation?

Remember

Schedules are not always needed on contracts, particularly smaller ones.

Schedules

A schedule is used to record repeated design information that applies to a range of components or fittings. Schedules are mainly used on bigger sites where there are multiples of several types of house (4-bedroom, 3-bedroom, 3-bedroom with dormers, etc.), each type having different components and fittings. The schedule avoids the wrong component or fitting being put in the wrong house. Schedules can also be used on smaller jobs, such as a block of flats with 200 windows, where there are six different types of window.

Find out

Think of a job or contract that would require a schedule and produce one for a certain part of that job, for example, doors or brick types.

The need for a schedule depends on the complexity of the job and the number of repeated designs there are. Schedules are mainly used to record repeated design information for:

- doors
- windows
- ironmongery
- joinery fitments
- sanitary components
- heating components and radiators
- lintels
- kitchens.

A schedule is usually used in conjunction with a range drawing and a floor plan.

On the following page are basic examples of these documents, using windows as an example.

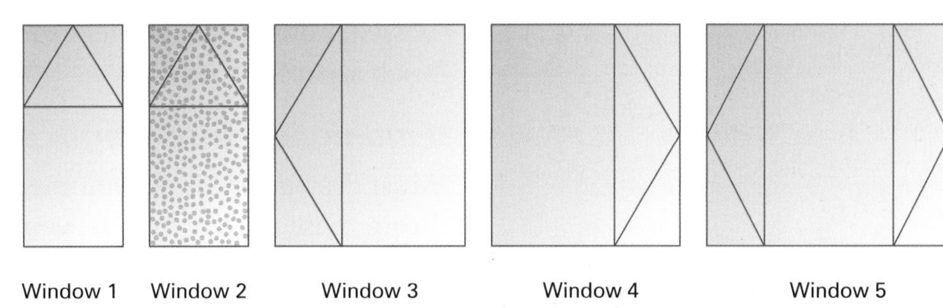

Figure 2.20 Range drawing

The schedule (see Figure 2.22) shows that there are five types of window, each differing in size and appearance; the range drawing (see Figure 2.20) shows what each type of window looks like; and the floor plan (see Figure 2.21) shows which window goes where. For example, the bathroom window is a type 2 window, which is 1200 × 600 mm with a top-opening sash and obscure glass.

Setting-out drawings

Setting-out drawings are as important as contract documents. You must be aware of how certain tasks are set out and what drawings can be created to aid in the setting-out process.

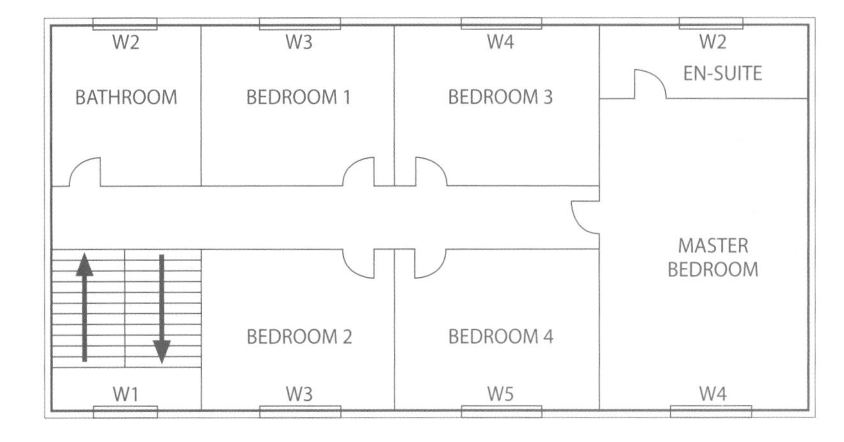

Figure 2.21 Floor plan

Setting-out drawings are most often needed on smaller jobs, where there is limited or no information from the architect in the form of contract document drawings. Setting-out drawings can also be used on larger sites where there has been an alteration or on oversight by the architect.

WINDOW	SIZE	EXTERIOR	INTERIOR	LOCATION	GLASS	FIXING
Window 1	600 x 1200 mm	Mahogany wood grain UPVC	White UPVC	Stairwell	22 mm thermal resistant double glazed units	Fixed with 100 mm frame fixing screws
Window 2	600 x 1200 mm	Mahogany wood grain UPVC	White UPVC	Bathroom En-suite	22 mm thermal resistant double glazed units with maple leaf obscure pattern	Fixed with 100 mm frame fixing screws
Window 3	1100 x 1200 mm	Mahogany wood grain UPVC	White UPVC	Bedroom 1 Bedroom 2	22 mm thermal resistant double glazed units	Fixed with 100 mm frame fixing screws
Window 4	1100 x 1200 mm	Mahogany wood grain UPVC	White UPVC	Bedroom 3 Master bedroom	22 mm thermal resistant double glazed units	Fixed with 100 mm frame fixing screws
Window 5	1500 x 1200 mm	Mahogany wood grain UPVC	White UPVC	Bedroom 4	22 mm thermal resistant double glazed units	Fixed with 100 mm frame fixing screws

Figure 2.22 Schedule

The most common forms of setting-out drawings are used:

- **in carpentry**, for cut roofing, where there may be no information on the true lengths of rafters
- **in joinery**, when setting out for stairs, where there may be no information on the individual rise, etc.
- **in bricklaying**, where you may come across setting-out drawings for arch centres, such as segmental or gothic arches (see Figure 2.23 on the next page).

Figure 2.23 Setting-out drawings are crucial for creating arches like these

We will now look at a brief example of how roofing and brick arches are set out.

Finding the true length of a common rafter

Most drawings will tell you the **span** and **rise** of the roof. From these measurements, you can create a drawing that will tell you the true length of the common rafter, and also what angle the ends of the rafter should be cut at.

This true length is the actual length that the rafter needs to be, and all the rafters can be cut to length from the setting-out drawing. The setting-out drawing for a roof is usually drawn on a sheet of plywood to a scale that fits the sheet.

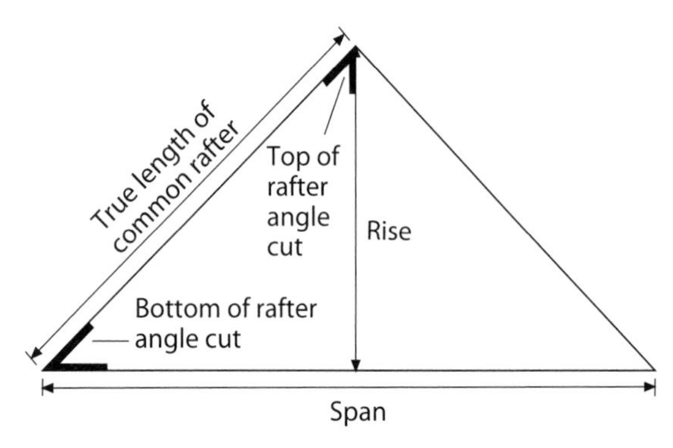

Figure 2.24 Finding common rafter true length

Key terms

Span – the distance measured in the direction of ceiling joists, from the outside of one wall plate to another, known as the overall (O/A) span.

Rise – the distance from the top of the wall plate to the roof's peak.

Setting out a segmental brick arch

Most drawings will show you the opening span of the arch, but some may not tell you the radius. Without the radius, you cannot build the arch correctly.

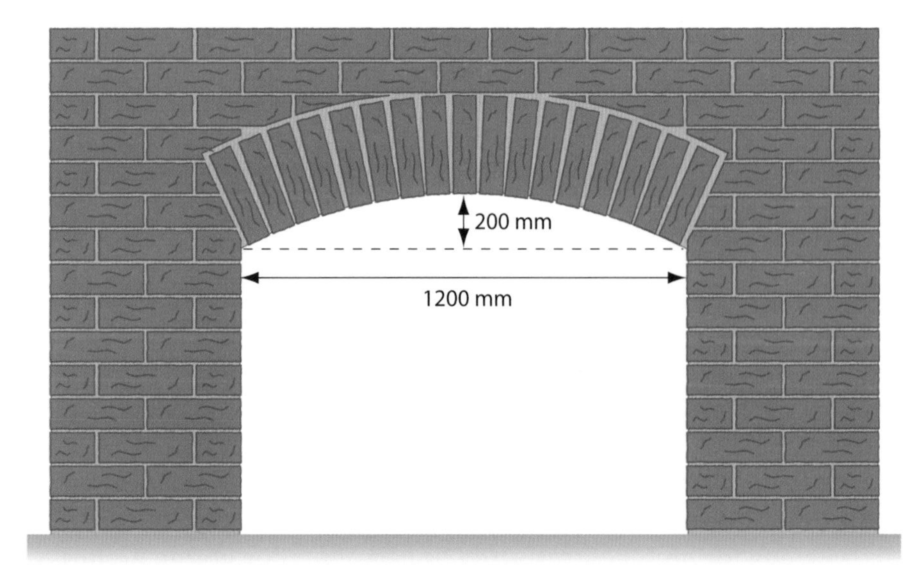

Figure 2.25 An example of a segmental arch

Find out

Using the Internet and other resources, find out what sort of scales are best to use for drawing up setting-out drawings for different components and builds.

We will now look at how setting-out drawings can aid you in setting out this arch.

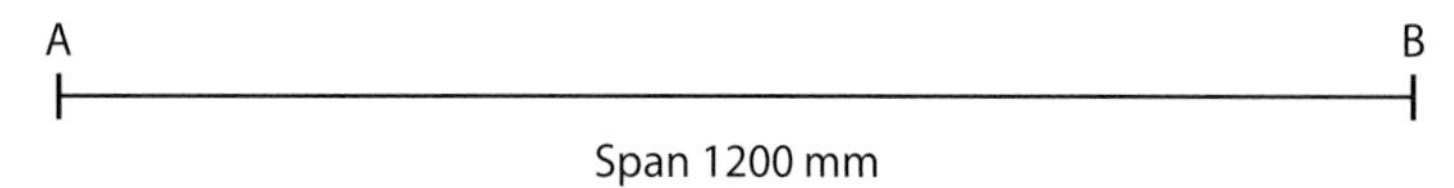

Figure 2.26 Establish the span (a length of 1200 mm has been used here, shown as A–B)

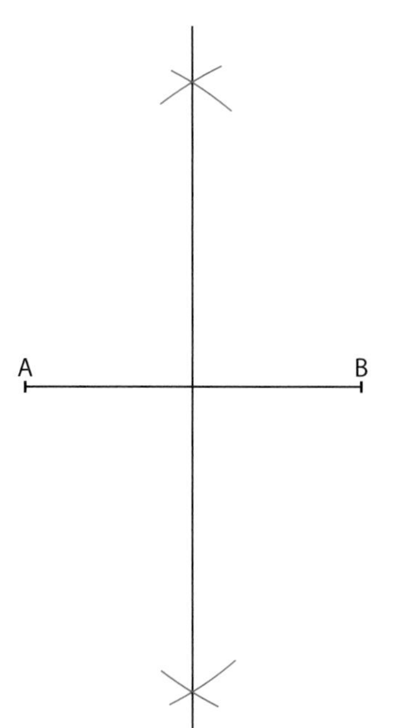

Figure 2.27 Bisect this line

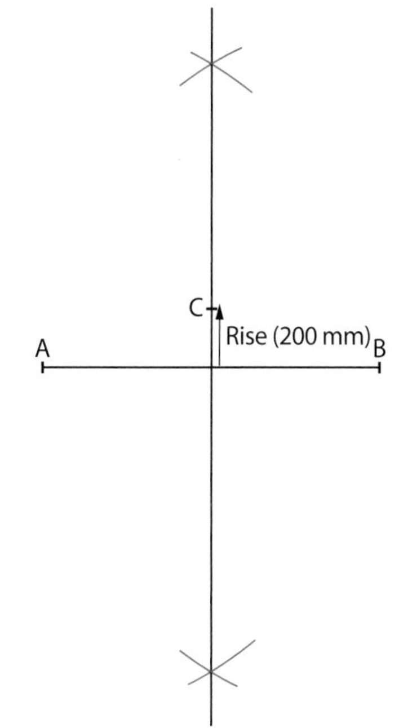

Figure 2.28 Establish the rise (the distance from the springing line (A–B) to the highest point of the soffit shown as C). The rise is normally one-sixth of the span so, in this case, the rise is shown as 200 mm

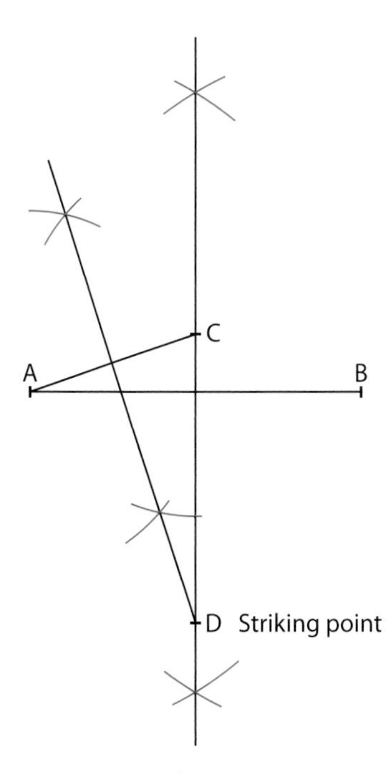

Figure 2.29 Draw a line from A to C and bisect this line. The point where this bisecting line crosses the bisecting line of the span will be the striking point for the arch (shown here as D). From striking point D, open out compass to A and draw an arc across to B. This will provide the intrados for the arch

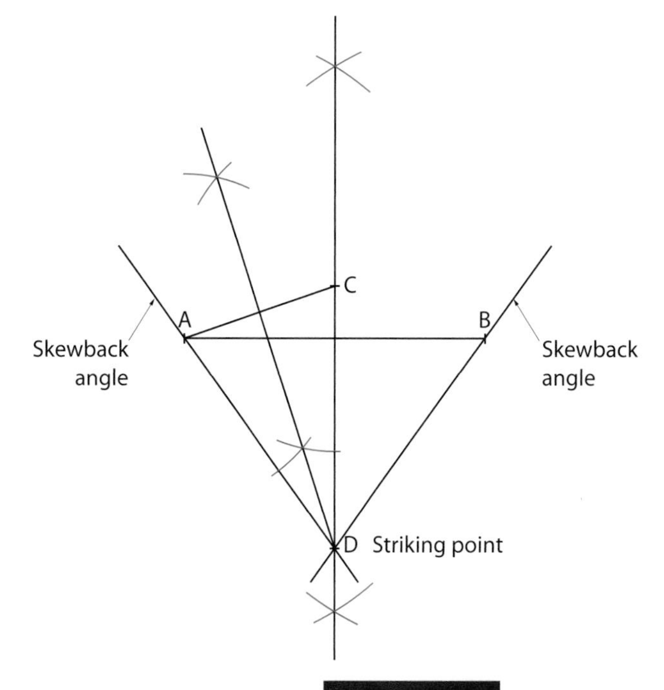

Figure 2.30 Draw a line from D through A and a line from D through B. These lines will provide the angle for the **skewbacks**

Key term

Skewbacks – the angle at the springing point at which the arch rings will be laid.

Setting out for a segmental arch can again be drawn out on a sheet of plywood but, in this case, it can be drawn full size, with the drawing being cut out and used as a template for the arch centre.

Reasons for the use of elevations and projections

Building, engineering and similar drawings aim to give as much information as possible in a way that is easy to understand. They frequently combine several views on a single drawing. These may be of two kinds:

- **Elevation** – the view we would see if we stood in front or to the side of the finished building.
- **Plan** – the view we would have if we were looking down on the finished building.

The view we see depends on where we are looking from. There are, then, different ways of 'projecting' what we would see onto the drawings. The three main methods of projection, used on standard building drawings, are orthographic, isometric and oblique.

Orthographic projection

Orthographic projection works as if parallel lines were drawn from every point on a model of the building on to a sheet of paper held up behind it (an elevation view), or laid out underneath it (plan view). There are different ways that we can display the views on a drawing. The method most commonly used in the building industry for detailed construction drawings is called 'first angle projection'. In this the front elevation is roughly central. The plan view is drawn directly below the front elevation and all other elevations are drawn in line with the front elevation. An example is shown in Figure 2.31.

Figure 2.31 Orthographic projection

Isometric projection

In isometric views, the object is drawn at an angle where one corner of the object is closest to the viewer. Vertical lines remain vertical but horizontal lines are drawn at an angle of 30° to the horizontal. This can be seen in Figure 2.32.

Oblique projection

Oblique projection is similar to an isometric view, with the object drawn at an angle where one corner of the object is closest to the viewer. Vertical lines remain vertical but horizontal lines are drawn at an angle of 45° to the horizontal. This can be seen in Figure 2.33.

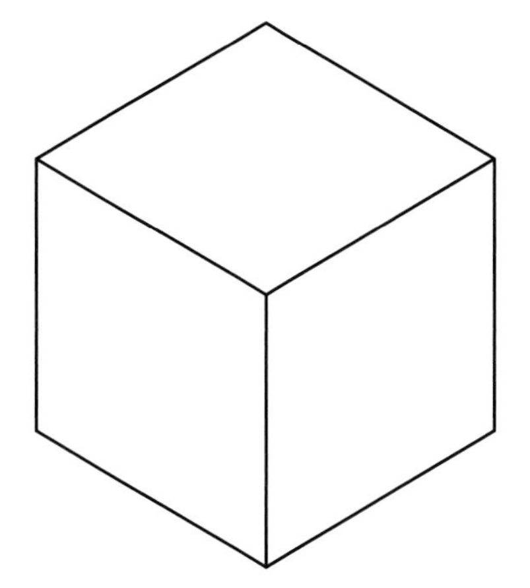

Figure 2.32 Isometric projection

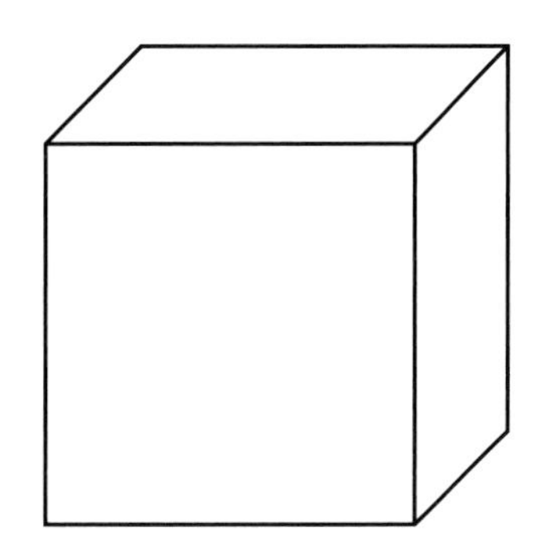

Figure 2.33 Oblique projection

Working life

You have been tasked with building a segmental brick arch, but there is minimal information on the drawing. You decide to just build the arch but soon run into problems with the radius.

What could have prevented the problems? You will need to think about the processes you could have followed to check information and who you could have consulted with about any problems. What should you do now? You will need to think about the impact any action could have not only on you but also on anyone else you may be working with on site. What effect could all this have on the building and profitability of the job?

Use of hatchings and symbols

All plans and drawings contain symbols and abbreviations, which are used to show the maximum amount of information in a clear and legible way. A range of well-known symbols and abbreviations are shown on the next page.

Find out

Research, using the Internet and other resources, how isometric projection is used to sketch components.

Functional skills

This task will allow you to practise **FE 2.3.1–2.3.5** Write documents, including extended writing pieces, communicating information, ideas and opinions effectively and persuasively.

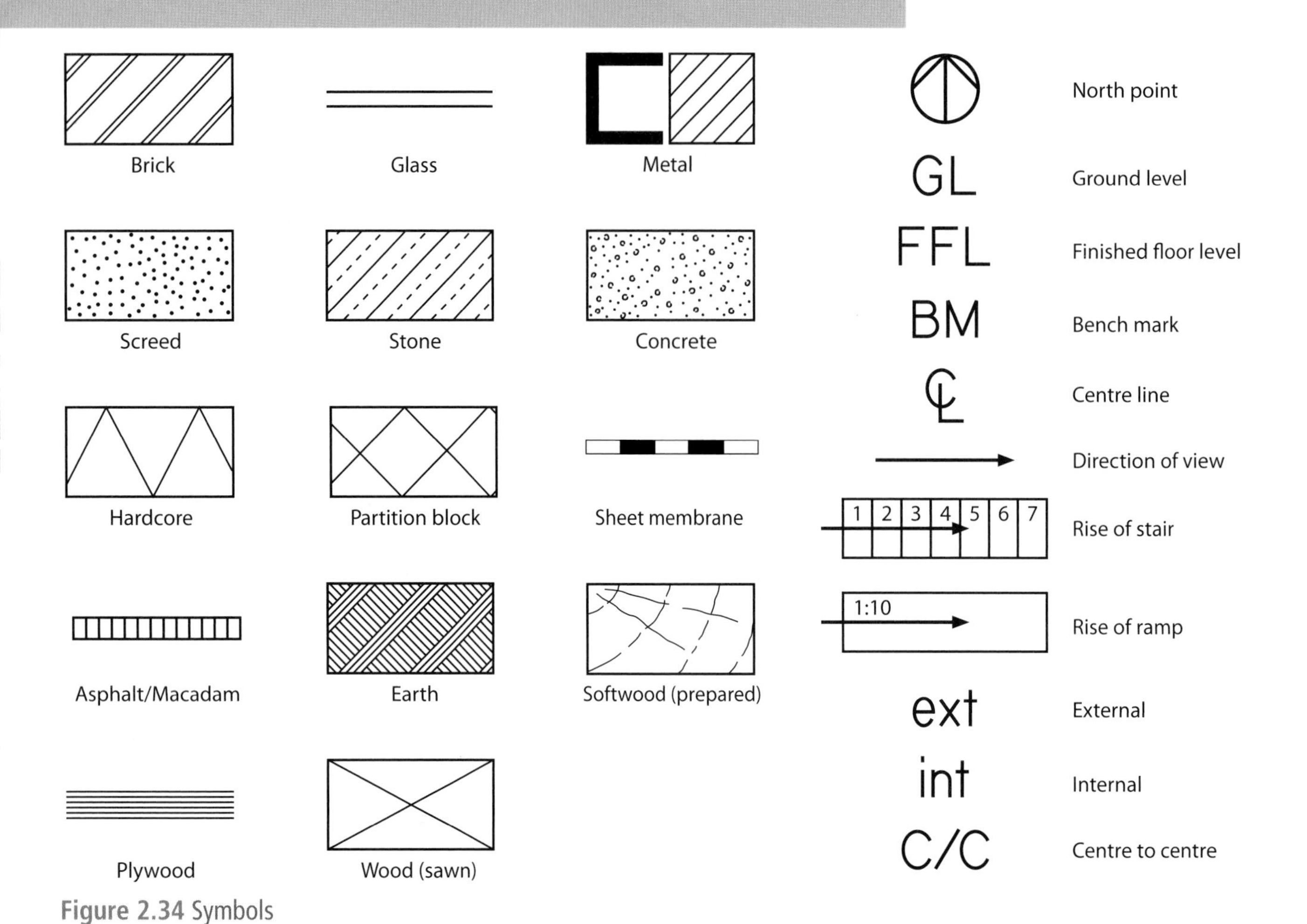

Figure 2.34 Symbols

Table 2.1 Abbreviations

Item	Abbreviation	Item	Abbreviation
Airbrick	AB	Hardcore (or hard fill)	Hc
Asbestos	abs	Hardwood	Hwd
Bitumen	bit	Insulation	Insul
Boarding	bdg	Joist	Jst
Brickwork	bwk	Mild steel	MS
Building	bldg	Plasterboard	Pbd
Cast iron	CI	Polyvinyl acetate	PVA
Cement	ct	Polyvinyl chloride	PVC
Column	col	Reinforced concrete	RC
Concrete	conc	Satin anodised aluminium	SAA
Cupboard	cpd	Satin chrome	SC
Damp-proof course	DPC	Softwood	Swd
Damp-proof membrane	DPM	Stainless steel	SS or SST
Drawing	dwg	Tongue and groove	T&G
Foundation	fnd	Wrought iron	WI
Hard board	hdbd		

Scale drawings

All building plans are drawn to scale by using symbols and abbreviations. To draw a building on a drawing sheet, its size must be reduced. This is called a scale drawing.

Remember

A scale is merely a convenient way of reducing a drawing in size.

Using scales

Each length on a building plan is in proportion to the real length. On a drawing that has been drawn to a scale of 1 mm represents 1 m:

- a length of 50 mm represents an actual length of 50 m
- a length of 120 mm represents an actual length of 120 m
- a length of 34 m is represented by a line 34 mm long.

Scales are often given as **ratios**. For example:

- a scale of 1:100 means that 1 mm on the drawing represents an actual length of 100 mm (or 0.1 m)
- a scale of 1:20 000 means that 1 mm on the drawing represents an actual length of 20 000 mm (or 20 m).

Table 2.2 shows some common scales used in the construction industry.

Remember

You can use a scale to:

- work out the actual measurement from a plan
- work out how long to draw a line on the plan to represent an actual measurement.

Key term

Ratio – one value divided by the other.

Table 2.2 Common scales used in the construction industry

Scale	Representation	Proportion
1:5	1 cm represents 5 cm	5 times smaller than actual size
1:10	1 cm represents 10 cm	10 times smaller than actual size
1:20	1 cm represents 20 cm	20 times smaller than actual size
1:50	1 cm represents 50 cm	50 times smaller than actual size
1:100	1 cm represents 100 cm = 1 m	100 times smaller than actual size
1:1250	1 cm represents 1250 cm = 12.5 m	1250 times smaller than actual size

Why different scales are used

The scales that are preferred for use in building drawings are shown in Table 2.3.

Table 2.3 Preferred scales for building drawings

Type of drawing	Scales
Block plans	1:2500, 1:1250
Site plans	1:500, 1:200
General location drawings	1:200, 1:100, 1:50
Range drawings	1:100, 1:50, 1:20
Detail drawings	1:10, 1:5, 1:1
Assembly drawings	1:20, 1:10, 1:5

Find out

With a little practice, you will easily master the use of scales. Try the following:

- On a scale of 1:50, 40 mm represents: _____
- On a scale of 1:200, 70 mm represents: _____
- On a scale of 1:500, 40 mm represents: _____

These scales mean that, for example, on a block plan drawn to 1:2500, 1 mm on the plan would represent 2500 mm (or 2.5 m) on the actual building. Some other examples are:

- on a scale of 1:50, 10 mm represents 500 mm
- on a scale of 1:100, 10 mm represents 1000 mm (1.0 m)
- on a scale of 1:200, 30 mm represents 6000 mm (6.0 m).

Accuracy of drawings

Printing or copying drawings introduces variations that can affect their accuracy. Hence, although measurements can be read from drawings using a rule with common scales marked (Figure 2.35), you should work to written instructions and measurements wherever possible.

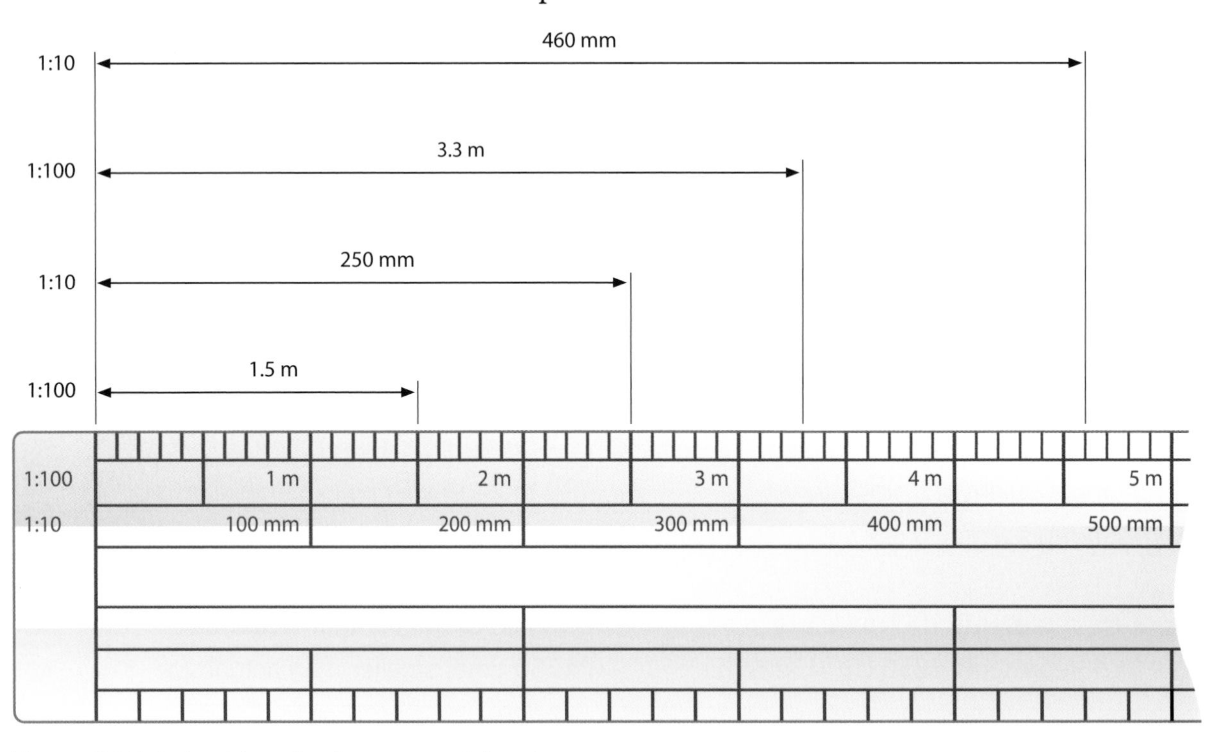

Figure 2.35 Rule with scales for maps and drawings

Working life

You have been issued a scale drawing for building internal walls, but some of the dimensions are missing. What should you do?

What complications could arise from scaling from the drawing as it is? Think about the importance the drawings have to the planning of the whole project. What could happen if the information you are using is unreliable or incorrect?

What effect could building a wall in the wrong place have? This could have an impact on a range of people – not just you, but also other craftspeople working on site and the client.

Now look at the following examples.

Example

A plan is drawn to a scale of 1:20.

On the plan, a wall is 45 mm long. How long is the actual wall?

1 mm on the plan = actual length 20 mm

45 mm on the plan = 45 × 20

= 900 mm or 0.9 m.

A window is 3 m tall. How tall is it on the plan?

3 m = 3000 mm

actual length 20 mm = 1 mm on the plan

actual length 3000 mm = 3000 ÷ 20

= 150 mm

Therefore, the window is 150 mm tall on the plan.

Remember

As we have seen, to make scale drawings architects use a scale rule. The different scales on the ruler give the equivalent actual length measurements for different lengths in cm for each scale.

K2. Know how to estimate quantities and price work

For all construction projects it is necessary to calculate the amounts of materials and other resources that will be needed. As part of this you will also need to be able to make a calculation on the expected cost of these materials. This is called an estimate.

Estimates are used on all construction projects when setting a budget for the work. For many projects, a client will look for tenders from a range of contractors. This means the potential contractors put together their own estimates for the work, with the client selecting the estimate that best meets their needs – usually the estimate that presents the best value.

This section will look at the information used to create an estimate and plan a project.

The tender process

Tendering is a competitive process where the contractor works with a specification and drawings from the client and submits a cost estimate for the work (including materials, labour and equipment). Tenders are often invited for large contracts, such as government contracts, with strict fixed deadlines for the tenders to be received.

An estimator will calculate the total cost in the tender. Using the information in the specification, the estimator calculates the amount of materials and labour needed to complete the work. The final tender is based on this estimation.

All the tenders for a contract will then present their case and costs to the client, who will then decide on one business to be offered the contract.

Functional skills

This exercise will allow you to practise **FM 2.3.1** Interpret and communicate solutions to multistage practical problems. If you give oral answers to questions from your tutor, you will be able to practise **FE 2.1.1–2.1.4** Make a range of contributions to discussions and make effective presentations in a wide range of contexts.

Working life

You have been invited to tender a bid for a large public contract. Business has been slow, and you really need this contract if you are to keep your business afloat and avoid redundancies.

Two of the other tenders concern you. One is priced so low that, if you match it, you may make a small loss. In the other, the contractor promises to recycle 45 per cent of materials, to use only sustainable materials and to employ 70 per cent of the local workforce – matching this may mean you have to lay off some workers and may only make a small profit.

What should you do? What stipulations could you introduce to help improve your bid? What could the consequences be of not getting the contract – or, indeed, of getting it?

Quoting

A quote is basically part of the tender process, but it will only contain pricing information on materials, labour, etc. The quote will state how much the job will cost without any additional information that may appear on a tender, such as using a percentage of the local workforce or recycling a certain amount of materials.

The quote is then used as part of the tender to give an idea of the potential cost of a job. Companies submitting tenders will look to make this quote as attractive as possible to the client.

Estimated pricing

Estimated pricing is used to create the quote. An estimator will look at what is required and provide an estimated price for it.

It may take many months for a successful tender to be selected, so an estimator who prices everything up exactly as it is now may be wrong in six months' time, as the price of labour or materials may have changed. This means that an estimator will instead give an estimated price, based on a calculation of how much the materials or labour may cost in the future.

The resources used for making an estimated price and then a final price include:

- materials
- purchase orders and invoices
- time study sheets, labour schedules, job sheets and site diaries
- building suppliers' price lists and equipment availability lead times.

Prime cost

This is the final total cost of all material costs, labour costs and expenses for the project. This sum will need to be agreed by all parties before work begins.

Provisional sums

A provisional sum describes work which has not yet been defined in terms of scope and extent. Neither party involved will try to create an accurate price for this work until a contract is agreed. The provisional sum is usually included in the contract price as an estimate. Many contracts include a clause allowing for the provisional sum to be omitted and replaced with the final figure.

Resources required for a construction task

The materials used for a job will usually depend on what the client wants. This is particularly the case for smaller jobs where the client may want certain fixtures or fittings. Where any structural or large alterations will be needed to accommodate the client's plans, a client may want to consult an architect or local planning authority, or even their contractor, as to what materials they must have to meet regulations.

If a client insists on arranging and organising any material requirements, then it is important to ensure that they are aware of exactly what type and size of materials are required and when they are required. Otherwise, not only can it hold up the job, it may also lead to a poor job being done with sub-standard materials, which can affect your reputation.

Other problems can develop when clients order materials themselves, as they may lack the technical knowledge needed for ordering materials. For example, they may think they are getting cheaper materials by ordering 3" × 2" timber studwork or cheaper common bricks. However, 3" × 2" may not be strong enough and 4" × 2" should be used, or particular bricks, such as engineering bricks, may be needed.

Did you know?

On smaller jobs the client may wish to order the materials themselves, as they may be able to get a better deal and save money.

Remember

As well as ordering possibly the wrong size, other problems arising from the client dealing with the materials can be sub-standard materials and delays in materials delivery, both of which can cause delays.

Larger jobs will be led by the client's wishes, carried out through an architect. This will ensure the correct materials are stated on the building documentation. The larger companies will usually have contracts in place with suppliers, which will allow them to purchase materials at cheap rates.

Any specialist materials will be resourced by a buyer. They will look at which companies provide the materials, what the cost is and what attributes the company has, such as whether they work with fair trade, etc.

Working life

You are working on a renovation project when your boss calls you to ask what materials you need for the next few weeks. You are caught a bit off-guard, and you rush around – giving your boss a list of materials over the phone. When the materials are delivered, there are some discrepancies: it is not what you said, as far as you can remember. You phone your boss to tell him and he gets cross, blaming you for the mistakes.

Who is to blame? What should have been done? You will need to think about the ideal process that should have been followed. What information could you have used? Where would you get this information? What would be the best way of getting this information to your boss?

Remember

Bills of quantities are used to help contractors provide a tender for a contract. A bill of quantities is put together for a task, including labour, materials, etc.

Bill of quantities

The bill of quantities is produced by the quantity surveyor. It gives a complete description of everything that is required to do the job, including labour, materials and any items or components, drawing on information from the drawings, specification and schedule. The same single bill of quantities is sent out to all prospective contractors so they can submit a tender based on the same information – this helps the client select the best contractor for the job.

Figure 2.36 Every item needed should be listed on the bill of quantities

All bills of quantities contain:

- **preliminaries** – general information such as the names of the client and architect, details of the work and descriptions of the site
- **preambles** – similar to the specification, outlining the quality and description of materials and workmanship, etc.
- **measured quantities** – a description of how each task or material is measured, with measurements in metres (linear and square), hours, litres, kilograms or simply the number of components required
- **provisional quantities** – approximate amounts where items or components cannot be measured accurately
- **costs** – the amount of money that will be charged per unit of quantity.

The bill of quantities may also contain:

- any costs that may result from using sub-contractors or specialists
- a sum of money for work that has not yet been detailed
- a sum of money to cover contingencies for unforeseen work.

Figure 2.37 shows an extract from a bill of quantities that might be sent to prospective contractors, who would then complete the cost section and return it as their tender.

Item ref no	Description	Quantity	Unit	Rate £	Cost £
A1	Treated 50 × 225 mm sawn carcass	200	M		
A2	Treated 75 × 225 mm sawn carcass	50	M		
B1	50 mm galvanised steel joist hangers	20	N/A		
B2	75 mm galvanised steel joist hangers	7	N/A		
C1	Supply and fit the above floor joists as described in the preambles				

Figure 2.37 Sample extract from a bill of quantities

To ensure that all contractors interpret and understand the bill of quantities consistently, the Royal Institution of Chartered Surveyors and the Building Employers Confederation produce a document called the Standard Method of Measurement of Building Works (SMM). This provides a uniform basis for measuring building work, for example stating that carcassing timber is measured by the metre whereas plasterboard is measured in square metres.

Advantages and disadvantages of purchasing and hiring plant

Plant hire is an important aspect of a construction job to be taken into consideration – usually during the tender stage. The hiring of plant is not essential on all jobs, but most jobs, and especially large ones, will require some plant to be hired in one way or another.

The type of plant that can be hired ranges from portable power tools to mobile tower scaffolding. It could also include cranes and diggers.

However, not all plant is hired and some items are bought outright by the company rather than hired. Most companies will have all the relevant trade-related power tools. For example, carpenters will have bought their own cordless drills or circular saws and bricklayers will have bought their own cement mixers.

Remember

If you are hiring plant such as cranes, you will also need to employ a qualified operator to use it.

If a certain item of plant is required and is not already owned, is it better to buy or hire? The final decision you make will depend on a variety of factors, but will mainly come down to cost. If something large like a crane is required then obviously the cheapest option is to hire one rather than buy it. But for something small it may still often be better to hire. Similarly, if you only need to use an item for a small length of time for a specialist job, it would be much better to hire it rather than buy it.

There are exceptions to this. For example, if a carpenter had ten kitchens to fit then it would be better to buy a worktop jig for a router at £80 rather than hire one on ten different occasions at £10 per hire.

Planning the sequence of material and labour requirements

We have already looked at how specifications and schedules are used to plan construction projects and establish the requirements of the project.

The main planning relating to the sequence of material and labour requirements will be taken into account when the programme of work is devised. This usually takes the form of a bar or progress chart, which is covered in more detail on the next page.

To plan sequences of material and labour requirements, you will also need to be familiar with some of the common methods of working used to ensure the smooth operation of materials and labour. This includes:

- work programmes and critical path analysis
- stock rotation systems
- lead times
- pricing systems.

Work programmes

Bar charts

The bar or Gantt chart is the most popular work programme as it is simple to construct and easy to understand. Bar charts have tasks listed in a vertical column on the left and a horizontal timescale running along the top.

Did you know?

The Gantt chart is named after the first man to publish it. This was Henry Gantt, an American engineer, in 1910.

	Time in days									
Activity	1	2	3	4	5	6	7	8	9	10
Dig for foundation and service routes										
Lay foundations										
Run cabling, piping, etc. to meet existing services										
Build up to DPC										
Lay concrete floor										

Figure 2.38 Basic bar chart

Each task is given a proposed time, which is shaded in along the horizontal timescale. Timescales can overlap as one task often overlaps another.

	Time in days									
Activity	1	2	3	4	5	6	7	8	9	10
Dig for foundation and service routes	■	■								
Lay foundations			■	■						
Run cabling, piping, etc. to meet existing services				■	■					
Build up to DPC						■	■			
Lay concrete floor								■	■	■

Key: proposed time ■

Figure 2.39 Bar chart showing proposed time for a contract

> **Did you know?**
>
> Bar charts are used to identify possible areas that have gone wrong. You will need to have an idea about the sort of things that can go wrong on a project and plan contingencies to deal with any problems.

The bar chart can then be used to check progress. Often the actual time taken for a task is shaded in underneath the proposed time (in a different way or colour to avoid confusion). This shows how what has been done matches up to what should have been done.

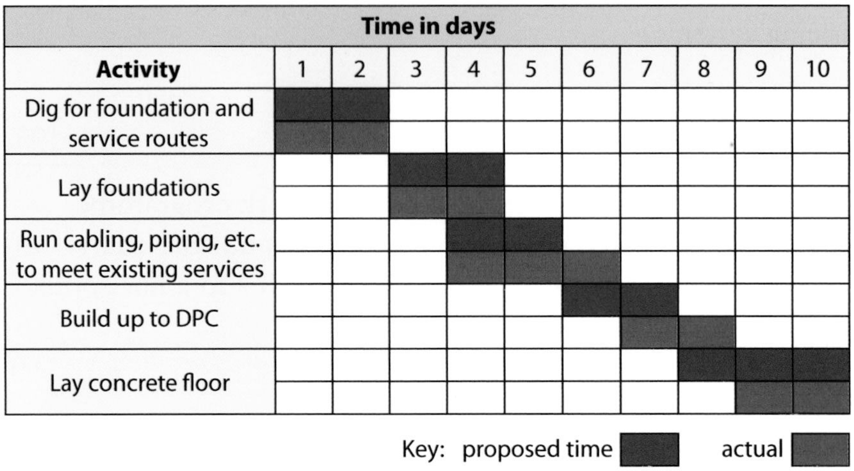

	Time in days									
Activity	1	2	3	4	5	6	7	8	9	10
Dig for foundation and service routes										
Lay foundations										
Run cabling, piping, etc. to meet existing services										
Build up to DPC										
Lay concrete floor										

Key: proposed time ■ actual ■

Figure 2.40 Bar chart showing actual time halfway through a contract

> **Did you know?**
>
> Bad weather is the main external factor responsible for delays on building sites in the UK. A Met Office survey showed that the average UK construction company experiences problems caused by the weather 26 times a year.

As you can see, a bar chart can help you plan when to order materials or plant, see what trade is due in and when, and so on. A bar chart can also tell you if you are behind on a job; if you have a penalty clause written into your contract, this information is vital.

When creating a bar chart, you should build in some extra time to allow for things such as bad weather, labour shortages, delivery problems or illness. It is also advisable to have contingency plans to help solve or avoid problems, such as:

- capacity to work overtime to catch up time
- a bonus scheme to increase productivity
- a penalty clause on suppliers to try to avoid late or poor deliveries
- a source of extra labour (e.g. from another site) if needed.

Good planning, with contingency plans in place, should allow a job to run smoothly and finish on time, leading to the contractor making a profit.

Critical paths

Another form of work programme is the critical path. Critical paths are rarely used these days as they can be difficult to decipher. The final part of this section of the unit will give a brief overview of the basics of a critical path, in case you should come across one.

A critical path can be used in the same way as a bar chart to show what needs to be done and in what sequence. It also shows a timescale, but in a different way to a bar chart: each timescale shows both the minimum and the maximum amount of time a task might take.

The critical path is shown as a series of circles called event nodes. Each node is split into three: the top third shows the event number, the bottom left shows the earliest start time, and the bottom right the latest start time.

The nodes are joined together by lines, which represent the tasks being carried out between those nodes. The length of each task is shown by the times written in the lower parts of the nodes. Some critical paths have information on each task written underneath the lines that join the nodes, making them easier to read.

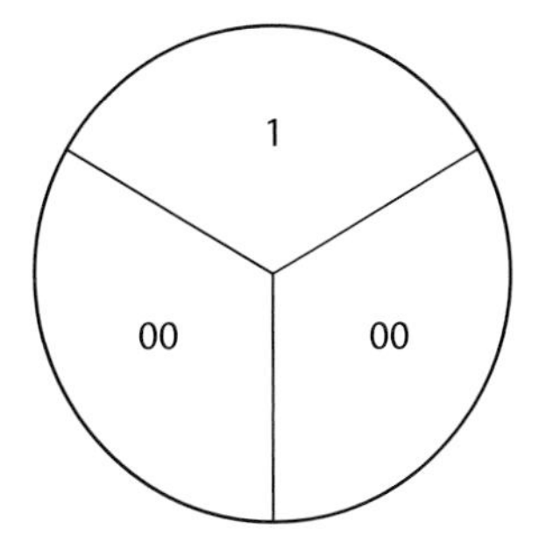

Figure 2.41 Single-event node

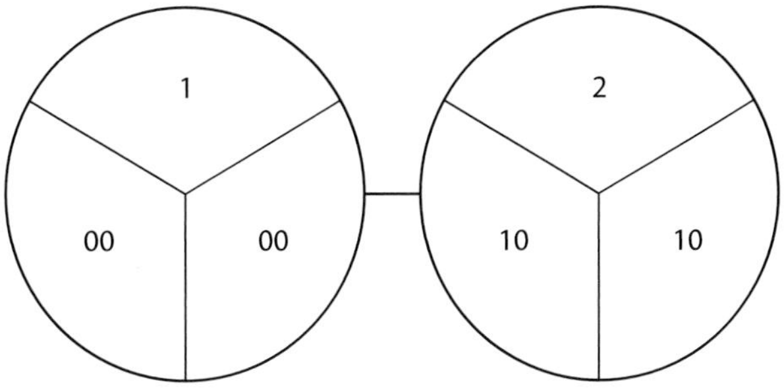

1
00
00
2
10
10

Figure 2.42 Nodes joined together

On a job, many tasks can be worked on at the same time, e.g. the electricians may be wiring at the same time as the plumber is putting in the pipes. To show this on a critical path, the path can be split.

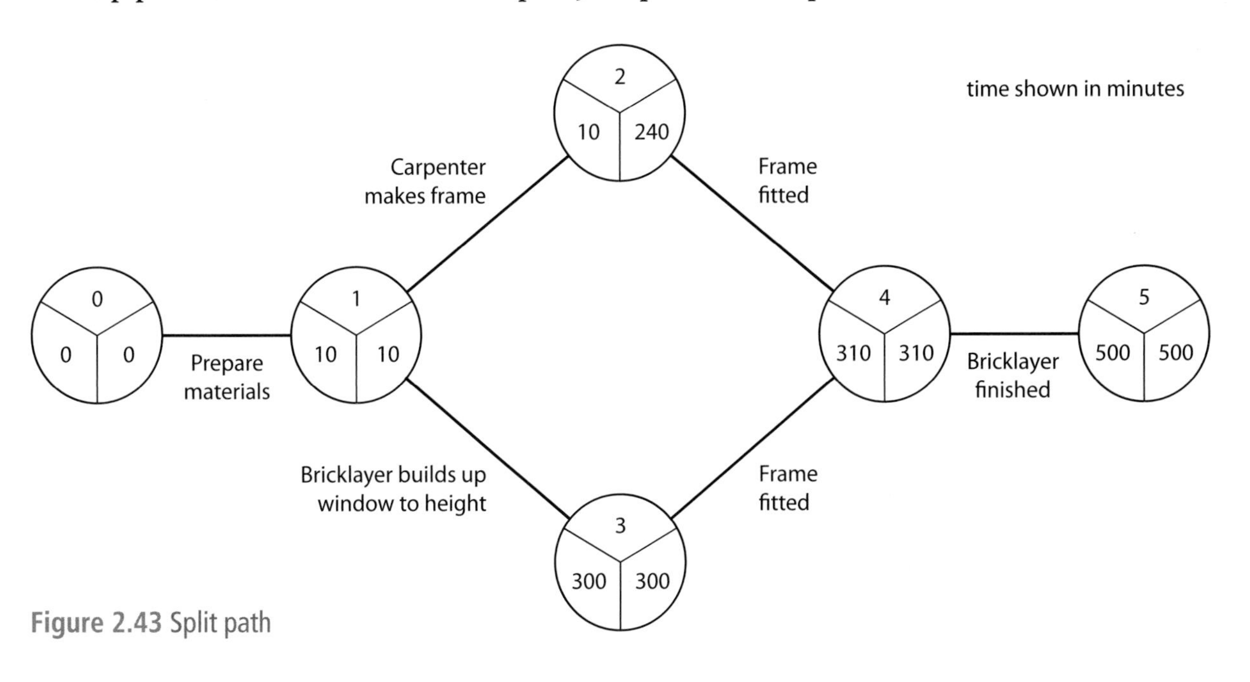

Figure 2.43 Split path

The example on the previous page shows how a critical path can be used for planning building in a window opening, with a carpenter creating a dummy frame.

The event nodes work as follows:

- **Node 0** – this is the starting point.
- **Node 1** – this is the first task, where the materials are prepared.
- **Node 2** – this is where the carpenter makes the dummy frame for the opening. Notice that the earliest start time is 10 minutes and the last start time is 240 minutes. This means that the carpenter can start building the frame at any time between 10 minutes and 240 minutes into the project. This is because the frame will not be needed until 300 minutes, but the job will only take 60 minutes. If the carpenter starts after 240 minutes, there is a possibility that the job may run behind.
- **Node 3** – this is where the bricklayer must be at the site, ready for the frame to be fitted at 300 minutes, or the job will run behind.
- **Node 4** – with the frame fitted, the bricklayer starts at 310 minutes and has until node 5 (500 minutes) to finish.
- **Node 5** – the job should be completed.

When working with a split path it is vital to remember that certain tasks have to be completed before others can begin. If this is not taken into account on the critical path, the job will run over (which may prove costly, both through penalty clauses and also in terms of the contractor's reputation).

On a large job, it can be easy to misread a critical path as there may be several splits, which could lead to confusion (see Figure 2.44).

Remember

Whichever way you choose to programme your work, your programme must be realistic, with clear objectives and achievable goals.

Functional skills

This exercise will allow you to practise **FM 2.3.1** Interpret and communicate solutions to multistage practical problems and **FM 2.3.2** Draw conclusions and provide mathematical justifications.

Working life

You have been tasked with designing a programme of work for a large contract involving the building of 20 houses. You have been told you need to plan all the work that needs to be carried out, as well as ensuring that all the materials required for the work are purchased and delivered to site on time and to schedule, so that they will be ready for work to continue without delay.

- What sort of thing should you check prior to starting?
- What should you do about plant, labour and materials?
- What sort of programme should you use (bar chart or critical path)?
- What amenities should you consider?

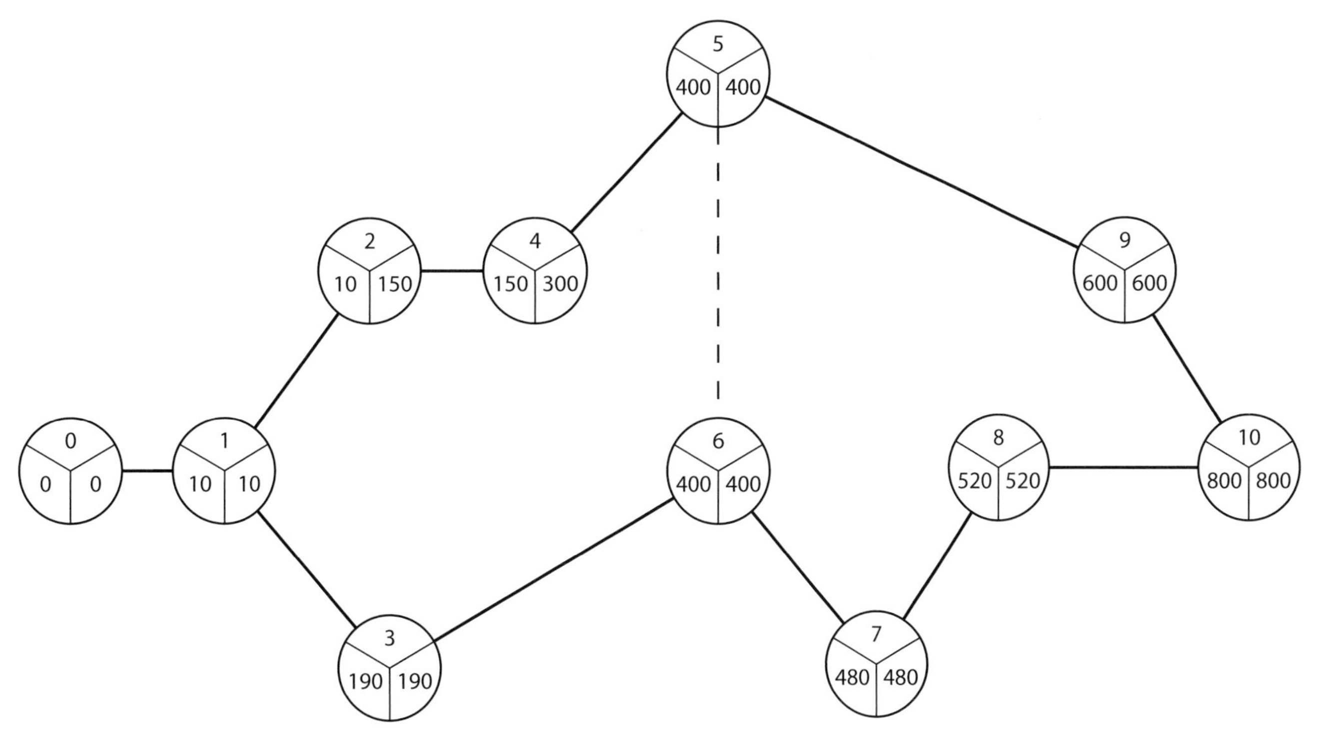

Figure 2.44 Critical path for a large job

Stock systems

Stock systems are used with mainly larger companies, suppliers and larger sites. A good stock system will ensure that all the materials required on site are available when needed and that no materials are damaged, either through overstocking which can lead to storage problems or by materials such as plaster going off.

When taking delivery of materials, the newest materials are placed at the back in storage. This will ensure that the older materials are used first, while they are still in date.

Remember

Materials such as plaster or cement have a use-by date on them; generally, such materials will set or go off at about this date. To prevent this, it is vital that the materials are used before their use-by date.

Lead times

Lead times are how long you should expect to wait for a new delivery of materials. Certain materials are not always readily available in a supplier's stock and even suppliers may have to order certain things. For example, oversized timbers would need to be ordered in, or steel beams which required special machining or manufacturing would need to be created to order. Where items like these are required, time is built into the planning stage to contact the suppliers and ask for the lead times for these items, to ensure that they are available when they are required.

Example

If hanging a door is priced at £40 per door, then a tradesperson may only do five doors in a day, as they feel then that they have earned enough for the day. If the price is £30 or even £20 per door, then the worker will try to do more, so that they have earned a good daily wage.

Did you know?

When estimating the labour costs for a job, it is easier to use the day work rate method as you can calculate how many hours the job will take.

Pricing systems

Pricing systems are also thought through at the tender and the planning stage to ensure that the price paid for labour during price work is not too much. Price work, which is covered in more detail below, is set to ensure that all labour tasks are given a price and that the price is set in a way which will get the job done on time.

Pricing tasks too high can see the job taking longer. The reverse of this is pricing jobs too low, which may see the company struggling to get anyone to do the work for a low price.

Calculating hours

The way that labour is paid for can be split into two different methods.

Day work/hourly rate

This rate is used when the tradesperson is paid a specific amount for every hour that they work. The amount will depend on where the work is being carried out, as the cost of living is different in each area.

Places where the cost of living is low may receive £10–£20 per hour. In areas with a high cost of living (such as London), the rate may be £20–£30 per hour or more. The experience of a worker will also affect the day work rate. Newly qualified Level 2 apprentices will not be paid the same as someone with 30 years of experience.

Price work

This rate is used when the tradesperson is paid for the work they carry out. Examples of this include a carpenter who receives £20 for every door they hang, or a painter who gets £300 for every flat they decorate. This method is often preferred, although it can mean that you may have to work harder. However, the more work you do the more you will earn. Again, prices will vary not only from area to area but even within trades.

Example

A carpenter may get paid £2000 to fit a truss roof but only £250 to fit a small kitchen in a flat. A painter may get paid £15 to paint the inside of a window compared with £20 to paint the outside. These differences in price are worked out prior to the job starting and take into account factors such as weather or hazards. The roof may look like the best job at £2000, but if it is raining heavily for a week, or alterations to the scaffold are required, then you may not get much work done. However, you may be able to fit seven kitchens in a week no matter what the weather.

The price work method is calculated by working out how many hours it will take to complete the task and then giving a certain price to that task, based on the day work rate.

For example, the day work rate may be £20 per hour and a roof should take 100 hours. This means a price of £2000 will be put forward.

Remember

The £2000 price is for the whole roof. If four people work on it they will get paid £500 each, not the full £2000.

Range of added costs for estimating which affect profitability

When estimating prices, there is a range of added costs that need to be considered. You will need to factor in these costs before you begin pricing a job.

These added costs can also affect the final profitability of a project.

Remember

Factors that can affect profitability include delays caused by weather, worker strike action and other external elements over which you may have no control.

Types of insurance

All companies that carry out work must be insured through public liability and if they employ others then they will also need employers' liability insurance.

- **Public liability insurance** – this will cover you if someone is accidentally injured by you or your business operation. It will also cover you if you damage third party property while on any work-related business. The cover should include any legal fees and expenses which can result from any claim by a third party. You should aim to have a level of insurance which covers you for at least £1,000,000. This may seem a lot but you could have several cases directed at you at any one time. **Premiums** for public liability can start from around £100 per year. Failure to have public liability insurance can see your company go bankrupt if a claim is made and you are not covered.
- **Employers' liability insurance** – this is required by law if you employ other people. If an employee should be injured at work, or become ill as a result of the work that you ask them to carry out, then employer's liability insurance gives you a minimum level of cover if you are sued. Cover for this should start at £5,000,000. Again, this may seem a lot but the premium can be as little as £100. Failure to have employers' liability insurance can lead to a fine of up to £2,500 for any day you operate without this insurance. Insurance policy certificates must be retained for 40 years, as illness that can occur may appear at a later stage and you need evidence of cover.

Key term

Premium – the amount you pay in order to be covered by an insurance company. The premium will be based on a quote given to you by the company and will be paid in one lump sum or, as is more common, in instalments over a year.

Did you know?

Combined policies for both of these types of insurance are usually the best way to operate and premiums start from around £180.

You will also need to make National Insurance contributions. These are paid to build up your entitlement to certain social security benefits, including state pensions. The amount of money you pay will depend upon your employment status (employed or self-employed). Employers will also be expected to make a contribution to each employee's National Insurance, with the amount depending on how much they earn.

Stage payments

Stage payments are used in contracts of any size but more so in large contracts, and they usually mean that a percentage of the total price for the job will be paid upon the construction reaching a certain stage. This can be beneficial to both parties, as some small contractors will like a small percentage paid up-front so that they can organise the delivery of materials or pay for other costs that have arisen during the job. If the job is half done, then 50 per cent of the payment can be made. Usually, with a stage payment, a small percentage is held back at the end of the job for a short period of time to allow for any faults or blemishes to be fixed.

VAT

VAT, or Value Added Tax, is a form of tax that is charged on most goods and services that VAT-registered businesses provide in the UK. The current rate of VAT in the UK is 20 per cent (since January 2011), but it can fluctuate. This happened after the financial crisis in 2009 when it was lowered to 15 per cent. The VAT amount means that services or goods have an extra percentage added to them as a tax. For example, a power tool may cost £100, but with the current rate of VAT added you will have to pay £120 for it.

Functional skills

This task will allow you to practise **FE 2.3.1–2.3.5** Write documents, including extended writing pieces, communicating information, ideas and opinions effectively and persuasively.

Find out

Using the Internet and other sources, find out the current rate of tax and show the impact it has on earnings.

PAYE

PAYE, or Pay As You Earn, is a method of paying income tax and National Insurance contributions. Your employer will deduct these amounts from your wages before you are paid. For example, you may earn £300 per week but, after deductions, this could fall to just over £200. With regard to PAYE, every employed person is given a tax code which relates to the amount of tax they pay. A single-person tax code could be 117L, which means that you could earn up to £1,170 per year before you pay tax, but any money you earn over this amount will be taxed. Employers also pay tax on each employee.

Self-employed workers do not follow the PAYE system, but are paid the full amount (called 'gross'). However, they are expected to keep a note of their earnings and expenses and once a year file a tax return in which they pay all the year's taxes in one lump sum (usually in January, six months in arrears and for the six months ahead, with another sum payable in July).

Travel expenses

Travel expenses are incurred when travelling to and from a job, and although they usually consist of fuel expenses it is also important to consider other things such as bridge tolls or congestion charges. It may not seem like much at the time but a job that lasts 20 or 30 days and which includes a £10 charge for these can easily see the profits start to fall, so it is worth considering this when pricing up a job.

Profit and loss

It is important that profit and loss balance sheets are kept by a company, as these will show how the company is performing over a period of time. The simplest form of profit and loss will show how much money a company has taken and deduct from it the amount spent. For example, a job may be priced at £10,000 and has taken four weeks to complete – during which £8,000 has been spent on materials, wages, taxes etc.; this means that the profit for this period is £2,000.

Suppliers' terms and conditions

All suppliers will have some form of terms and conditions which will outline what restrictions are in place for the use of their goods and services. These are used as a form of insurance by the company and will include things such as payments, the customer's responsibilities and the company's liability.

Wastage

Wastage can have a massive effect on profitability as the more waste there is, with regard to materials and other ancillaries, the more money is lost out of your profits. A simple way to keep a tab on wastage is to monitor what is ordered against what is needed. For example, if 2.5 m lengths of timber are required then you should only order the next size up, which would be 2.7 m, as ordering 3.0 m lengths would create waste and expense.

Building up a price

As has been shown previously, building up a price is not simply about calculating what materials, equipment and labour are required but also involves other factors such as insurances, taxes, expenses and so on. If you do not take these things into account, you can quickly see your profits evaporate.

K3. Know how to ensure good working relationships

Good working relationships are absolutely vital when working on site. It is important to have good relationships not only with those who you are working with directly, but also with other trades and professionals you come into contact with. In addition, you need to have a good relationship with the client, who is the overall 'boss' of the entire project.

There are a number of methods that can be used to achieve good working relationships on site. These include the following:

- **Good planning** – it has been mentioned time and again that good planning is vital on a site, but never more so than when ensuring that good working relationships are maintained. Planning that the correct trades are in when needed will avoid problems between them, as having the wrong trades in or having them arrive in the wrong order can lead to work having to be re-done, which will cause conflict.
- **Regular site meetings** – these can be vital in preventing conflict, as each trade should be represented and they can give updates on progress or possible conflicts.
- **Ensuring that competent tradespeople are employed** – working with, or after, workers who have not done a good job will lead to conflicts, as the work they do may have to be re-done. This will cause problems for the people who have to put the work right and can lead to delays, which will create conflict with the trades who are waiting to get on with their work.
- **Leaving your work area clean and clear** – this may sound simple but leaving a mess for the following tradespeople will upset them, as they will either have to clean up themselves or wait until it is clean. This will delay their work and, if on price work, will cost them money.
- **Working safely** – again, a simple point but this is crucial, as not working safely will lead to hazards and possible injuries, which can cause conflicts.

Maintaining trust and confidence in colleagues

One of the main components of a continued good working relationship is trust. Just as in everyday life, having trust and confidence in a colleague or friend can help with the relationship, in the same way that having no trust or confidence can break it. By doing simple things such as arriving when you say, doing what you say and by being open and honest, you can start to bring trust into the relationship. Being professional, helpful and working to the best of your ability will instill confidence in your colleagues. This is important, as having no confidence in work colleagues can create problems and conflict.

Explain the need for accurate communication throughout the stages of construction

Accurate communication is vital for efficient relations between everyone who may be involved in a business, from the employer and employees through to clients and suppliers.

Most of the crucial moments when you will need to use good, clear and effective communication relate to decisions that will have a wider effect on the business and those working around you.

Some examples of these include the following:

- **Alterations to drawings** – it is important to communicate any changes to these to everyone involved, as all the planning, estimating, material orders and work programmes will be based in part on these drawings. Not communicating changes could lead to mistakes in all these areas.
- **Variations to contracts** – the contract with the client is the crucial document that dictates all decisions that are made on a worksite. Changes to this document must be made known throughout a business.
- **Risk assessments** – the results of these assessments have a direct impact on the safety of workers on site, and should be made known to everyone.
- **Work restrictions** – these should be communicated to everyone, as a restriction is put in place for a specific reason. A restriction may be put in place for safety reasons. This would mean the area is unsafe, so everyone who may be affected needs to be told.

FAQ

How do I know what scale a drawing is at?

The scale should be written on the title panel (the box included on a plan or drawing giving basic information such as who drew it, how to contact them, the date and the scale).

How do I know if I need a schedule?

Schedules are only really used in large jobs where there is a lot of repeated design information. If your job has a lot of doors, windows, etc. it is a good idea to use one.

Which type of programme should I use: bar chart or critical path?

It is up to the individual which programme they use – both have their good points – but a bar chart is the easiest to set up and work from.

What if it rains for the entire 20-day duration of a job?

The job would be seriously behind schedule. You can't plan for the weather in the UK, but it would be unwise to start some jobs during a rainy season. There are companies that can provide scaffolding with a fitted canopy to protect the work area, which would be ideal for a job of this size. Larger jobs have longer programmes, and when they are drawn up they are made more flexible to allow for a lot of rainy days.

Check it out

1. Describe the main advantages of using a CAD system and use it to create a 3-D wireframe and dumb solids program.
2. Produce a component drawing for an item that you are familiar with.
3. Produce a detail drawing of a component you are familiar with.
4. Using a suitable scale, create a setting-out drawing for a segmental brick arch with an opening span of 1.8 m, so that the rise and radius can be identified.
5. Using a suitable scale, create a setting-out drawing for a rafter with a span of 3.5 m and a rise of 1.4 m, so that the true length of the rafter and angles of cuts can be identified.
6. Explain in detail the process followed in the client–architect consultation on projects.
7. Describe the purpose of a bill of quantities and then put together your own example using a job you are familiar with.
8. Draw a flowchart explaining the different steps involved in the tender process, both for the client and for those companies who are submitting tenders.
9. Put together your own programme for the work on one of the jobs you have carried out. Put together two versions of this programme: a bar chart and a critical path using event nodes.
10. Take a task that you are familiar with as part of your work. Think about all the possible costs and implications that are connected with this job. Try to work out a price for it, including all the ancillaries discussed in this unit, such as taxes.
11. Draw up a method statement that describes the best working practices for communicating with other trades on site. Make a clear note of the important information that you will need to make sure is conveyed from one person to another while working.

Getting ready for assessment

The information contained in this unit, as well as continued practical assignments that you will carry out in your college or training centre, will help you with preparing for both your end of unit test and the diploma multiple-choice test. It will also aid you in preparing for the work that is required for the synoptic practical assignments.

Working with contract documents such as drawings, specifications and schedules is something that you will be required to do within your apprenticeship and even more so after you have qualified. Similarly, when working professionally you will need to be able to build up a price accurately and correctly.

You will need to be familiar with:

- producing drawn information
- determining quantities of material and price estimates
- how to ensure good working relationships.

A particular focus of this unit has been the estimating of quantities and materials needed to build up a price. Learning outcome two has shown you how to analyse the resources required for a construction task, as well as the advantages and disadvantages that exist in hiring or purchasing equipment. You will need to be able to assess and evaluate the material needs of a project and then complete a plan for the sequence of labour and materials. You will also need to be able to identify the added costs that can contribute to the final price you build for a project.

To get all the information you require, you will need to build on the maths and arithmetic skills that you learnt at school. These skills will give you the understanding and knowledge you will need to complete many of the practical assignments, which will require you to carry out calculations and measurements, including estimations of price for tasks to be completed.

You will also need to use your English and reading skills. These skills will be particularly important, as you will need to make sure that you are following all the details of any instructions you receive. This will be the same for the instructions you receive for the synoptic test, as it will for any specifications you might use in your professional life.

Communication skills have also been a focus of this unit, and of learning outcome three. This unit has shown the importance not only of communicating clearly and consistently with everyone on site, but also of building up relationships that are built on trust and confidence.

Teamwork is a very important part of all construction work. It can help work to run smoothly and ensure people's safety. Relationships are a vital part of all teamwork and you will need to be able to demonstrate how the key personnel should communicate effectively within this team. The communication skills that are explained within the unit are also vital in all tasks that you will undertake throughout your training and in life.

Good luck!

CHECK YOUR KNOWLEDGE

1. What scale are block plans usually drawn at?
   - a 1:1250
   - b 1:500
   - c 1:200
   - d 1:10
2. What stage is CAD normally used in a project?
   - a Final planning
   - b Drafting
   - c Pricing
   - d On-site plans
3. What is the British Standard code number for structural timber?
   - a BS EN 12390-3:2009
   - b BS 5268-2:2002
   - c BS EN 1097-8:2009
   - d BS EN 12350-1:2009
4. What information will a specification contain?
   - a Site description
   - b Services
   - c Workmanship
   - d All of the above
5. What does a line drawn 45 mm long at a scale of 1:20 represent?
   - a 90 mm
   - b 900 mm
   - c 9000 mm
   - d 45 mm
6. A line drawn 25 mm long at a scale of 1:250 represents what?
   - a 6000 mm
   - b 6250 mm
   - c 6500 mm
   - d 6750 mm
7. When is employers' liability insurance required?
   - a If you are employed
   - b If you employ others
   - c If you do any work
   - d If you have public liability insurance
8. What must an employee pay?
   - a VAT
   - b PAYE
   - c Public liability insurance
   - d All of the above
9. When might a stage payment be used?
   - a When a small contractor may need some payment up front
   - b When a job is to be completed in stages
   - c When a contractor does not have all the money ready
   - d For all large contracts
10. Which of these is the most important for building good working relationships?
    - a Good planning
    - b Working safely
    - c Regular communication
    - d All of the above

Functional skills

In answering the Check it out and Check your knowledge questions, you will be practising **FE 2.2.1** Select and use different types of texts to obtain relevant information, **FE 2.2.2** Read and summarise succinctly information/ideas from different sources and **FE 2.2.3** Identify the purposes of texts and comment on how effectively meaning is conveyed.

You will also cover **FM 2.3.1** Interpret and communicate solutions to multistage practical problems and **FM 2.3.2** Draw conclusions and provide mathematical justifications.

Unit 3003

Knowledge of building methods and construction technology 3

Many of us are aware of the growing concerns around global warming and the current focus on minimising our 'carbon footprint'. This unit deals with some of the factors that may be contributing to climate change, and the way in which modern construction methods and design can help to tackle it by creating a more sustainable environment – one that provides for today's needs without compromising the needs of future generations.

As such it is important that, when looking at building methods and construction technology, the use of sustainable building methods and materials is considered. You need to be familiar with how to use these in connection with the basic elements of a building. Remember, all buildings have common elements that must be included.

This unit contains material that supports the following NVQ units:

- VR 209 Confirm work activities and resources for the work
- VR 211 Confirm the occupational method for work

This unit will cover the following learning outcomes:

- Know about new technology and methods used in construction
- Know about energy efficiency in new construction buildings
- Know about sustainable methods and materials in construction work

K1. Know about new technology and methods used in construction

One of the most controversial issues affecting new technologies and methods used in construction is climate change. Climate change has been at the centre of a number of disagreements between experts.

However, it is hard to ignore the changes in weather patterns that we have witnessed in recent years – not just in the UK but also around the world. In the UK, these changes have resulted in extensive flooding, with winter months becoming warmer and summer months becoming wetter. Around the world, the ice caps are melting at worrying speeds, sea levels are rising and adverse weather systems have resulted in more frequent tornados and tropical storms.

Figure 3.01 Area of land under flood

The changing climate will certainly impact on the way we design buildings, in order to cope with some of the possible effects, which include:

- rising temperatures
- rising sea levels
- rising levels of rainfall
- higher humidity levels.

Although not fully proven, a number of factors are thought to contribute to global warming. The main causes are considered to be burning fossil fuels and the high levels of carbon dioxide (CO_2) being emitted into the air. CO_2 is emitted in many different ways, including through car emissions, aerosol gases and burning untreated waste products.

Many practices have been introduced in an attempt to reduce the **carbon footprint**, and many more initiatives are being planned. These practices and initiatives include:

- recycling packaging, such as plastics, paper, cardboard, metals and glass
- the use of alternative fuels to power cars and industrial machinery
- the use of natural resources such as wind to supply electricity
- transport sharing to reduce the number of cars on the road, thus reducing CO_2 emissions.

Energy efficiency and the reduction of waste are now goals for every building project, and must be considered carefully from the design stage right through the building process.

New legislation has been introduced to make sure that the industry works towards the design of energy-efficient buildings, but at the same time the construction industry is becoming more conscious of the need to address climate change itself, by introducing its own initiatives and practices to minimise its carbon footprint:

- The construction industry is putting greater emphasis on reducing building waste, and is being proactive in recycling many materials that would once have been disposed of to the detriment of the environment.
- There have been significant changes in the design of both domestic and industrial buildings.
- More natural resources are being used in the construction of buildings which, when the building is demolished, can be re-used on a new development.
- Buildings now have improved insulation properties, thus reducing fuel usage for heating.
- There are alternative methods of providing energy for a building, including the use of solar panels to provide natural heat.

There are many, many ways of making buildings eco-friendly and we will look at some of these later in this unit.

Key term

Carbon footprint – total amount of CO_2 emissions produced by individuals and industry.

Find out

Identify other initiatives and practices currently being used or developed in an attempt to reduce people's carbon footprint.

Different structures used in commercial and domestic dwellings

There are several different construction methods that can be used both commercially and domestically. The table on the next page shows the types of construction that are used in these different structures for the key elements of the building.

We will cover the different structures in greater detail later in this unit.

Table 3.01 Comparison between domestic and commercial construction techniques

Element	Domestic	Commercial
Foundations	• Traditional strip footings and stepped foundations • Raft foundations on poor load-bearing ground • Mass-fill trench foundations	• Piled foundations with pile caps • Ground beams • Raft foundations with edge beam • Pad foundations • Reinforced wide strip
Structural frame	• Traditional cavity walls • Timber-framed construction • Load-bearing insulated formwork	• Steel portal frame • Steel columns and beams • Composite construction • Concrete frameworks
Cladding	• Facing brickwork • Rendering • Timber cladding • Clay tiles	• Insulated composite cladding panels • Steel powder-coated panels • Brickwork facing skin dwarf walls • Timber cladding
Flooring	• Solid concrete floors • Beam and block • Traditional timber flooring joists	• Solid reinforced concrete floors, power-floated with edge beams
Roofing	• Roof trusses with interlocking roof tiles • Traditional roof timbering with clay/concrete roof tiles	• Insulated cladding panels

Find out

What is a steel portal frame used for in commercial building insulation?

Good environmental design

The design of a building or structure should now take on elements that make it environmentally friendly, both in its construction and in the way the occupants use it. The design should also take into account the needs of the local community, in terms of the infrastructure that will be needed to support its eventual construction and use.

Environmental considerations may include some or all of the following elements:

- **The design brief** – this should aim to create a design that lowers pollutants to the atmosphere, reduces waste in its construction, reduces noise pollution, and gives the local community something that it can enjoy.
- **Recycling materials** – the design should specify the inclusion of recycled materials into the structure and aesthetics of the building project: for example, the use of crushed hardcore from demolition on site, or the re-use of slate roofing materials.

- **Energy efficiency** – the amount of **embedded energy** and used energy must be carefully considered for each element of the design, from boiler management systems to highly engineered aerated concrete thermal blocks.
- **Sustainability** – the design must contain elements that will meet the needs of future generations. A long, maintenance-free lifespan for the building is essential. Spending more at the initial cost stage can pay dividends in the future by reducing the cost of energy use and maintenance later on.
- **Green materials** – the use of green materials is a vital environmentally friendly way of ensuring minimum impact on the local and global environment. For example, cedar timber boarding is a sustainable timber product that does not require chemical treatment or painting.

Key term

Embedded energy – the amount of energy that has been used to create and manufacture the material and transport it to site for inclusion in the structure.

Did you know?

In a good environmental design, you can include a rainwater harvesting system that collects the rainwater from guttering and uses it to flush toilets and to wash down. This is known as 'grey water'.

K2. Know about energy efficiency in new construction buildings

Whatever type of building you may be involved in constructing, there are certain elements that must be included and certain principles that must be followed. For example, a block of flats and a warehouse will both have foundations, a roof, and so on.

At Level 2, you learned about the basic elements of a building. For this learning outcome, you will look in greater depth at the main elements and principles of building work and the materials used.

Accurate setting out of foundations

Any building work will start with the foundations. The design of any foundation will depend on a number of factors, including ground conditions, soil type, the location of drains and trees in relation to the building, and any loads that may be generated, either by the structure or naturally.

The purpose of foundations

The foundations of a building ensure that all **dead** and **imposed loads** are absorbed safely and transmitted through to the natural foundation or subsoils on which the building is constructed. Failure to adequately absorb and transmit these loads will result in the stability of the building being compromised, and will undoubtedly cause structural damage.

Key terms

Dead load – the weight of the structure.

Imposed load – the additional weight/loading that may be placed on the structure itself.

Foundations must also be able to allow for ground movement, brought about by the soil shrinking as it dries out and expanding

Figure 3.02 A building damaged through subsidence

as it becomes wet. The severity of shrinkage or expansion depends on the type of soil you are building on.

Frost may also affect ground movement, particularly in soils that hold water for long periods. When this retained water freezes, it can make the subsoil expand.

Tree roots and future excavations can also cause movement that affects the subsoil.

Types of soil

As you can imagine there are many different types of soils. For foundation design purposes, these have been categorised as:

- rock
- gravel
- clay
- sand
- silt.

Each of these categories of subsoil can be broken down even further, for example:

- clay which is sandy and very soft in its composition
- clay which is sandy but very stiff in its composition.

This information will be of most interest to the architect, but nonetheless is of the utmost importance when designing the foundation.

Key term

Load-bearing capacity of subsoil – the load that can be safely carried by the soil without any adverse settlement.

A number of calculations are used to determine the size and make-up of the foundation. These calculations take into account the **load-bearing capacity of the subsoil**. Calculations for some of the more common types of foundations can be found in the current Building Regulations. However, these published calculations cannot possibly cover all situations. Ultimately it will be down to the expertise of the building design teams to accurately calculate the bearing capacity of soils and the make-up of the foundation.

In the early stages of the design process, before any construction work begins, a site investigation will be carried out to ascertain any conditions, situations or surrounding sites which may affect the proposed construction work. A great deal of data will need to be established during site investigations, including:

- the position of boundary fences and hedges
- the location and depth of services, e.g. gas, electricity, water, telephone cables, drains and sewers

- existing buildings which need to be demolished or protected
- the position, height, girth and spread of trees
- the types of soil and the depths of these various soils.

The local authorities will normally provide information relating to the location of services, existing buildings, planning restrictions, preservation orders and boundary demarcation. However, all of these will still need to be identified and confirmed through the site investigation. In particular, hidden services will need to be located with the use of modern electronic surveying equipment.

Find out

Look at the different methods and equipment used to locate and identify various hidden services.

Soil investigations are critical. Samples of the soil are taken from various points around the site and tested for their composition and for any contamination. Some soils contain chemicals that can seriously damage the foundation concrete. These chemicals include sodium and magnesium sulphates. The effects of these chemicals on the concrete can be counteracted with the use of sulphate-resistant cements.

Many different tests can be carried out on soil. Some are carried out on site; others need to be carried out in laboratories. Tests on soil include:

Find out

How are different soil tests carried out?

- **penetration tests** – to establish the density of the soil
- **compression tests** – to establish shear strength of the soil or its load-bearing capacity
- **various laboratory tests** – to establish particle size, moisture content, humus content and chemical content of the soil.

Once all site investigations have been completed and all necessary information and data has been established in relation to the proposed building project, site clearance can take place.

Working life

You have been tasked with building a garage. You decide not to go with a soil survey as this is an extra expense, and you just put in a standard strip foundation.

What could go wrong? You will need to think about the implications of not finding out as much as you can about the type of soil you are building on.

What could the cost implications be? You may have saved money avoiding the soil survey, but what impact could this have on your future spending? What effect could this have on your business?

What other implications could there be? Consider how not only the finances but also the stability and long-term life of the building could be affected.

Did you know?

Site investigations or surveys also establish the contours of the site. This will identify where certain areas of the site may need to be reduced or increased in height. An area of the site may need to be built up in order to mask surrounding features outside the boundaries of the proposed building project.

Site clearance

The main purpose of site clearance is to remove existing buildings, waste, vegetation and, most importantly, the surface layer of soil referred to as topsoil. It is necessary to remove this layer of soil, as it is unsuitable to build on. This surface layer of soil is difficult to compact down due to the high content of vegetable matter, which makes the composition of the soil soft and loose. The topsoil also contains various chemicals that encourage plant growth, which may adversely affect some structures over time.

Find out

How can plant growth affect some structures?

The process of removing the topsoil can be very costly, in terms of both labour and transportation. The site investigations will determine the volume of topsoil that needs to be removed.

In some instances, the excavated topsoil may not be transported off site. Where building projects include garden plots, the topsoil may just need to be stored on site, thus reducing excessive labour and transportation costs. However, where this is the case, the topsoil must be stored well away from areas where buildings are to be erected or materials are to be stored, to prevent contamination of soils or materials.

Once the site clearance is complete, excavations for the foundations can start.

Figure 3.03 Removing soil from a site

Trench excavation

In most modern-day construction projects, trenches are excavated by mechanical means. Although this is an expensive method, it reduces labour time and the risks associated with manual excavation work. Even with the use of machines to carry out excavations, an element of manual labour will still be needed to clean up the excavation work: loose soil from both the base and sides of the trench will have to be removed, and the sides of the trench will have to be finished vertically.

Manual labour is still required for excavating trenches on some projects where machine access is limited and where only small strip foundations of minimum depths are required.

Trenches to be excavated are identified by lines attached to and stretched between profiles. This is the most accurate method of ensuring trenches are dug to the exact widths.

Excavation work must be carefully planned, as workers are killed or seriously injured every year while working in and around trenches. Thorough risk assessments need to be carried out and method statements produced before any excavation work is started.

Potential hazards are numerous and include possible collapse of the sides of the trench, hitting hidden services, plant machinery falling into the trench and people falling into the trench.

One main cause of trench collapse is the poor placement of materials near to sides of the trench. Not only can materials cause trench collapse, but they may also fall into the trench onto workers. Materials should not be stored near to trenches. Where there is a need to place materials close to the trench for use in the trench itself, always ensure these are kept to a minimum, stacked correctly and used quickly and, most importantly, ensure the trench sides are supported.

Did you know?

The Health and Safety Executive (HSE) has produced detailed documents that deal exclusively with safety in excavations. These documents can be downloaded from the HSE website or obtained directly from the HSE on request.

Regulations relating to safety in excavations are set out in the Construction Regulations and these must be strictly adhered to during the work.

Trench support

The type and extent of support required in an excavated trench will depend predominantly on the depth of the trench and the stability of the subsoil.

Traditionally, trench support was provided just by using varying lengths and sizes of timber, which can easily be cut to required lengths. However, timber can become unreliable under certain loadings, pressures and weather conditions and can fail in its purpose.

More modern types of materials have been introduced, as they are less costly and time-consuming methods of providing the required support. These materials include steel sheeting, rails and props. Trench support can be provided with a mixture of both timber and steel components.

On the next page you can see the methods of providing support in trenches using timber (see Figure 3.05) and a combination of timber and steel (see Figure 3.06).

The amount of timber or other materials required to provide adequate temporary support will be determined by the characteristics of the soil and the soil's ability to remain stable during the time over which the work is carried out.

Figure 3.04 Trench being excavated prior to support

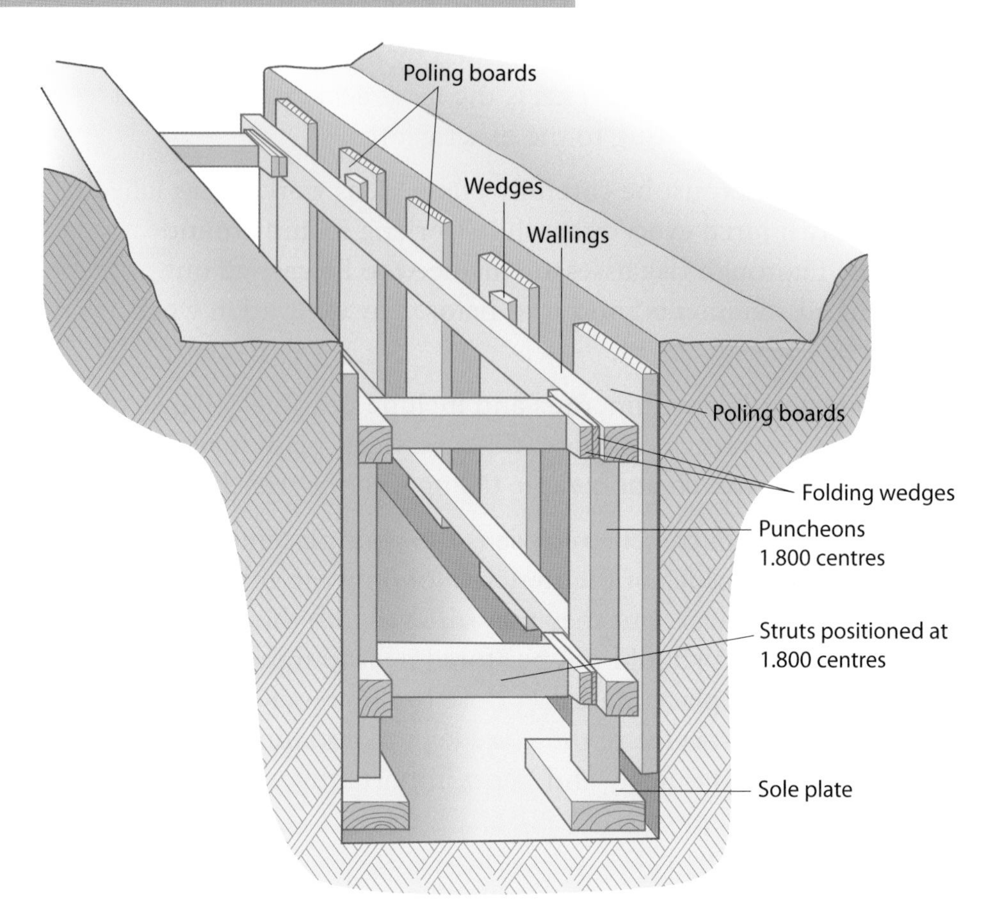

Figure 3.05 Timber used in trench support

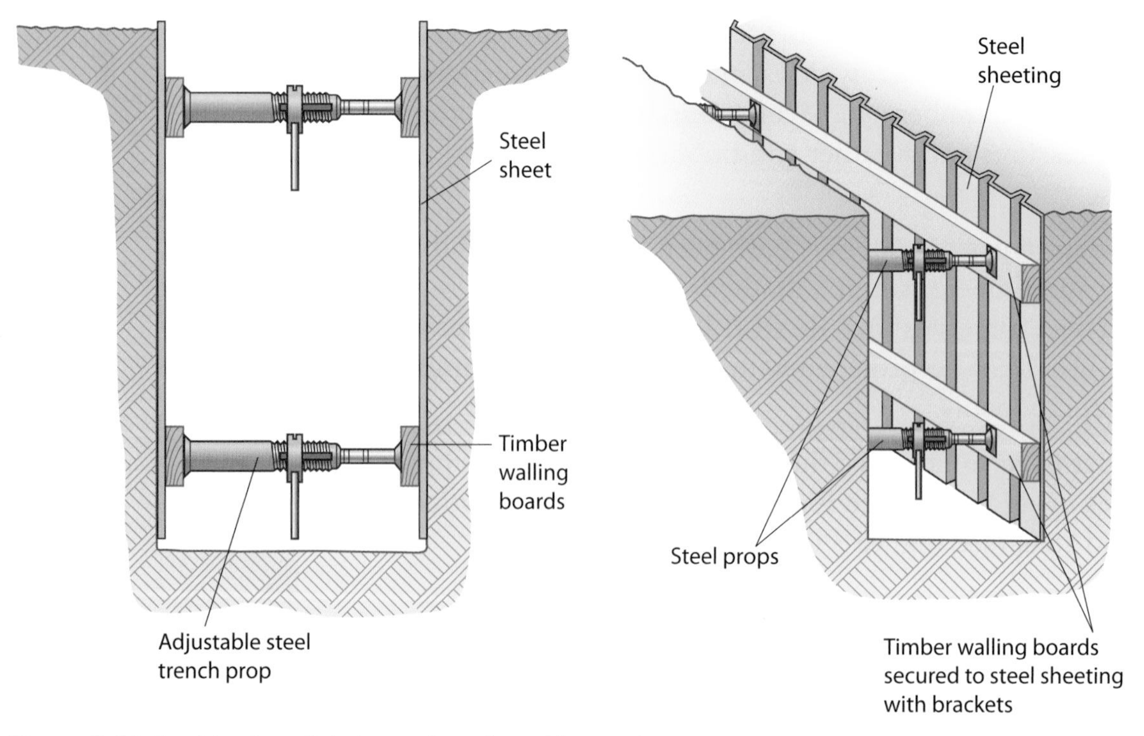

Figure 3.06 Combination of timber and steel used in trench support

The atmospheric conditions will also affect the soil's ability to remain stable. The longer the soil is exposed to the natural elements, the greater chance of the soil shrinking or expanding.

Without support, soil will have a natural angle of repose (in other words, the angle at which the soil will rest without collapsing or moving). Again, this will be affected by the natural elements to which the soil is exposed. It is virtually impossible to accurately establish the exact angle at which a type of soil will settle, so it is always advisable to provide more support than is actually required.

Site engineers will carry out calculations in relation to the support requirements for trenches.

Temporary barriers or fences should also be provided around the perimeters of all trenches, to prevent people falling into the trenches and also to prevent materials from being knocked into them. Good trench support methods will incorporate extended trench side supports, which provide a barrier – similar to a toe board on a scaffold – to prevent materials being kicked or knocked into the trench. Where barriers or fences are impractical, then trenches should be covered with suitable sheet materials.

In addition to the supports already mentioned, any services which run through the excavated trenches (in particular drains and gas pipes) need to be supported, especially where the ground has to be excavated underneath them (see Figure 3.07).

Where trenches have to be excavated close to existing buildings, it may be necessary to provide support to the elevation adjacent to the excavation. This is due to the fact that, as ground is taken away from around the existing foundations, the loads will not be adequately and evenly distributed and absorbed into the natural or sub-foundation, possibly causing the structure to collapse. This support is known as shoring (see Figure 3.08).

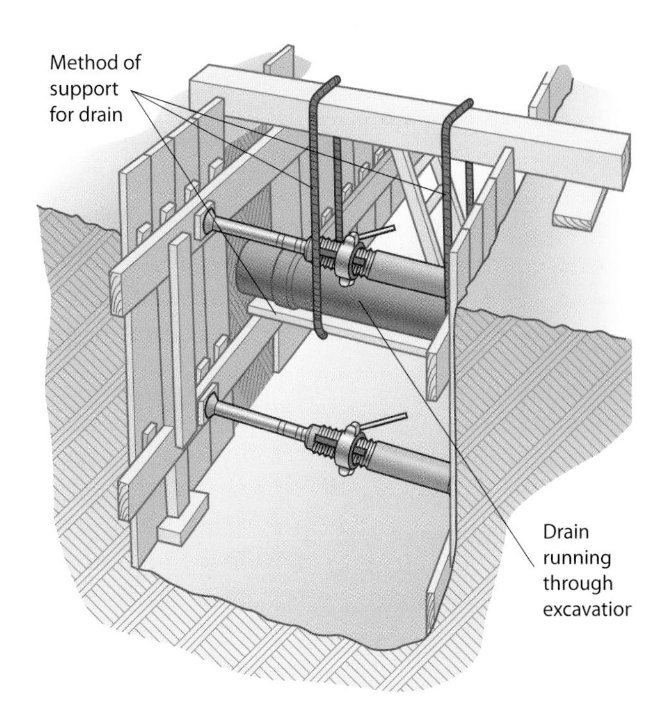

Figure 3.07 Support for drains running through an excavation

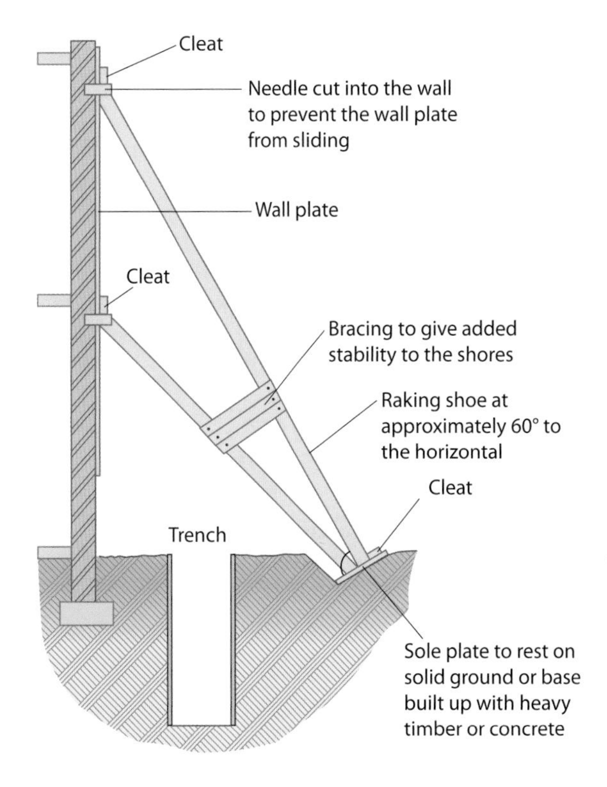

Figure 3.08 Raking shores used to support an existing building

One other factor that can affect the safety of workers in excavations and the stability of the soil is surface water. Surface water can be found at varying levels within the soil and, depending on the depth, trenches can easily cause flooding. Where this occurs, water pumps will need to be used to keep the trench clear. Failure to keep the trench free of water during construction will not only make operations difficult, but may also weaken and loosen the support systems due to soil displacement.

Functional skills

This exercise will allow you to practise **FM 2.3.1** Interpret and communicate solutions to multistage practical problems and **FM 2.3.2** Draw conclusions and provide mathematical justifications.

Working life

You have been asked by your boss to enter an excavation to clean out some loose soil. The excavation is 1.5 m deep, and has been excavated for a foundation. There are no supports on the excavation sides and overnight there has been a considerable amount of rainfall. Your boss shouts at you to get on with it as the concrete for the foundation will be here in 10 minutes.

What should you do? What could the implications be if you do it? What could the implications be if you don't?

Types of foundation

As previously stated, the design of a foundation will be down to the architect and structural design team. The final decision on the suitability and depth of the foundation, and on the thickness of the concrete, will rest with the local authority's building control department.

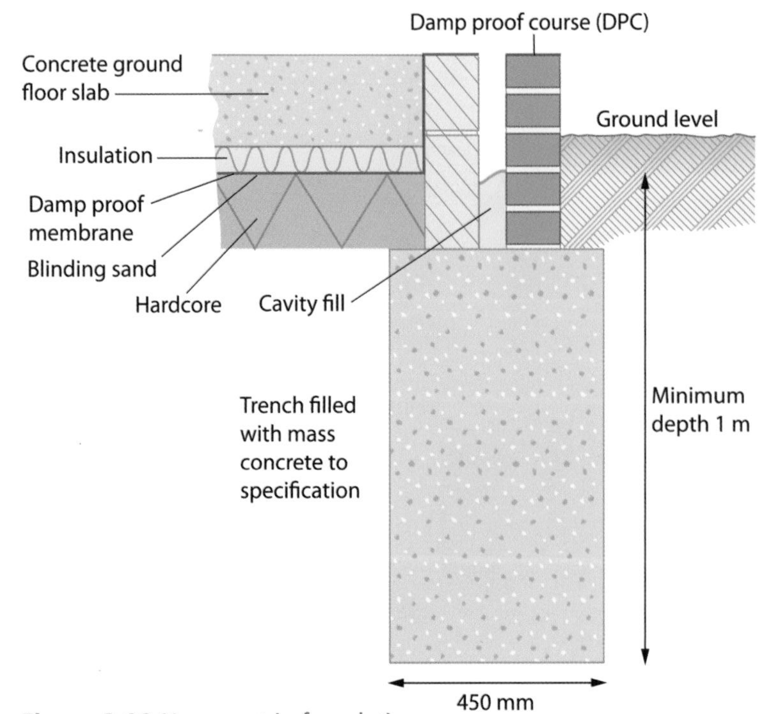

Figure 3.09 Narrow strip foundation

Strip foundations

The most commonly used strip foundation is the 'narrow strip' foundation, which is used for small domestic dwellings and low-rise structures. Once the trench has been excavated, it is filled with concrete to within 4–5 courses of the ground level damp-proof course (DPC). The level of the concrete fill can be reduced in height, but this makes it difficult for the bricklayer due to the confined area in which to lay bricks or blocks.

The depth of this type of foundation must be such that the subsoil acting as the natural foundation cannot be affected by the weather. This depth would normally not be less than 1 m.

The narrow strip foundation is not suitable for buildings with heavy structural loading, or where the subsoil is weak in terms of supporting the combined loads imposed on it. Where this is the case, a wide strip foundation is needed.

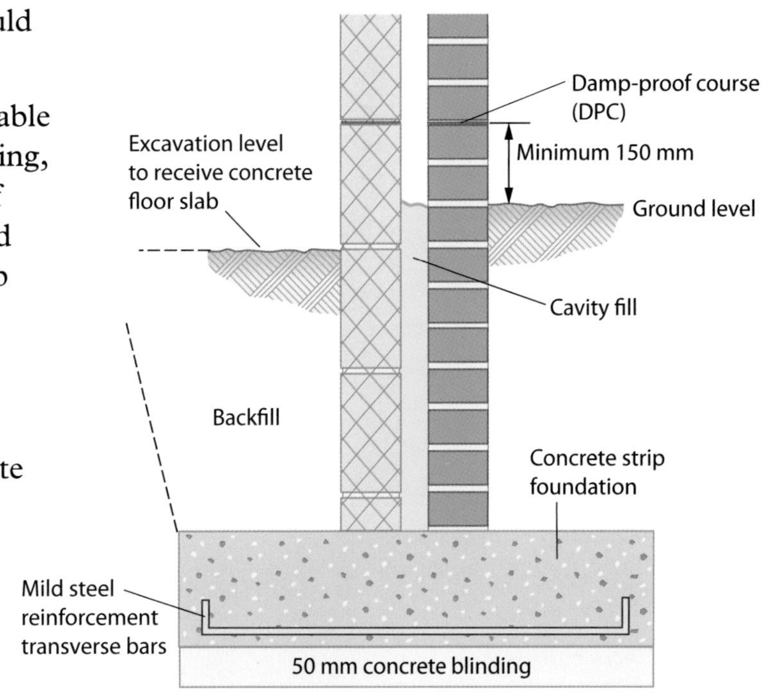

Figure 3.10 Wide strip foundation

Wide strip foundations

Wide strip foundations consist of steel reinforcement placed within the concrete base of the foundation. This removes the need to increase the depth considerably in order to spread the heavier loads adequately.

Raft foundations

These types of foundation are used where the soil has poor bearing capacity, making the soil prone to settlement. A raft foundation consists of a slab of reinforced concrete covering the entire base of the structure. The depth of the concrete is greater around the edges of the raft in order to protect the load-bearing soil directly beneath the raft from further effects of moisture taken in from the surrounding area.

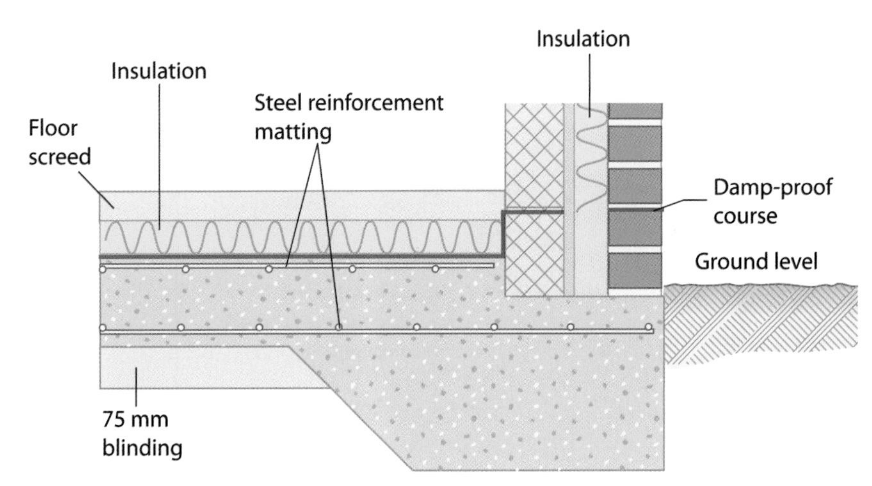

Figure 3.11 Raft foundation

Pad foundations

Pad foundations are used where the main loads of a structure are imposed at certain points. An example would be where brick or steel columns support the weight of floors or roof members, and walls between the columns are of non-load bearing cladding panels. The simplest form of pad foundation is where individual concrete pads are placed at various points around the base of the structure, and concrete ground beams span across them. The individual concrete pads will absorb the main imposed loads, while the beams will help support the walls.

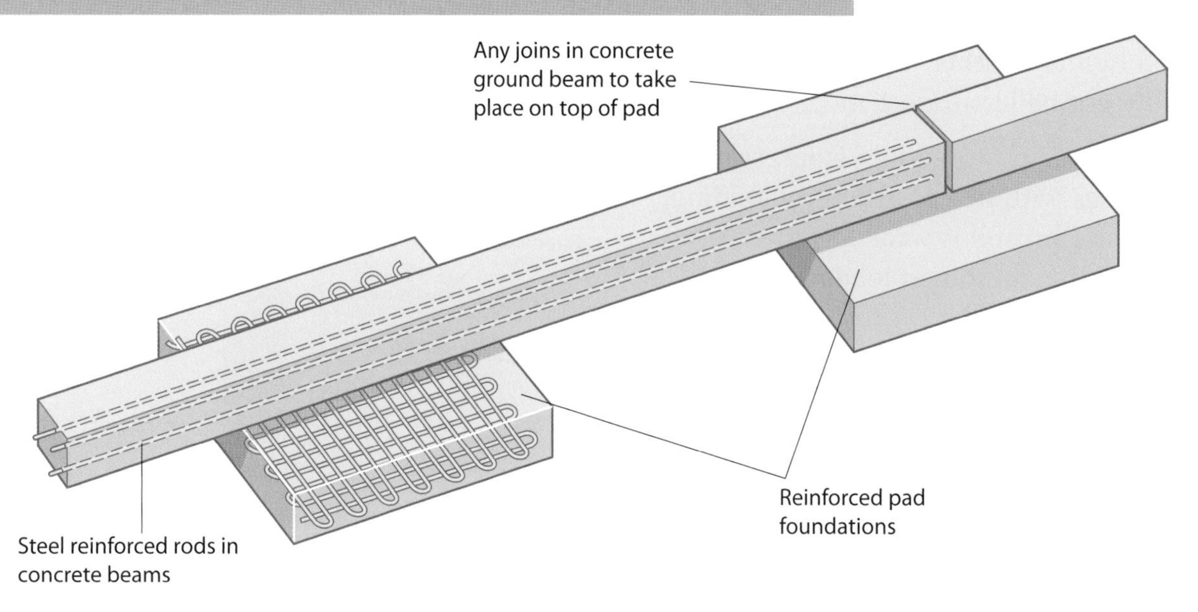

Figure 3.12 Square pad foundation with spanning ground beams

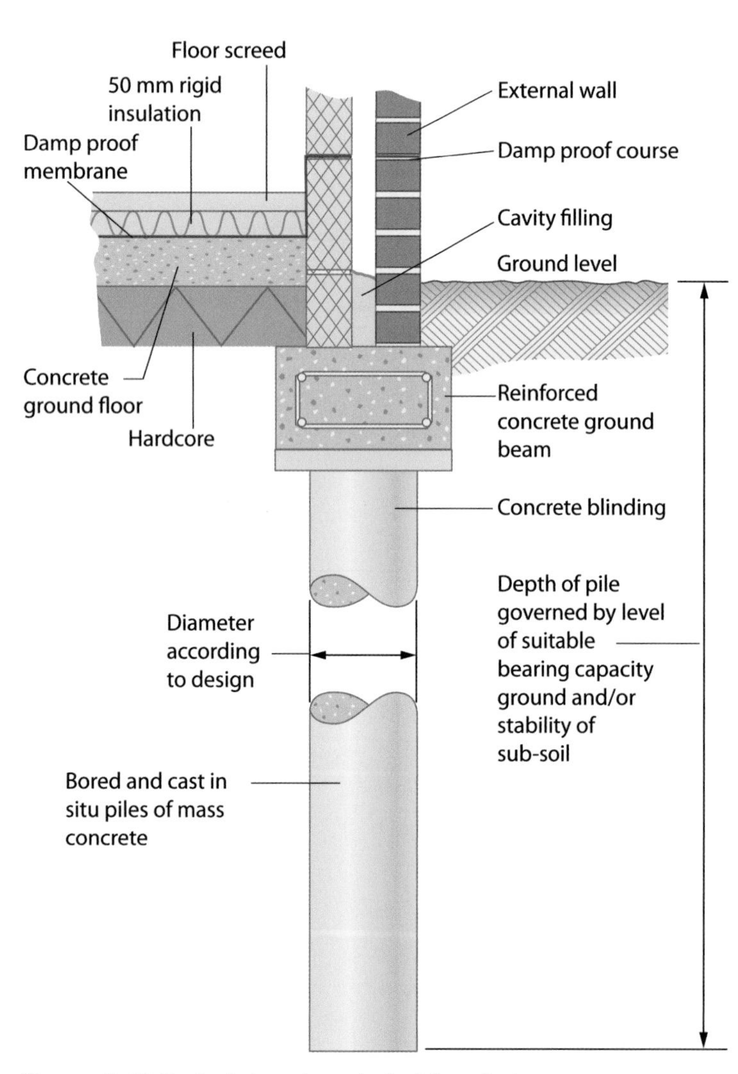

Figure 3.13 Typical short bored piled foundation

The depth of a pad foundation will depend on the load being imposed on it; in some instances, it may be necessary to use steel reinforcement to prevent excessive depths of concrete. This type of pad foundation can reduce the amount of excavation work required, as trenches do not need to be dug out around the entire base of the proposed structure.

Piled foundations

There are a large number of different types of piled foundations, each with an individual purpose in relation to the type of structure and ground conditions.

Short bored piled foundations are the most common piled foundations. They are predominantly used for domestic buildings where the soil is prone to movement, particularly at depths below 1 m.

A series of holes are bored, by mechanical means, around the perimeter of the base of the proposed building. The diameter of the bored holes will normally be between 250 and

350 mm and can extend to depths of up to 4 m. Once the holes have been bored, shuttering is constructed to form lightweight reinforced concrete beams, which span across the bored piles. The bored holes are then filled with concrete, with reinforcement projecting from the top of the pile concrete, so it can be incorporated into the concrete beams that span the piles.

As with the pad foundation, short bored piled foundations can significantly reduce the amount of excavated soil, because there is no need to excavate deep trenches around the perimeter of the proposed structure.

Stepped foundations

A stepped foundation is used on sloping ground. The height of each step should not be greater than the thickness of the concrete, and should not be greater than 450 mm. Where possible, the height of the step should coincide with brick course height, in order to avoid oversized mortar bed joints and eliminate the need for split brick courses. The overlap of the concrete to that below should not be less than 300 mm or less than the thickness of the concrete.

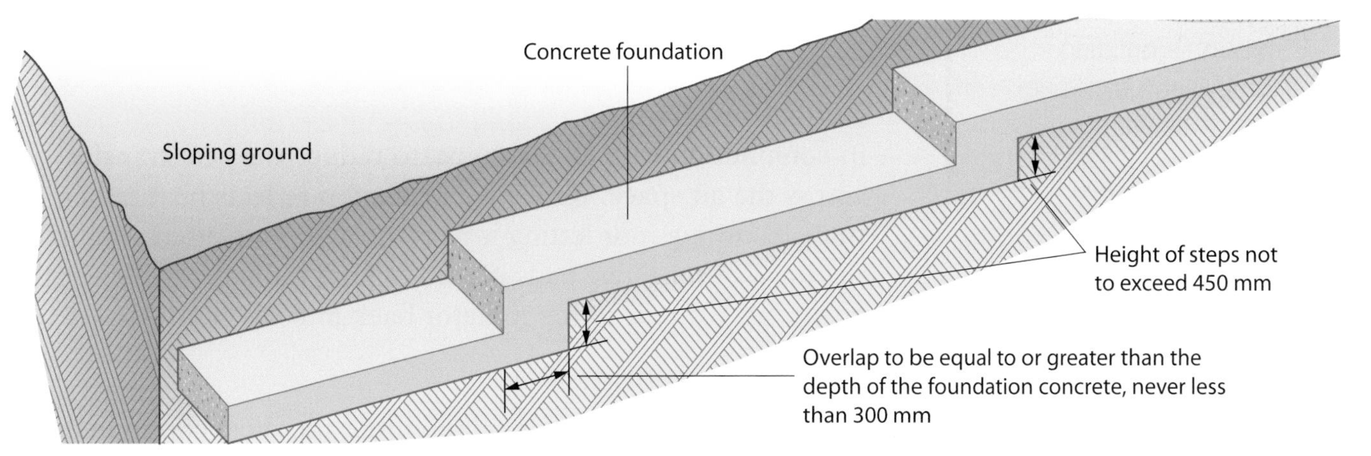

Figure 3.14 Typical stepped foundation

Different methods used to insulate against heat loss and gain

The term 'building insulation' refers broadly to any object in a building used as insulation for any purposes – the term applies to acoustic insulation and fire insulation as well as thermal insulation. Often an insulation material will be chosen for its ability to perform several of these functions at once.

Insulating buildings is vital: maintaining an acceptable temperature in buildings (by heating or cooling) uses a large proportion of a building's total energy consumption, and insulation can help to reduce this energy use.

Key term

R-value – the standard way of describing how effective insulation is. The higher the R-value, the more effective the insulation will be.

The effectiveness of insulation is commonly related to its **R-value**. However, the R-value does not take into account the quality of construction or local environmental factors for each building. Construction quality issues include poor vapour barrier and problems with draught-proofing. Local environmental factors are simply about where the building is located. With cold climates, the main aim is to reduce heat flow from the building; with hot climates, the main aim is to reduce the heat from entering the building, usually through solar radiation, which can enter a building through the windows.

Remember

The R-value is the inverse of the **U-value**.

Key term

U-value – a measure of thermal transmittance through a building component, usually a roof, wall, window, door or floor.

Materials used in insulating homes

There are essentially two types of building insulation – bulk insulation and reflective insulation:

- **Bulk insulation** blocks conductive heat transfer and convective flow, either into or out of a building. The denser the material is, the better it will conduct heat. As air has such a low density, it is a very poor conductor and therefore a good insulator. This is why air trapped between two materials is often used as an insulator, as in cavity walling.
- **Reflective insulation** works by creating a radiant heat barrier in conjunction with an air space, to reduce the heat transfer across the air space. Reflective insulation reflects heat rather than absorbing it or letting it pass through. It is often seen in the form of reflective pads placed behind radiators to reflect the heat from the rear of the radiator back into the room.

Forms of insulation

There are various forms of insulation, but the most common are mineral wool/Rockwool and fibreglass:

- **Mineral wool/Rockwool** products are made from molten rock spun on a high-spinning machine at a temperature of about 1600°C, similar to the process of making candy floss, with the product being a mass of fine, intertwined fibres.
- **Fibreglass** is made in a similar way to Rockwool, but using molten glass.

Mineral wool usually comes in sheets that are cut to fit between the rafters or studs. Fibreglass is similar, although it can come in rolls that can be cut to fit. Both materials are available in a variety of thicknesses to suit where they are going to be laid.

Other forms of insulation include:

- **polystyrene sheets** – polystyrene is a thermoplastic substance that comes in sheets which can be cut to size, again available in different thicknesses
- **loose-fill insulation** – this is particularly used where there is no insulation in the cavity walls: holes are drilled in the exterior wall and glass wool insulation is blown into the holes until the cavity is full.

Where to insulate

Where insulation should be used depends on the climate and particular living needs, but generally insulation should be placed:

- in the roof space, between rafters or trusses and between joists
- on all exterior walls
- between the joists at every floor level, including ground floor
- in solid ground floor construction
- in partition walls
- around any ducts and pipes.

Over the next few pages you will see how insulation is used and applied in the different elements of the structure that you will be working with.

Types of floor construction and components

Floors have a number of standard components, including the following:

- **DPC** – this is the damp-proof course that is inserted into both skins of the external cavity wall construction. It should be a proprietary product that is tested and has a long life expectancy.
- **DPM** – this is the damp-proof membrane, which is placed in large sheets within the floor structure so it can resist the passage of moisture and rising damp. This keeps the floors dry. The DPM should be taken up vertically and tucked into the DPC to form a complete seal. All DPMs should be lapped by at least 300 mm, with any joints taped.
- **Screeds** – floor screeds are considered in the solid concrete floors section (see page 86). They provide a finish to the concrete surface, cover up services and provide a level for floor finishes to be applied to. They also provide falls to floors, for example, in a wet room for the shower waste.

- **Wall plate** – the wall plate is on top of the sleeper wall that supports the floor joists. It has a DPC underneath it to prevent damp penetrating the timber. Wall plates should be treated.

The most common types of flooring used are:

- suspended timber floors
- floating floors
- beam and block floors
- solid concrete floors.

Suspended timber floors

Suspended timber floors can be fitted at any level, from top floor to ground floor. In the next few pages, you will look at:

- basic structure
- joists
- construction methods.

Basic structure

Suspended timber floors are constructed with timbers known as joists, which are spaced parallel to each other spanning the distance of the building. Suspended timber floors are similar to traditional roofs in that they can be single or double, a single floor being supported at the two ends only (see Figure 3.15), and a double floor supported at the two ends and in the middle by way of a honeycombed sleeper/dwarf wall, steel beam or load-bearing partition (see Figure 3.16).

All floors must be constructed to comply with the Building Regulations, in particular Part C, which is concerned with damp. The bricklayer must insert a **damp-proof course (DPC)** between the brick or block work when building the walls, situated no less than 150 mm above ground level. This prevents moisture moving from the ground to the upper side of the floor. No timbers are allowed below the DPC. Airbricks, which are built into the external walls of the building, allow air to circulate round the underfloor area, keeping the moisture content below the dry rot limit of 20 per cent, thus preventing dry rot.

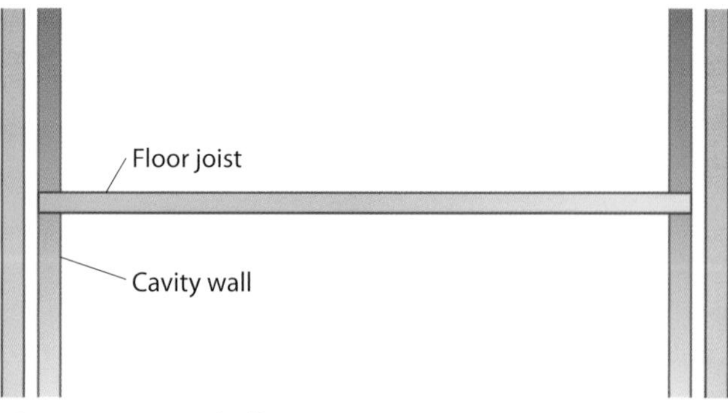

Figure 3.15 Single floor

Joists

In domestic dwellings, suspended upper floors are usually single floors, with the joist supported at each end by the structural walls but, if support is required, a load-bearing partition is used. The joists that span from one side of the building to the other are called bridging joists, but any joists that are affected by an opening in the floor such as a stairwell or chimney are called trimmer, trimming and trimmed joists (see Figure 3.17).

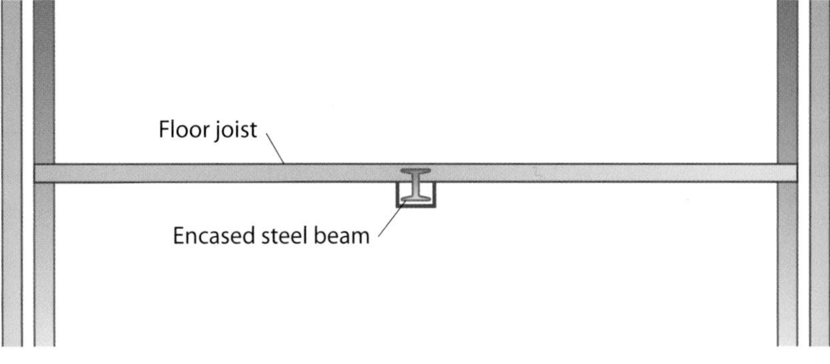

Figure 3.16 Double floor

Types of joist

As well as the traditional method of using solid timber joists, there are now alternatives available. Laminated and I-type joists are the most common.

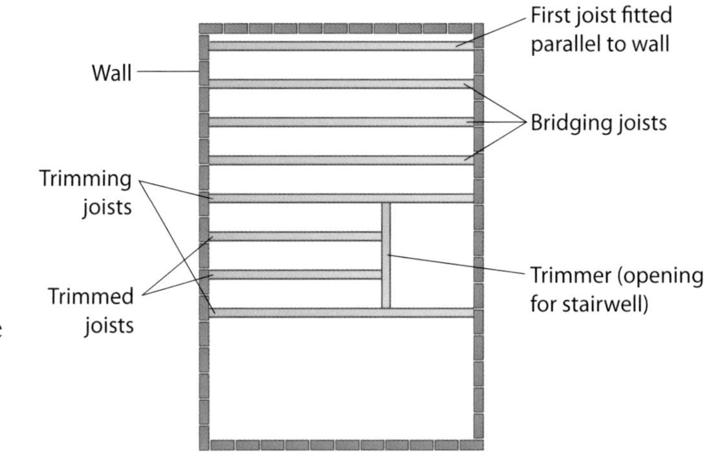

Figure 3.17 Joists and trimmers

Laminated joists

These were originally used for spanning large distances, as a laminated beam could be made to any size, but now they are more commonly used as an environmental alternative to solid timber – recycled timber can be used in the laminating process. They are more expensive than solid timber, as the joists have to be manufactured.

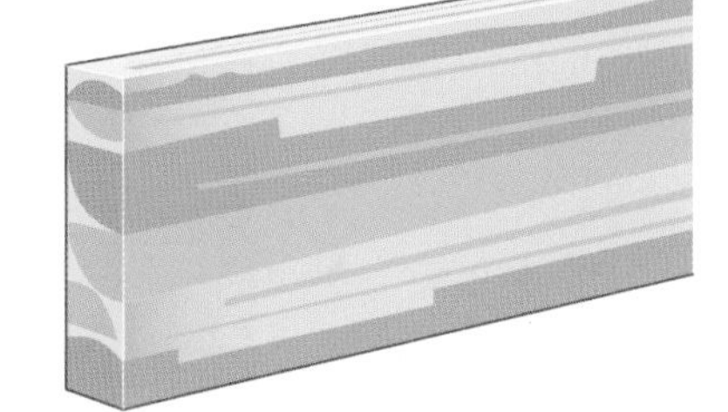
Figure 3.18 Laminated joist

I-type joists

These are now some of the most commonly used joists in the construction industry: they are particularly popular in new builds and are the only joists used in timber kit house construction. I-type beam joists are lighter and more environmentally friendly as they use a composite panel in the centre, usually made from oriented strand board, which can be made from recycled timber.

The following construction methods show how to fit solid timber joists but, whichever joists you use, the methods are the same.

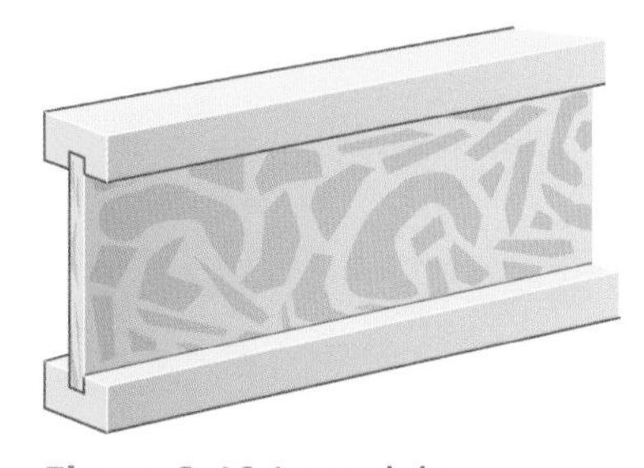
Figure 3.19 I-type joist

Construction methods

A suspended timber floor must be supported at either end. Figures 3.20 and 3.21 on the next page show ways of doing this.

If a timber floor has to trim an opening, there must be a joint between the trimming and the trimmer joists. Traditionally, a **tusk tenon joint** was used between the trimming and the trimmer joist (even now, this is sometimes preferred) (see Figure 3.22). If the joint is formed correctly a tusk tenon is extremely strong, but making one is time-consuming. A more modern method is to use a metal framing anchor or timber-to-timber joist hanger (see Figure 3.23).

Key term

Tusk tenon joint – a kind of mortise and tenon joint that uses a wedge-shaped key to hold the joint together.

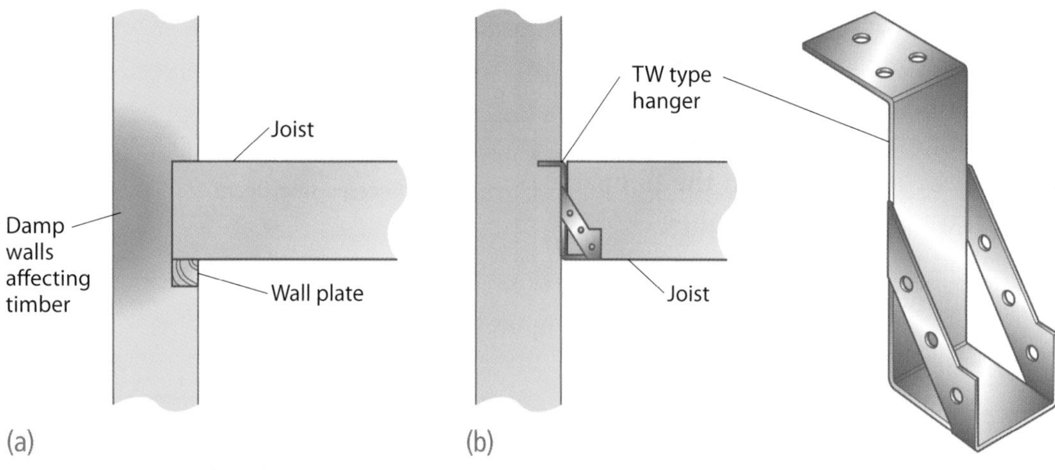

Figure 3.20 Solid floor bearings: (a) Old practice, (b) New practice

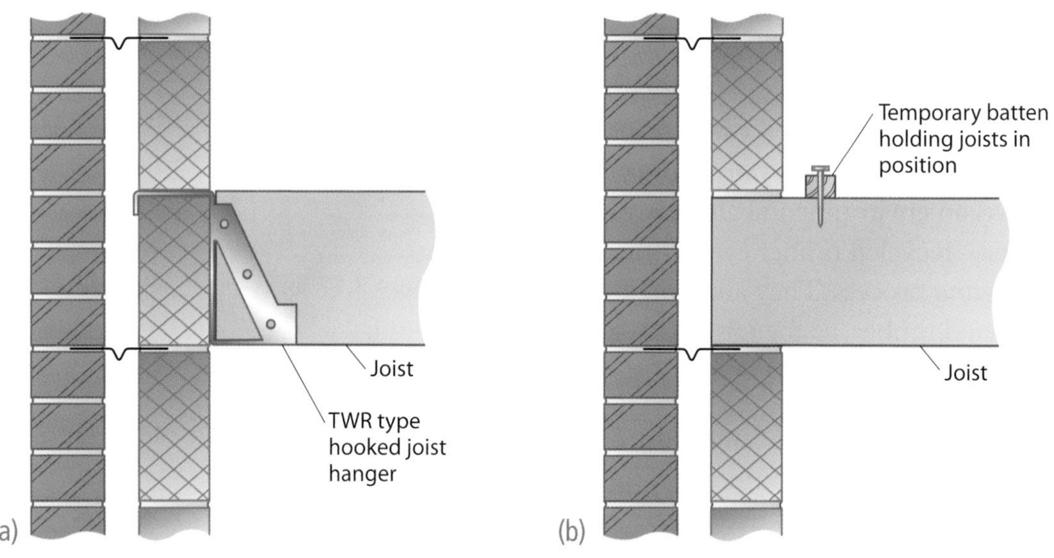

Figure 3.21 Cavity wall bearings: (a) Joist hanger bearing, (b) Built-in bearing

Figure 3.22 Traditional tusk tenon joint

Figure 3.23 Joist hanger

Fitting floor joists

Before the carpenter can begin constructing the floor, the bricklayer needs to build the honeycomb sleeper walls. This type of walling has gaps in each course to allow the free flow of air through the underfloor area. It is on these sleeper walls that the carpenter lays the timber wall plate, which will provide a fixing for the floor joists.

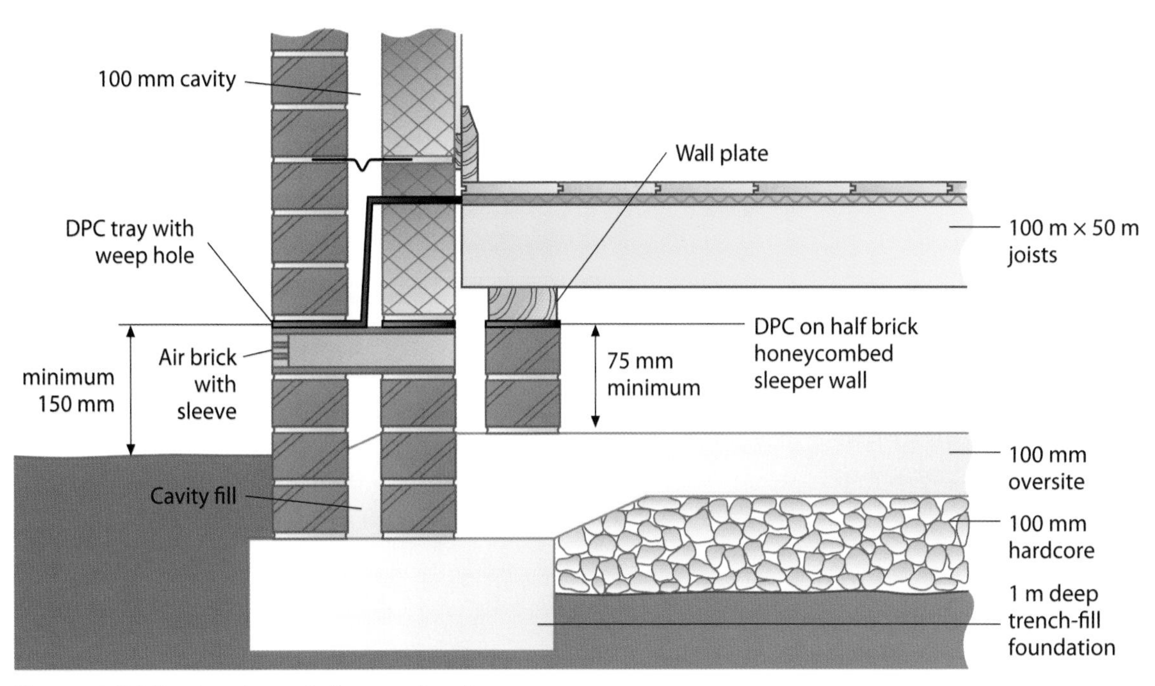

Figure 3.24 Section through floor and wall

Step-by-step: Fitting floor joists

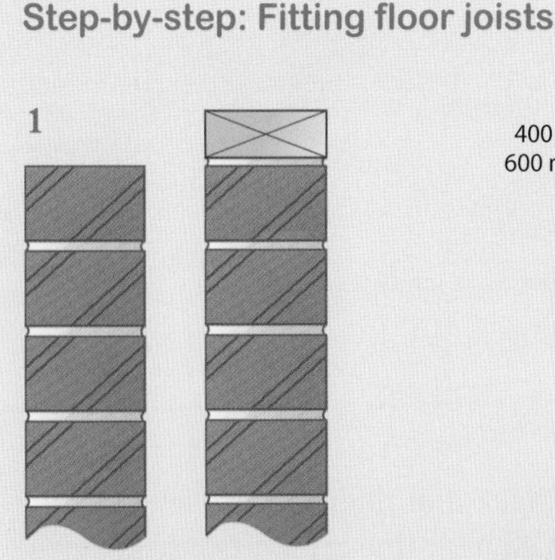

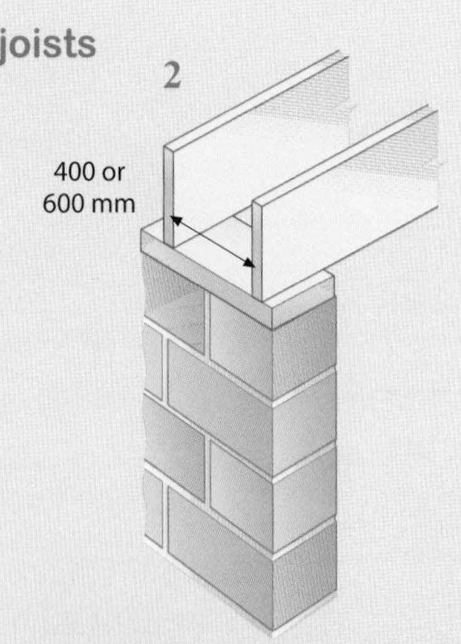

Step 1 Bed and level the wall plate onto the sleeper wall with the DPC under it.

Step 2 Cut joists to length and seal the ends with a coloured preservative. Mark out the wall plate with the required centres, space the joists out and fix temporary battens near each end to hold the joists in position. Ends should be kept away from walls by approximately 12 mm. It is important to ensure that the camber is turned upwards.

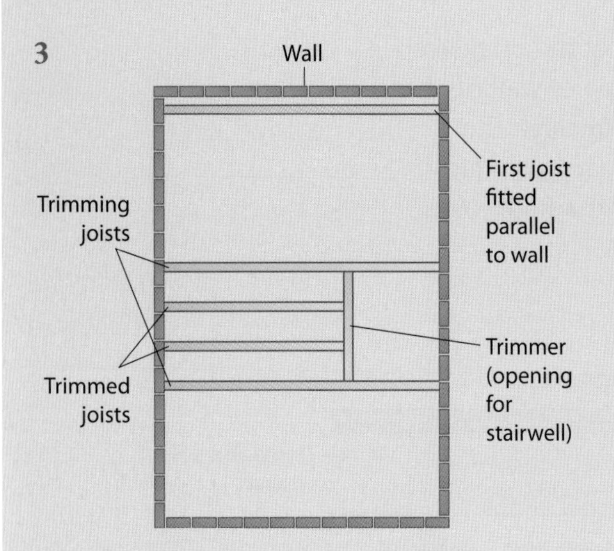

Step 3 Fix the first joist parallel to the wall with a gap of 50 mm. Fix trimming and trimmer joists next to maintain the accuracy of the opening.

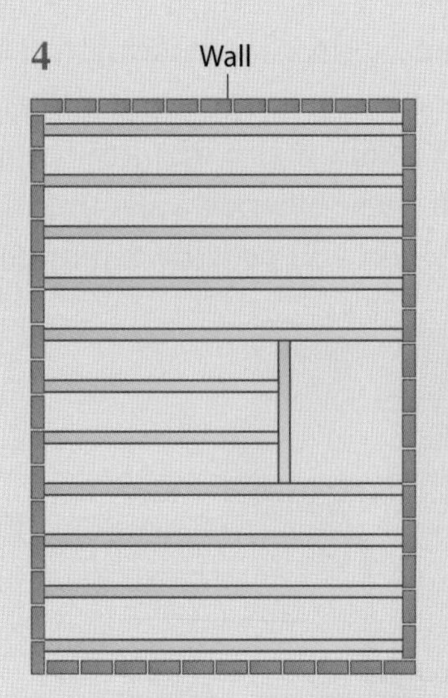

Step 4 Fix subsequent joists at the required spacing as far as the opposite wall. Spacing will depend on the size of joist and/or floor covering, but usually 400 mm to 600 mm centres are used.

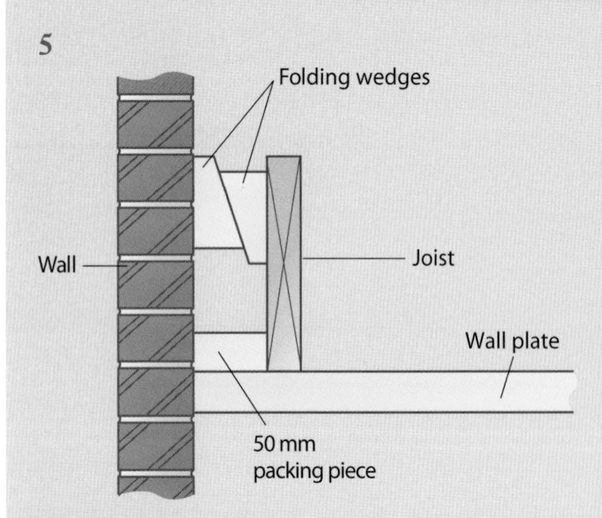

Step 5 Fit folding wedges to keep the end joists parallel to the wall. Overtightening is to be avoided in case the wall is strained.

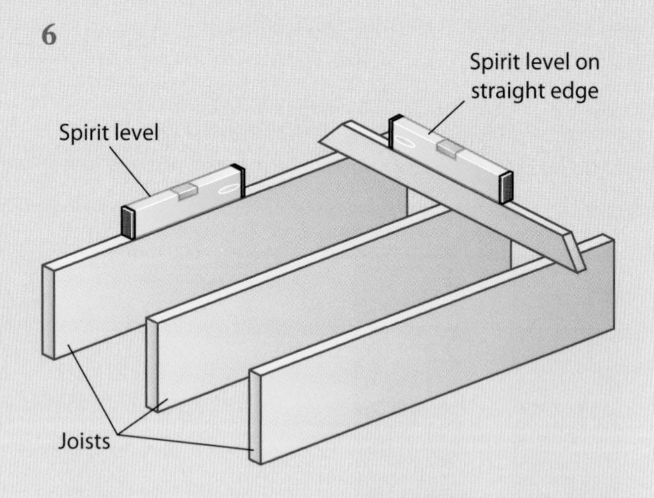

Step 6 Check that the joists are level with a straight edge or line and, if necessary, pack with slate or DPC.

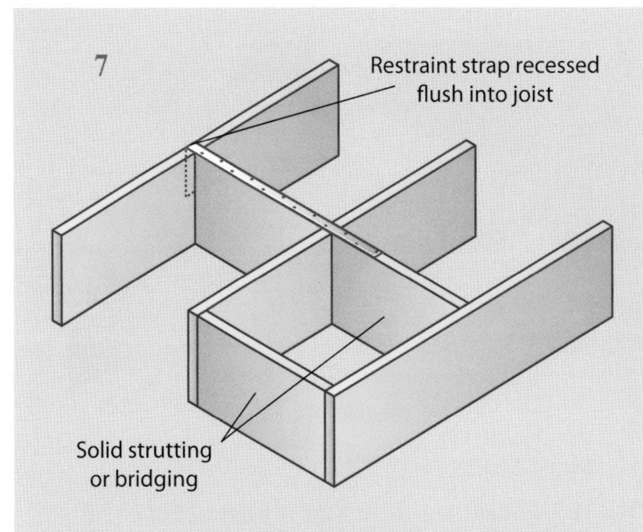

Step 7 Fit restraining straps and, if the joists span more than 3.5 m, fit strutting and bridging, described in more detail next.

Strutting and bridging

When joists span more than 3.5 m, a row of struts must be fixed midway between each joist. Strutting or bridging stiffens the floor in the same way that noggins stiffen timber stud partitions, preventing movement and twisting, which is useful when fitting flooring and ceiling covering. A number of methods are used, but the main ones are solid bridging, herringbone strutting and steel strutting.

Solid bridging

For solid bridging, timber struts the same depth as the joists are cut to fit tightly between each joist and **skew-nailed** in place. A disadvantage of solid bridging is that it tends to loosen when the joists shrink.

Herringbone strutting

In herringbone strutting, timber battens (usually 50 × 25 mm) are cut to fit diagonally between the joists. A small saw cut is put into the ends of the battens before nailing to avoid the battens splitting. This will remain tight even after joist shrinkage. The steps on the next page describe the fitting of timber herringbone strutting.

Remember

It is very important to clean the underfloor area before fitting the flooring, as timber cuttings or shavings are likely to attract moisture.

Key term

Skew-nailed – nailed with the nails at an angle.

Figure 3.25 Solid bridging

Space joists

Step 1 Nail a temporary batten near the line of strutting to keep the joists spaced at the correct centres.

Mark joist depths

Step 2 Mark the depth of a joist across the edge of the two joists, then measure 12 mm inside one of the lines and re-mark the joists. The 12 mm less than the depth of the joist ensures that the struts will finish just below the floor and ceiling level (as shown in step 5).

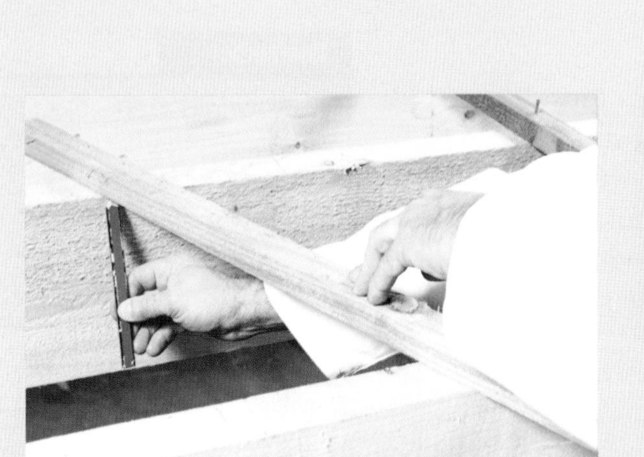

Lay struts across two joists at a diagonal

Step 3 Lay the strut across two joists at a diagonal to the lines drawn in Step 2.

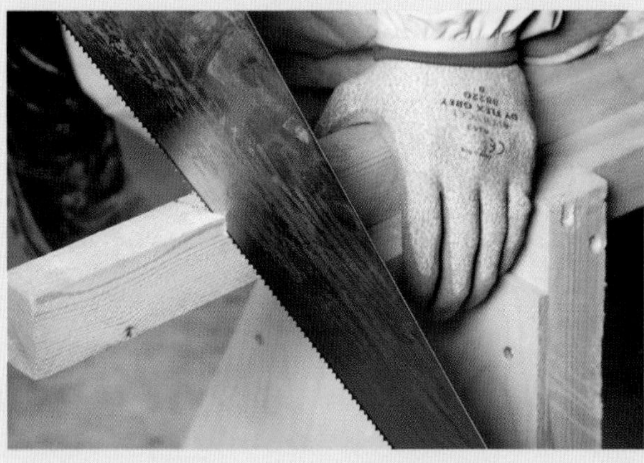

Cut to the mark

Step 4 Draw a pencil line underneath, as shown in Step 3, and cut to the mark. This will provide the correct angle for nailing.

Fix the strut

Step 5 Fix the strut between the two joists. The struts should finish just below the floor and ceiling level. This prevents the struts from interfering with the floor and ceiling if movement occurs.

Steel strutting

There are two types of galvanised steel herringbone struts available.

The first has angled lugs for fixing with the minimum 38 mm round head wire nails.

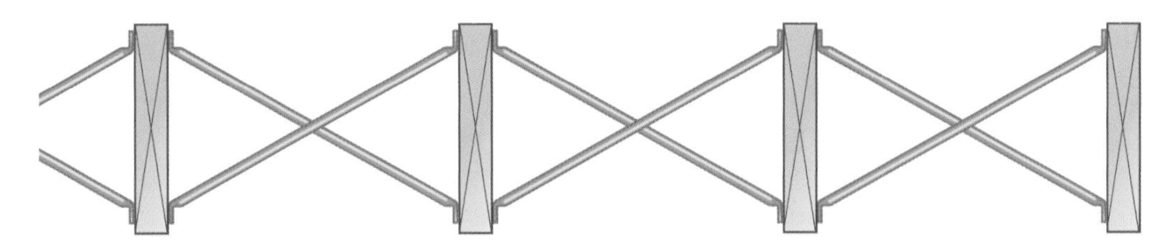

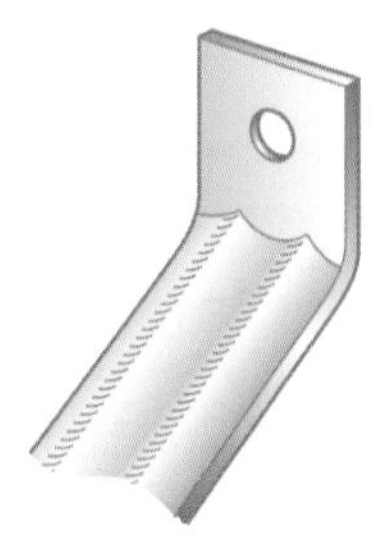

Figure 3.26 Catric® steel joist struts

The second has pointed ends, which bed themselves into joists when forced in at the bottom and pulled down at the top. Unlike other types of strutting, this type is best fixed from below.

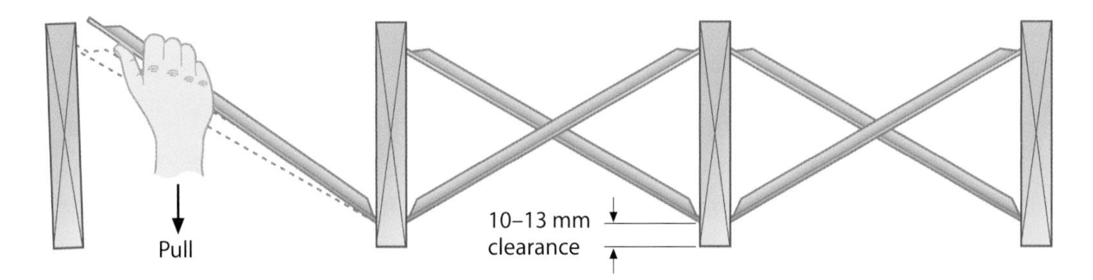

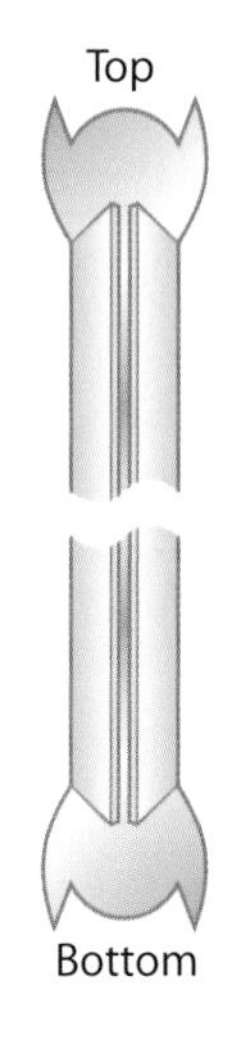

Figure 3.27 Batjam® steel joist struts

The disadvantage of steel strutting is that it only comes in set sizes, to fit centres of 400, 450 and 600 mm. This is a disadvantage as there will always be a space in the construction of a floor that is smaller than the required centres.

Restraint straps

Anchoring straps, normally referred to as restraint straps, are needed to restrict any possible movement of the floor and walls due to wind pressure. They are made from galvanised steel, 5 mm thick for horizontal restraints and 2.5 mm for vertical restraints, 30 mm wide and up to 1.2 m in length. Holes are punched along the length to provide fixing points.

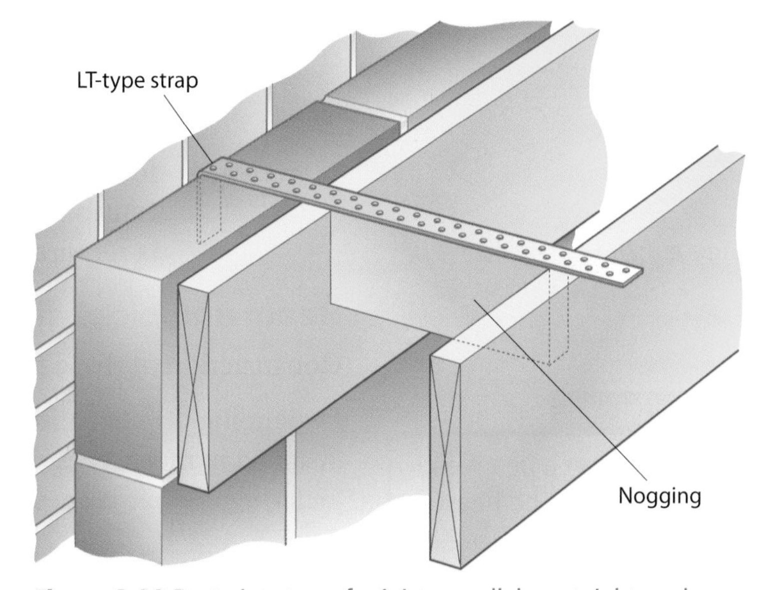

Figure 3.28 Restraint straps for joists parallel or at right angles to the wall

When the joists run parallel to the walls, the straps will need to be housed into the joist to allow the strap to sit flush with the top of the joist, keeping the floor even. The anchors should be fixed at a maximum of 2 m centre to centre. More information can be found in schedule 7 of the Building Regulations.

Functional skills

This exercise will allow you to practise **FM 2.3.1** Interpret and communicate solutions to multistage practical problems and **FM 2.3.2** Draw conclusions and provide mathematical justifications.

Working life

You have been tasked with joisting an upper floor. You fit the herringbone structure but, when it comes to fitting the floor and plasterboarding the ceiling, the strutting is interfering with both the floor and ceiling.

What could have caused this to happen? Think about all the stages you worked through, and the points where this problem could have come up. What could you do to rectify this situation? You will need to think about the impact any actions you take could have on the whole project.

Floating floors

These are basic timber floor constructions that are laid on a solid concrete floor. The timbers are laid in a similar way to joists, although they are usually 50 mm thick maximum as there is no need for support. The timbers are laid on the floor at pre-determined centres, and are not fixed to the concrete base (hence floating floor); the decking is then fixed on the timbers. Insulation or underfloor heating can be placed between the timbers to enhance the thermal and acoustic properties.

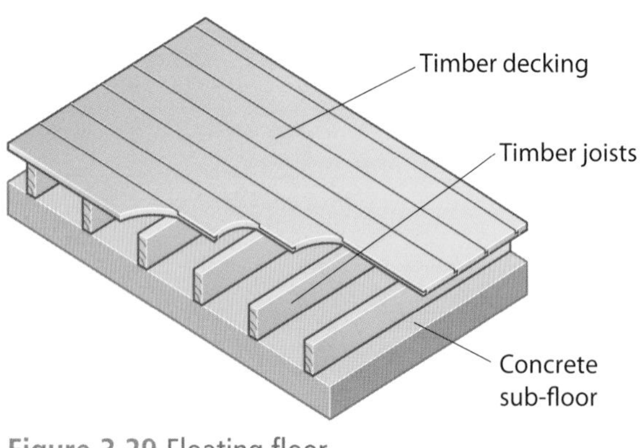

Figure 3.29 Floating floor

This type of floor 'floats' on a cushion of insulation. Floating floors are normally manufactured from chipboard, either standard or moisture-resistant for use in bathrooms. This type of floor is ideal for refurbishment work and where insulation upgrading is required.

Beam and block floors

Construction of these floors is generally quite simple.

Safety tip

The beams in a beam and block floor can be heavy, so you may need a crane to position them on the external walls safely.

Depending on site conditions, beam and block floors can be installed in most types of weather and by using a variety of methods, including mobile cranes or other site lifting plant or by hand.

Standard beam and block floors consist of 150 mm and 225 mm beams, with standard 100 mm deep concrete building blocks inserted between the beams.

The beams are pre-cast away from the site environment and are pre-stressed with high tensile steel wires suited to the environment and purpose for which they are to be used.

Once the blocks have been placed in position, the floor should be grouted with sand/cement grout consisting of four parts sand to one part cement. The grout should be brushed into the joints between the beams and the blocks in order to stabilise the floor.

Insulation can be slung underneath the beams or over the blocks where it is to be covered with a suitable finish.

Did you know?

Unlike solid floor construction, beam and block floors can be fitted at both ground floor level and upper floor levels.

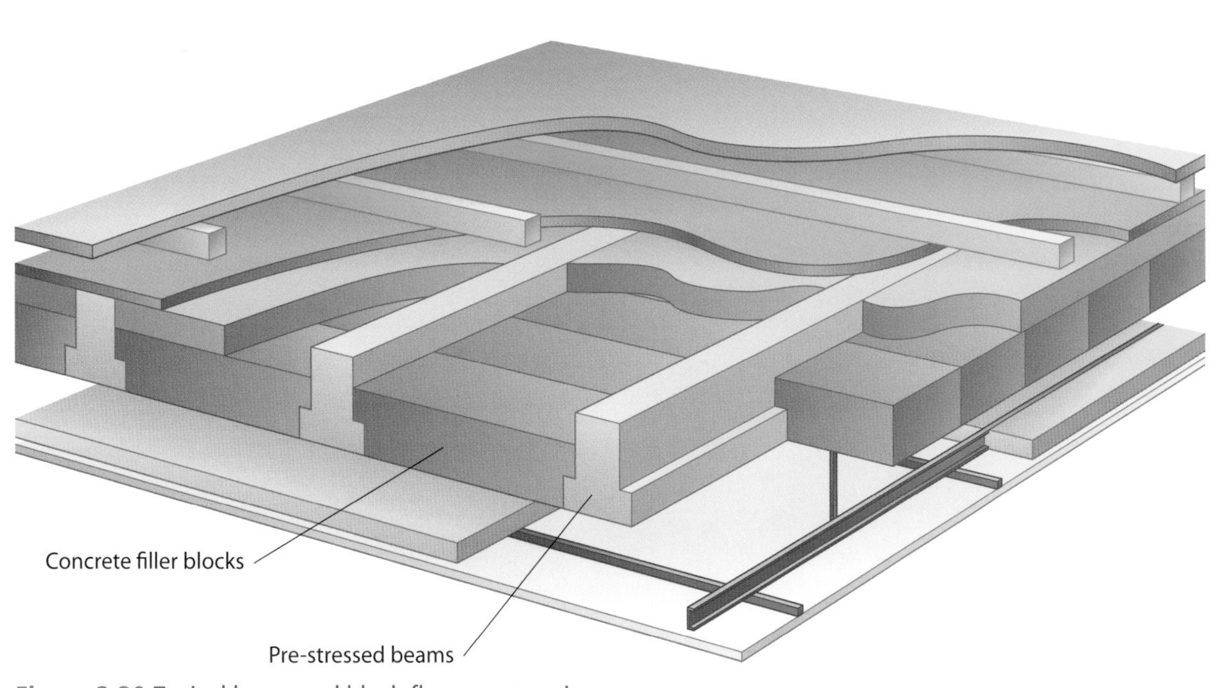

Figure 3.30 Typical beam and block floor construction

Solid concrete floors

Solid concrete floors are more durable than suspended timber floors. They are constructed on a sub-base incorporating hardcore, damp-proof membranes and insulation.

In the next few pages, you will look at:

- formwork for concrete floors
- reinforcement
- compacting of concrete
- surface finishes
- curing.

Remember

The depth of the hardcore and concrete will depend on the nature of the building, and will be set by the Building Regulations and the local authority.

Formwork

Any concreting job has to be supported at the sides to prevent the concrete just running off, and this support is known as formwork.

Floors for buildings such as factories and warehouses have large areas and would be difficult to lay in one slab. Floors of this type are usually laid in alternative strips up to 4.5 m wide, running the full length of the building (see Figure 3.31). The actual formwork is similar to that used for paths.

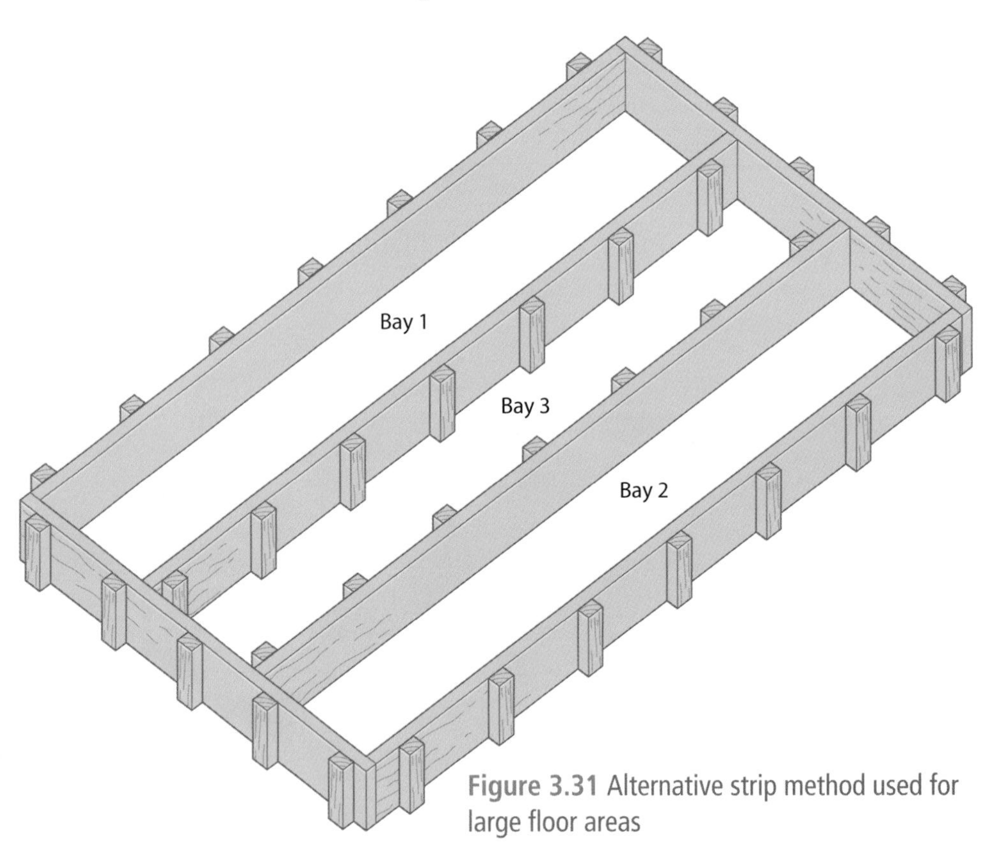

Figure 3.31 Alternative strip method used for large floor areas

Key terms

Compression – being squeezed or squashed together.

Tension – being stretched.

Reinforcement

Concrete is strong in **compression** but weak in **tension** so, to prevent concrete from being 'pulled' apart when under pressure, steel reinforcement is provided (see Figure 3.32). The type and position of the reinforcement will be specified by the structural engineer.

The reinforcement must always have a suitable thickness of concrete cover to prevent the steel from rusting if exposed to moisture or air. The amount of cover required depends upon the location of the site with respect to exposure conditions, and ranges from 20 mm in mild exposure to 60 mm for very severe exposure to water.

To prevent the reinforcement from touching the formwork, spacers made from concrete, fibre cement or plastic are used. They are available in several shapes and various sizes to give the correct cover.

Figure 3.32 Steel reinforcement of concrete

Compacting

When concrete has been placed, it contains trapped air in the form of voids. To get rid of these voids, we must compact the concrete. The more workable the concrete, the easier it is to compact, but if the concrete is too wet, the excess water will reduce the strength of the concrete.

Failure to compact concrete results in:

- reduction in the strength of the concrete
- water entering the concrete, which could damage the reinforcement
- visual defects, such as honeycombing on the surface.

The method of compaction depends on the thickness and the purpose of the concrete. For oversite concrete, floors and pathways up to 100 mm thick, manual compaction with a tamper board may be sufficient. This requires slightly overfilling the formwork and tamping down with the tamper board (see Figure 3.33).

Did you know?

For larger spans the tamper board may be fitted with handles.

For slabs up to 150 mm thick, a vibrating beam tamper should be used (see Figure 3.34 on the next page). This is simply a tamper board with a petrol-driven vibrating unit bolted on. The beam is laid on the concrete with its motor running and is pulled along the slab.

For deeper structures, such as retaining walls for example, a poker vibrator is required (see Figure 3.35 on the next page). The poker vibrator is a vibrating tube at the end of a flexible drive connected to a petrol motor. The pokers are available in various diameters from 25 mm to 75 mm.

Figure 3.33 Tamper board with handles

The concrete should be laid in layers of 600 mm, with the poker in vertically and penetrating the layer below by 100 mm. The concrete is vibrated until the air bubbles stop, and the poker is then lifted slowly and placed 150 to 1000 mm from this incision, depending on the diameter of the vibrator.

Key term

Screeding off – levelling off concrete by adding a final layer.

Surface finishes

Surface finishes for slabs may include the following:

Figure 3.34 Vibrating beam tamper

- **Tamped finish** – simply using a straight edge or tamper board when compacting the concrete will leave a rough finish to the floor, ideal for a path or drive surface, giving grip to vehicles and pedestrians. This finish may also be used if a further layer is to be applied to give a good bond.
- **Float and brush finish** – after **screeding off** the concrete with a straight edge, the surface is floated off using a steel or wooden float and then brushed lightly with a soft brush (see Figure 3.36). Again, this would be suitable for pathways and drives.
- **Steel float finish** – after screeding off using a straight edge, a steel float is applied to the surface. This finish attracts particles of cement to the surface, causing the concrete to become impermeable to water, but also very slippery when wet. This is not very suitable for outside but ideal for use indoors for floors, etc.
- **Power trowelling/float** – three hours after laying, a power float is applied to the surface of the concrete (see Figure 3.37). After a further delay to allow surface water to evaporate, a power trowel is then used. A power float has a rotating circular disc or four large flat blades powered by a petrol engine. The edges of the blades are turned up to prevent them digging into the concrete slab. This finish is most likely to be used in factories where a large floor area is needed.
- **Power grinding** – this is a technique used to provide a durable wearing surface without further treatment. The concrete is laid, compacted and trowel finished. After 1 to 7 days the floor is ground, removing the top 1–2 mm, leaving a polished concrete surface.

Figure 3.35 Poker vibrator in use

Figure 3.36 Brushed concrete finish

Remember

Make sure you always clean all tampers and tools after use.

Did you know?

The success of surface finishes depends largely on timing. You need to be aware of the setting times in order to apply the finish.

Surface treatment for other surfaces may include the following:

- **Plain smooth surfaces** – after the formwork has been struck, the concrete may be polished with a carborundum stone, giving a polished, water-resistant finish.
- **Textured and profiled finish** – a simple textured finish may be made by using rough sawn boards to make the formwork. When struck, the concrete takes on the texture of these boards. A profiled finish can be made by using a lining inside the formwork. The linings may be made from polystyrene or flexible rubber-like plastics, and gives a pattern to the finished concrete.
- **Ribbed finish** – this finish is made by fixing timber battens to the formwork (see Figure 3.38).
- **Exposed aggregate finish** – the coarse aggregate is exposed by removing the sand and cement from the finished concrete with a sand blaster. Another method of producing this finish is by applying a chemical retarder to the formwork, which prevents the cement in contact with it from hardening. When the formwork is removed, the mortar is brushed away to uncover the aggregate in the hardened concrete.

Figure 3.37 Power float

Figure 3.38 Ribbed finish

Remember

In the areas where you previously created the formwork, you will need to pour, compact and finish concrete. Mistakes at any of these stages can lead to long-term problems in the strength of the floor.

Remember

If the water is allowed to evaporate from the mix shortly after the concrete is placed, there is less time for the cement to 'go off'.

Curing

When concrete is mixed, the quantity of water is accurately added to allow for hydration to take place. To allow the concrete to achieve its maximum strength, this chemical reaction must be allowed to keep going for as long as possible. To do this we must 'cure' the concrete. This is done by keeping the concrete damp and preventing it from drying out too quickly.

Curing can be done by following these steps:

1. Spray the concrete with a chemical sealer, which dries to leave a film of resin to seal the surface and reduces the loss of moisture.
2. Spray the concrete with water, which replaces any lost water and keeps the concrete damp. This can also be done by placing sand or hessian cloth or other similar material on the concrete and dampening.
3. Cover the concrete with a plastic sheet or building paper, preventing wind and sun from evaporating the water into the air. Any evaporated moisture due to the heat will condense on the polythene and drip back on to the concrete surface.

Remember

In hot weather the concrete must be placed quickly and not left standing for too long.

Safety tip

Take precautions in hot weather against the effect of the sun on your skin, for example wear sun block and a T-shirt.

Concreting in hot weather

When concreting in temperatures over 20°C, there is a reduction in workability due to the water being lost through evaporation. The cement also tends to react more quickly with water, causing the concrete to set rapidly.

To remedy the problem of the concrete setting quickly, a retarding mixture may be used. This slows down the initial reaction between the cement and water, allowing the concrete to remain workable for longer.

Extra water may be added at the time of mixing so that the workability is correct at the time of placing.

Water must not be added during the placing of the concrete, to make it more workable, after the initial set has taken place in the concrete.

Concreting in cold weather

Water expands when freezing. This can cause permanent damage if the concrete is allowed to freeze when freshly laid or in hardened concrete that has not reached enough strength (5 N/mm^2, which takes 48 hours).

Concreting should not take place when the temperature is 2°C or less. If the temperature is only slightly above 2°C, mixing water should be heated.

After being laid, the concrete should be kept warm by covering it with insulating quilts, which allows the cement to continue its reaction with the water and prevents it from freezing.

Working life

You are working for a construction company and have been asked to complete a quote for the building of a garage. The client is unsure what type of floor to have and they ask your opinion.

What type of information would you need to know in order to make a decision? What type of floor do you think you are likely to pick? You will need to think about what the floor is likely to be used for. Why would you select that type of floor?

Floor coverings

Softwood flooring

Softwood flooring can be used at either ground or upper floor levels. It usually consists of 25 × 150 mm tongued and grooved (T&G) boards. The tongue is slightly off-centre to provide extra wear on the surface that will be walked upon.

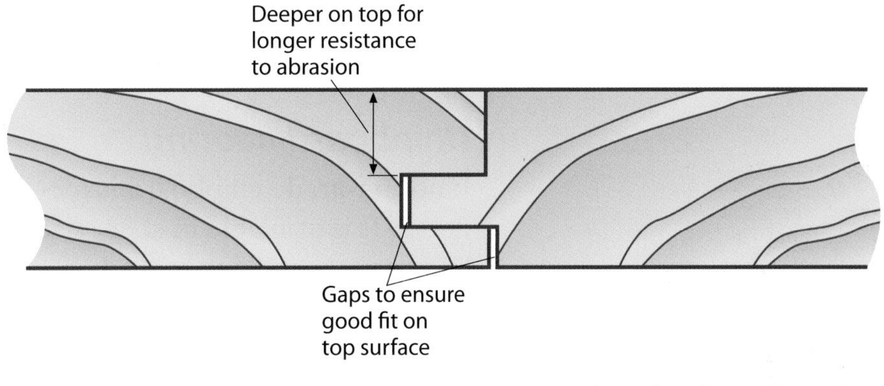

Figure 3.39 Section through softwood covering

When boards are joined together, the joints should be staggered evenly throughout the floor to give it strength. They should never be placed next to each other, as this prevents the joists from being tied together properly. The boards are either fixed with floor brads nailed through the surface and punched below flush, or secret nailed with lost head nails through the tongue. The nails used should be 2½ times the thickness of the floorboard.

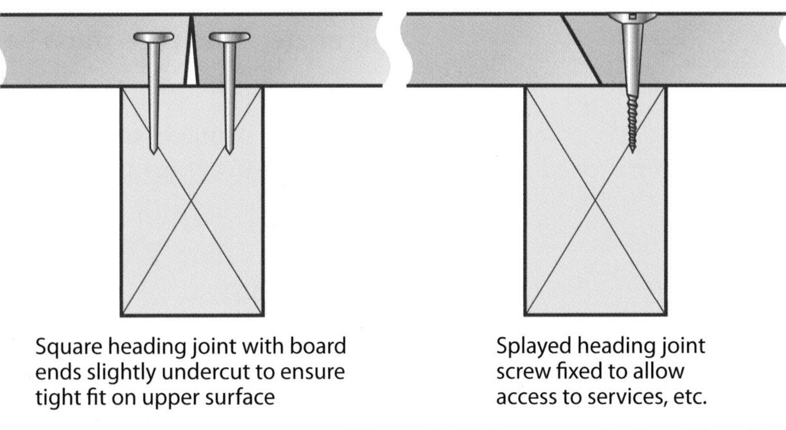

Figure 3.40 Square and splayed heading

The first board is nailed down about 10–12 mm from the wall. The remaining boards can be fixed four to six boards at a time, leaving a 10-12 mm gap around the perimeter to allow for expansion. This gap will eventually be covered by the skirting board.

There are two methods of clamping the boards before fixing. These are shown in Figure 3.41.

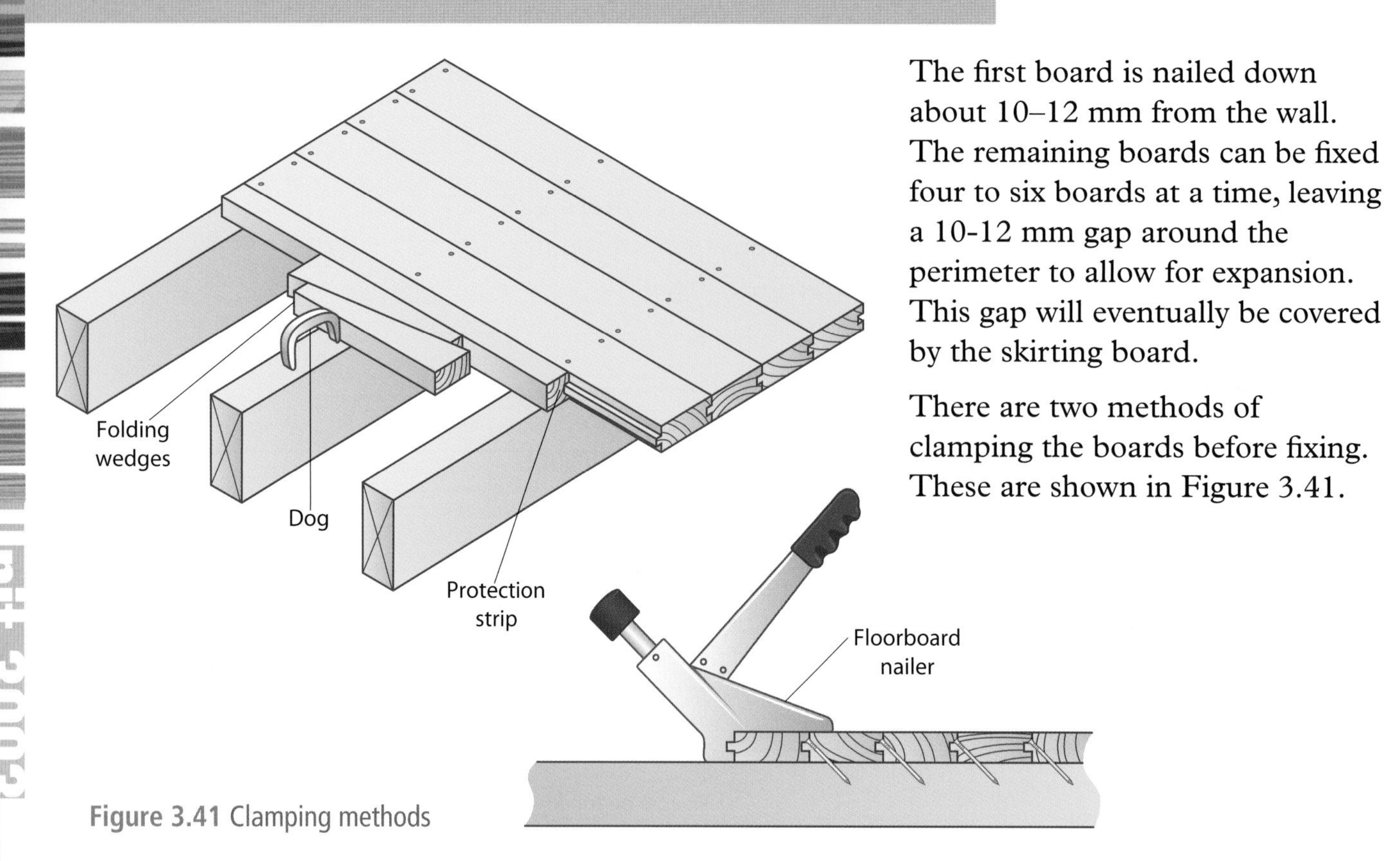

Figure 3.41 Clamping methods

Chipboard flooring

Flooring-grade chipboard is increasingly being used for domestic floors. It is available in sheet sizes of 2440 × 600 × 18 mm and can be square edged or tongued and grooved on all edges, the latter being preferred. If square-edged chipboard is used it must be supported along every joint.

Tongued and grooved boards are laid end to end, at right angles to the joists. Cross-joints should be staggered and, as with softwood flooring, expansion gaps of 10–12 mm left around the perimeter. The ends must be supported.

When setting out the floor joists, the spacing should be set to avoid any unnecessary wastage. The boards should be glued along all joints and fixed using either 50–65 mm annular ring shank nails or 50–65 mm screws. Access traps must be created in the flooring to allow access to services such as gas and water (see Figure 3.42).

Insulation in flooring

Flooring insulation in timber flooring usually takes the form of a quilted insulation such as mineral or Rockwool, or solid insulation such as polystyrene sheeting or insulation boards. Insulation in solid concrete flooring or slabs usually takes the form of insulation

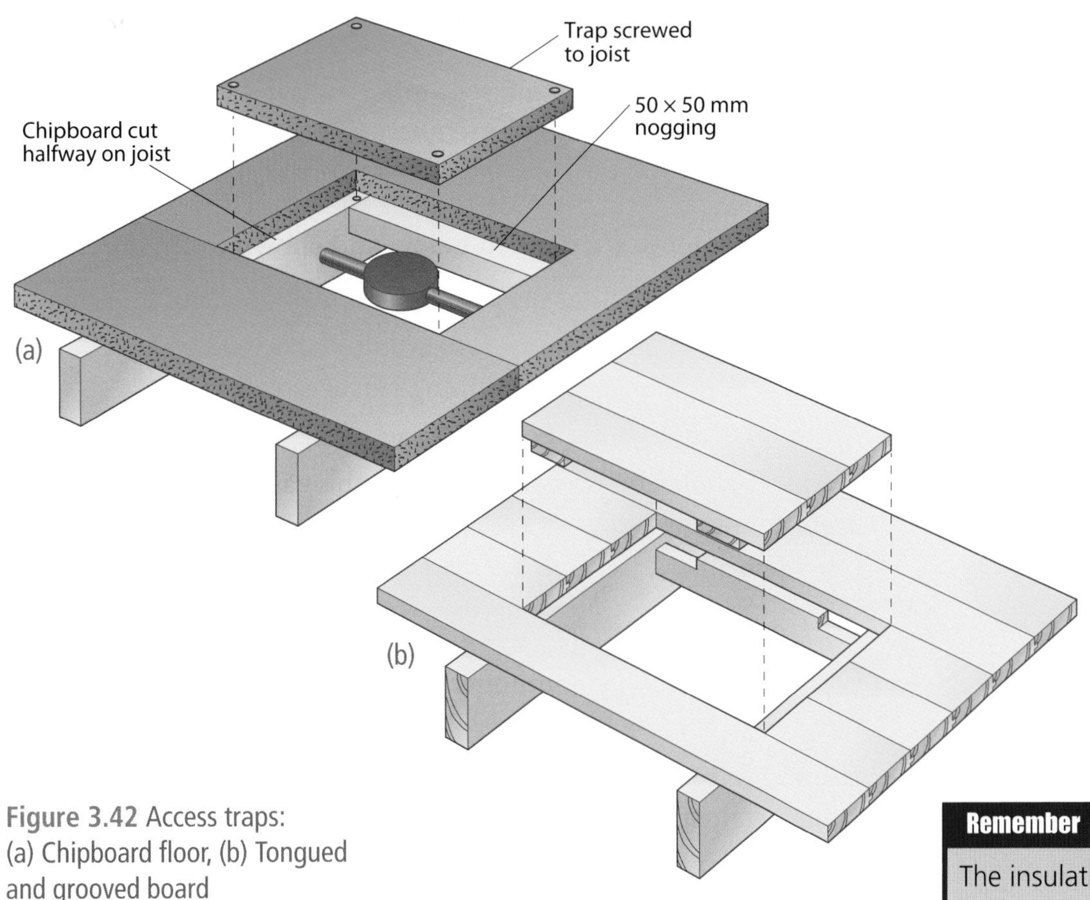

Figure 3.42 Access traps:
(a) Chipboard floor, (b) Tongued and grooved board

Remember

The insulation manufacturer will be able to help you with typical installation details for placing the thermal insulation into a design.

boards (which need to be a minimum of 50 mm thick to ensure the target U-value is reached). These should be staggered to avoid movement and can be placed under the slab or between the slab and floor screed, with a damp-proof membrane being placed under the insulation to prevent any punctures in the boarding.

Solid floors

Concrete floors are constructed as shown in Figure 3.43.

The insulation can be placed under or over the concrete, and must be of the right specification to take the loadings from the floor.

Hollow floors

Beam and block floors require specialist thermal insulation that clips under the beams, below the level of the blocks.

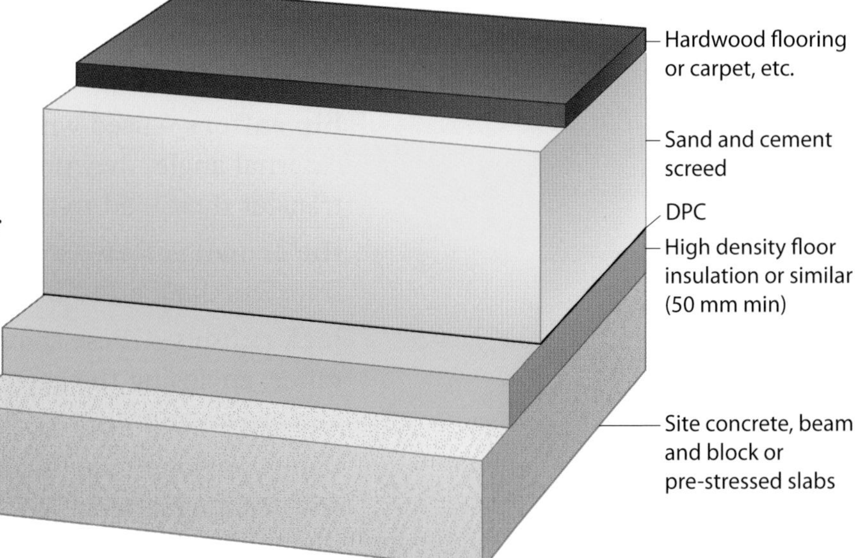

Figure 3.43 Solid floor construction

Roofing components

Roofs are made up of a number of different parts called elements. These in turn are made up of members or components.

Elements

The main elements are defined below and shown in Figure 3.44:

- **Gable** – the triangular part of the end wall of a building that has a pitched roof.
- **Hip** – where two external sloping surfaces meet.
- **Valley** – where two internal sloping surfaces meet.
- **Verge** – where the roof overhangs at the gable.
- **Eaves** – the lowest part of the roof surface where it meets the outside walls.

Members or components

The main members or components are defined below and shown in Figure 3.44. They include:

- **the ridge board** – a horizontal board at the apex acting as a spine, against which most of the rafters are fixed
- **the wall plate** – a length of timber placed on top of the brickwork to spread the load of the roof through the outside walls and to give a fixing point for the bottom of the rafters
- **purlins** – horizontal beams that support the rafters midway between the ridge and wall plate
- **the rafter** – a piece of timber that forms the roof, of which there are several types:
  - **common rafters** – the main load-bearing timbers of the roof
  - **hip rafters** – used where two sloping surfaces meet at an external angle, they provide a fixing for the jack rafters and transfer their load to the wall
  - **the crown rafter** – the centre rafter in a hip end that transfers the load to the wall
  - **jack rafters** – these span from the wall plate to the hip rafter, enclosing the gaps between common and hip rafters, and crown and hip rafters
  - **valley rafters** – like hip rafters but forming an internal angle, acting as a spine for fixing cripple rafters
  - **cripple rafters** – similar to a jack rafter, these enclose the gap between the common and valley rafters.

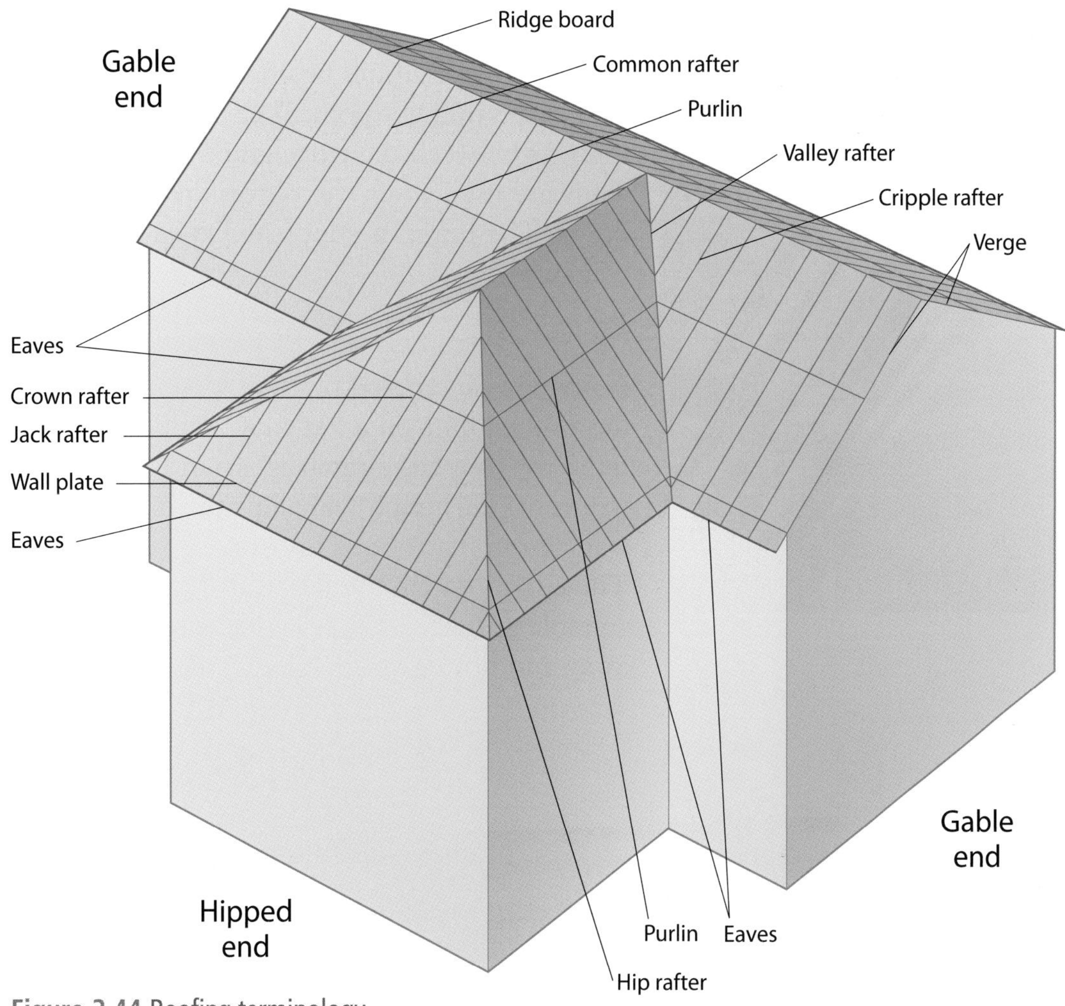

Figure 3.44 Roofing terminology

There are several different components that combine to form roofs. Some of the most common are covered briefly below.

These components combine to form three main types of roofs:

- flat roofs
- pitched roofs
- hipped roofs.

Ridge

A ridge is a timber board that runs the length of the roof and acts as a type of spine. It is placed at the apex of the roof structure. The uppermost ends of the rafters are then fixed to this. This gives the roof central support and holds the rafters in place.

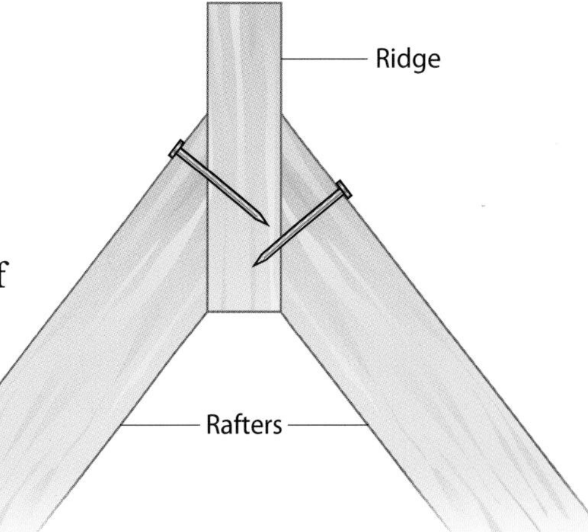

Figure 3.45 Ridge

Figure 3.46 Purlins

Purlins

Purlins are horizontal beams that support the roof at the midway point. They are placed midway between the ridge and the wall plate. They are used when the rafters are longer than 2.5 m. The purlin is supported at each end by gables.

Firrings

A firring is an angled piece of wood, laid on top of the joints on a flat roof. They provide a fall. This supplies a pitch of around 10° or less to a roof. This pitch will allow the draining of water from flat roofs. This drainage is vital. The edge – where the fall leads to – must have suitable guttering to allow rainwater to run away and not down the face of the wall.

Did you know?

A flat roof without drainage will have a lot of strain on it when rainwater collects there. This could lead the roof to collapse.

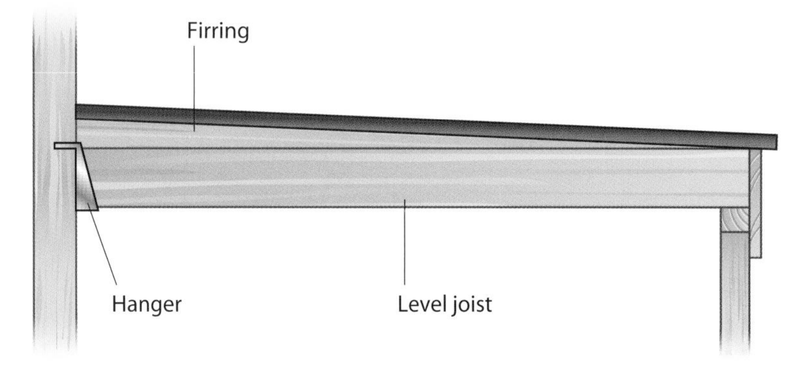

Figure 3.47 Firring

Battens

Battens are wood strips that are used to provide a fixing point for roof sheet or roof tiles. The spacing between the battens depends on the type of roof and they can be placed at right angles to the trusses or rafters. This makes them similar to purlins.

Some roofs use a grid pattern in both directions. This is known as a counter-batten system.

Figure 3.48 Battens

Did you know?

You may have heard the phrase 'batten down the hatches'. This comes from sailing when, in storms, the captain would order the crew to prevent water getting in through the hatches by sealing them with wooden strips – battening them down.

Wall plates

Wall plates are timber plates laid flat and bedded on mortar. These plates run along the wall to carry the feet of all trusses, rafters and ceiling joists. They can serve a similar role to lintels, but their main purpose is to bear and distribute the load of the roof across the wall. Because of the vital role that wall plates play in supporting the roof and spreading the pressure of its weight across the whole wall, it is important that they stay in place.

Wall plates are held in place by restraint straps along the wall (see Figure 3.50). These anchor the wall plates in place to prevent movement.

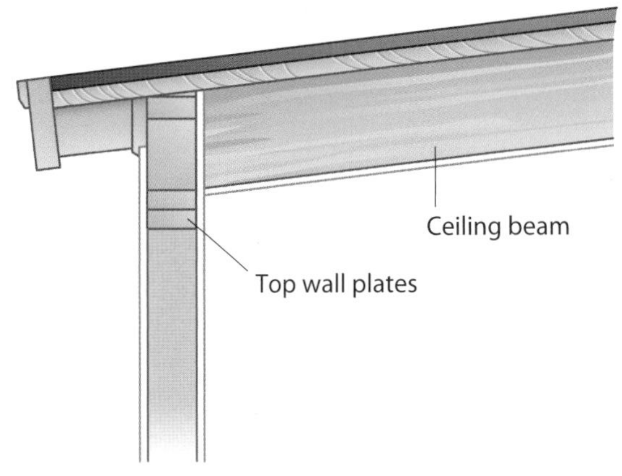

Figure 3.49 Wall plate

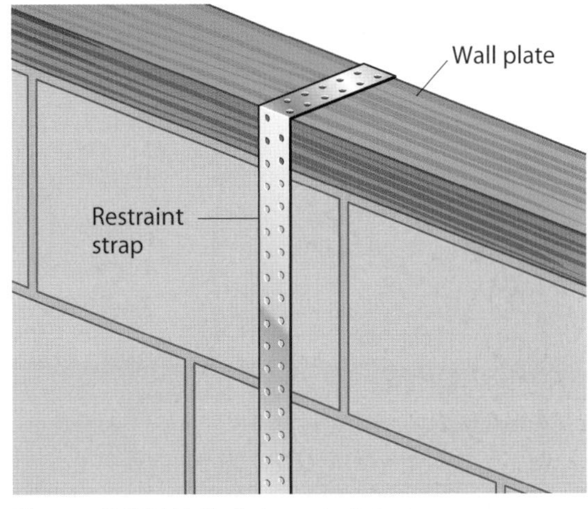

Figure 3.50 Wall plate restraint straps

Hangers and clips

Hangers, or clips, are galvanised metal clips that are used to fix trusses or joists in place on wall plates (see Figure 3.51). This anchors the trusses and joists to the wall plate, which is in turn firmly anchored to the wall. This gives the roof a stable and firm construction and helps it to avoid the pressures of wind and weather.

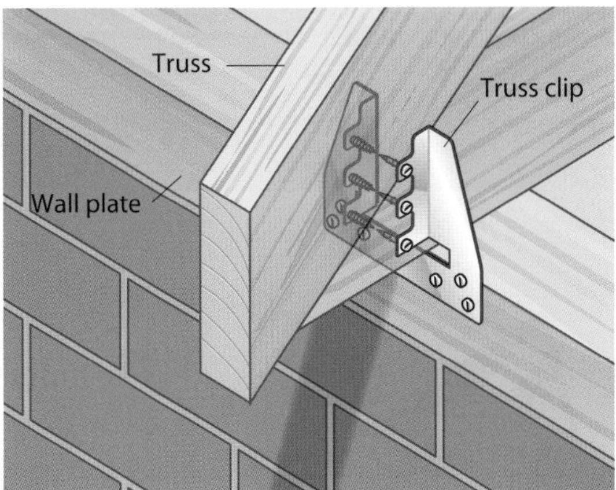

Figure 3.51 Truss clips

Bracing

Bracing are lengths of timber attached along the trusses. This holds them in place and helps to prevent any movement in high winds. Combined with the wall plates and truss clips, bracing is a major part of ensuring the stability of the roof in all conditions.

Figure 3.52 Bracing

Figure 3.53 Roofing felt

Felt

Felt is rolled over the top of the joists to provide a waterproof barrier. It is then fastened down to provide a permanent barrier. Overlapping the felt strips when placing them will help make this barrier even more effective, as will making sure that there are no air bubbles in the felt.

Slate and tiles

Slate is flat and easy to stack. The supplier will recommend the spacing between the battens on the roof. The slate is then laid onto the battens, with the bottom of each slate tile overlapping the top of the one below. The top and bottom rows are made up of shorter slates to provide this lapping for the slate below/above. This provides waterproofing to the roof. Slate is often nailed into place.

Did you know?

Some roofing tiles are made from slate.

Roofing tiles are made from concrete or clay. They are moulded or formed into a shape that allows them to overlap each other. This provides weatherproofing to the roof, similar to the technique used for slate.

Figure 3.54 Roofing slate

Figure 3.55 Roofing tiles

Figure 3.56 Roof flashing

Flashings

Flashings are made from aluminium or lead. They are used to provide water resistance around openings in a roof, such as chimneys or roof windows, or when a roof butts up to an existing wall. The joints can be sealed evenly by applying a mastic finish.

Metal flashing can be purchased in rolls or in sections specifically designed for certain roles – such as around a chimney.

Although there are several different types of roofing, all roofs will technically be either a flat roof or a pitched roof.

Flat roofs

A flat roof is any roof which has its upper surface inclined at an angle (also known as the fall, slope or pitch) not exceeding 10°.

A flat roof has a fall to allow rainwater to run off, preventing puddles forming as they can put extra weight on the roof and cause leaks. Flat roofs will eventually leak, so most are guaranteed for only 10 years (every 10 years or so the roof will have to be stripped back and re-covered). Today, **fibreglass** flat roofs are available that last much longer, so some companies will give a 25-year guarantee on their roof. Installing a fibreglass roof is a job for specialist roofers.

Flat roofs use bitumen to provide a waterproof seal. This is a black, sticky substance, a type of pitch or tar that turns into a liquid when it is heated. It is then used on flat roofs to provide a waterproof seal. It is a flexible DPC.

The amount of fall should be sufficient to clear water away to the outlet pipe(s) or guttering as quickly as possible across the whole roof surface. This may involve a single direction of fall or several directional changes of fall, such as:

- single fall into guttering
- double fall into guttering
- double fall to internal funnel outlet
- double fall to corner funnel outlet.

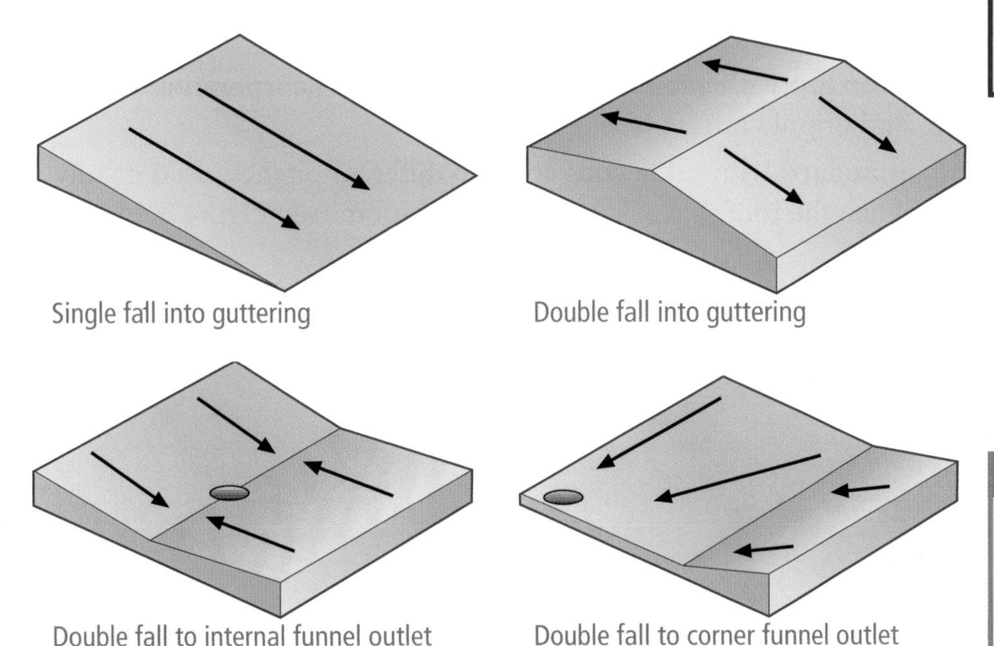

Figure 3.58 Falls on a flat roof and direction of fall

Remember

A roof with an angle of *more* than 10° is classified as a pitched roof.

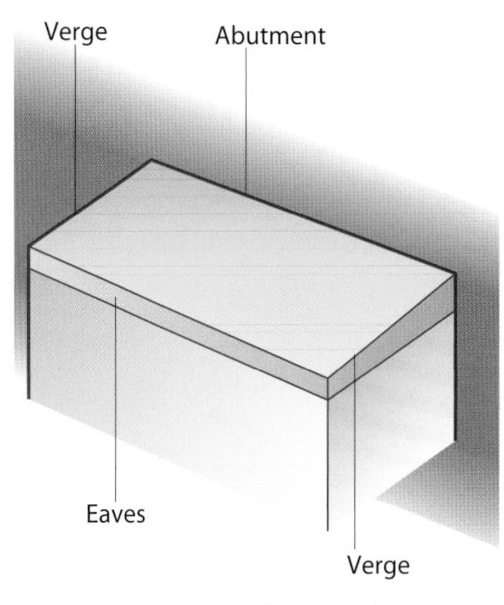

Figure 3.57 Flat roof terminology

Key term

Fibreglass – a material made from glass fibres and a resin that starts in liquid form and then hardens to become very strong.

Did you know?

If a puddle forms on a roof and is not cleared away quickly, over a period of years the water will eventually work its way through.

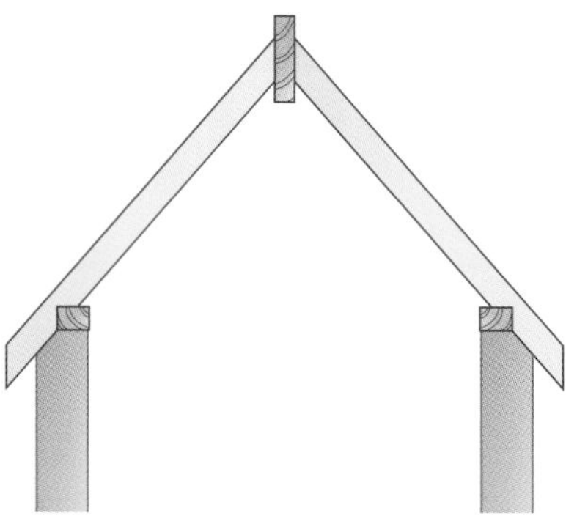

Figure 3.59 Single roof

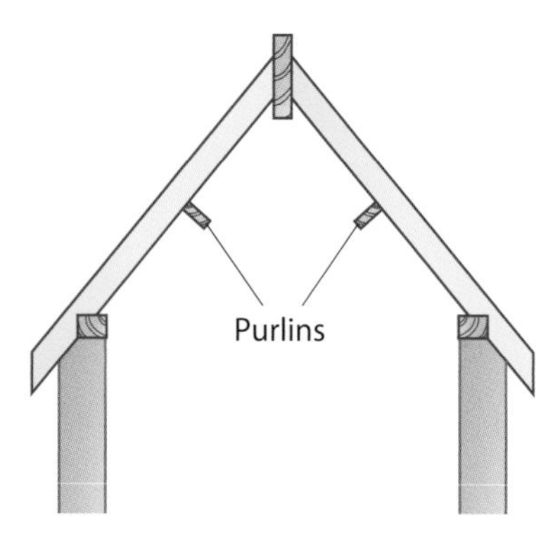

Figure 3.60 Double roof

Safety tip

Roofing requires working at height, so always use the appropriate access equipment (i.e. a scaffold or at least properly secured ladders, trestles or a temporary working platform).

Pitched roofs

There are several different types of pitched roof, but most are constructed in one of two ways:

- **Trussed roof** – a prefabricated pitched roof, specially manufactured prior to delivery on site, saving timber as well as making the process easier and quicker. Trussed roofs can also span greater distances without the need for support from intermediate walls.
- **Traditional roof** – a roof entirely constructed on site from loose timber sections using simple jointing methods.

Roof types

A pitched roof can be constructed either as a single roof, where the rafters do not require any intermediate support (see Figure 3.59), or a double roof where the rafters are supported (see Figure 3.60). Single roofs are used over a short span, such as a garage; double roofs are used to span a longer distance, such as a house or factory.

There are many different types of pitched roof, including:

- **mono pitch** with a single pitch (see Figure 3.61)
- **lean-to** with a single pitch, which butts up to an existing building (see Figure 3.62)
- **duo pitch** with gable ends (see Figure 3.63)
- **hipped** with hip ends incorporating crown, hip and jack rafters (see Figure 3.64)
- **over hip** with gable ends, hips and valleys incorporating valley and cripple rafters (see Figure 3.65)
- **mansard** with gable ends and two different pitches, used mainly when the roof space is to be used as a room (see Figure 3.66)

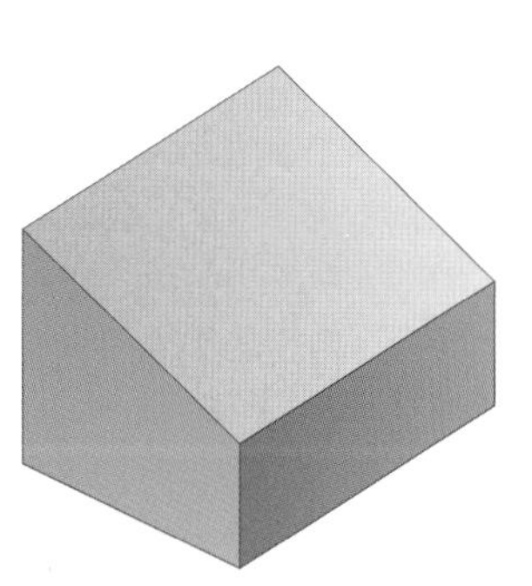

Figure 3.61 Mono pitch roof

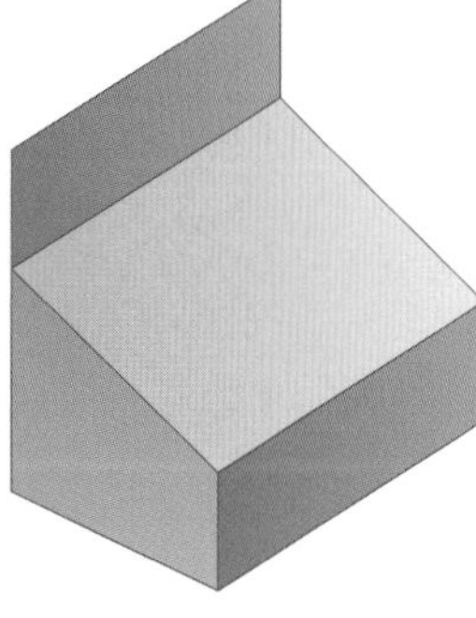

Figure 3.62 Lean-to roof

Figure 3.63 Duo pitch roof with gable ends

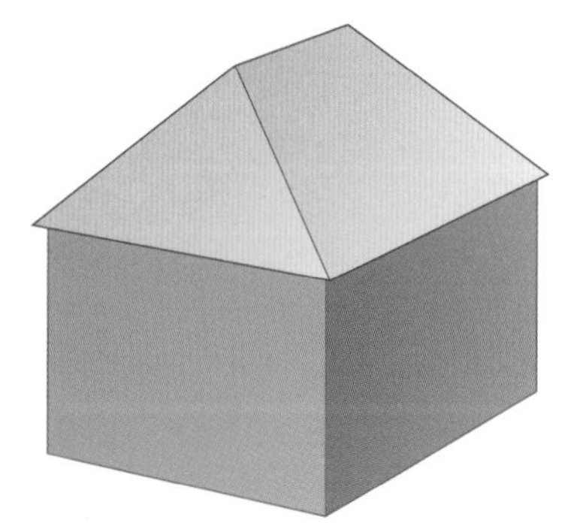

Figure 3.64 Hipped roof

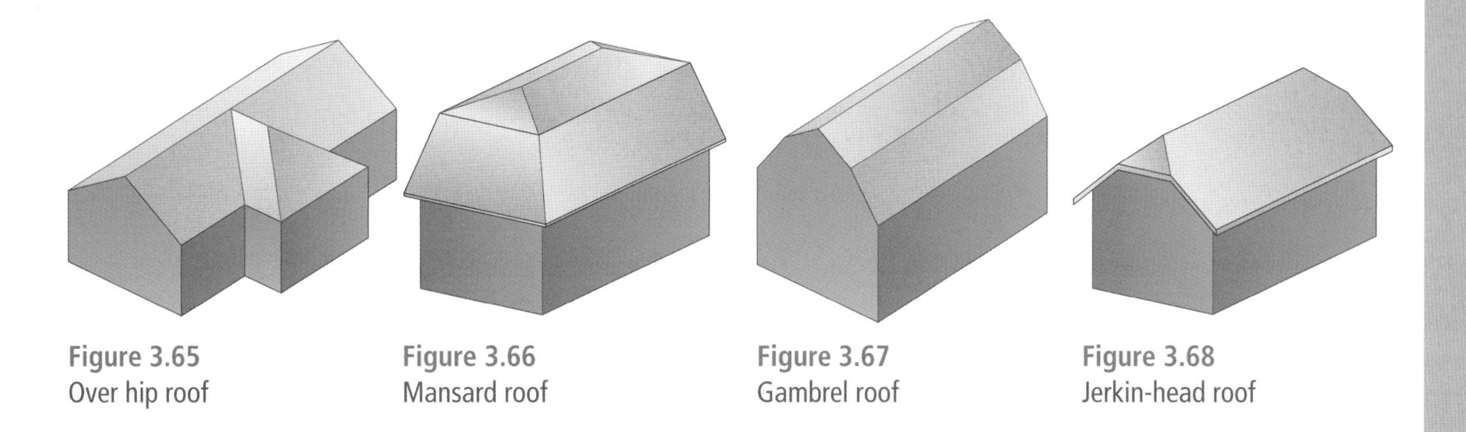

Figure 3.65
Over hip roof

Figure 3.66
Mansard roof

Figure 3.67
Gambrel roof

Figure 3.68
Jerkin-head roof

- **gambrel or gable hip** – double-pitched roof with a small gable (gablet) at the ridge and the lower part a half-hip (see Figure 3.67)
- **jerkin-head or barn hip** – a double-pitched roof hipped from the ridge part-way to the eaves, with the remainder gabled (see Figure 3.68).

The type of roof to be used will be selected by the client and architect.

Trussed rafters

Most roofing on domestic dwellings now comprises factory-made trussed rafters (see Figure 3.69 on the next page). These are made of stress graded, **PAR** timber to a wide variety of designs, depending on requirements. All joints are butt jointed and held together with fixing plates, face fixed on either side. These plates are usually made of galvanised steel and either nailed or factory pressed. They may also be **gang-nailed** gusset plates made of 12 mm resin bonded plywood.

One of the main advantages of this type of roof is the clear span achieved, as there is no need for intermediate, load-bearing partition walls. Standard trusses are strong enough to resist the eventual load of the roofing materials. However, they are not able to withstand pressures applied by lateral bending. Hence, damage is most likely to occur during delivery, movement across site, site storage or lifting into position.

Wall plates are bedded as described on page 95. Following this, the positions of the trusses can be marked at a maximum of 600 mm between centres along each wall plate. The sequence of operations then varies between gable and hipped roofs.

Key terms

PAR – a term used for timber that has been 'planed all round'.

Gang-nailed – galvanised plate with spikes used to secure butt joints.

Remember

Never alter a trussed rafter without the structural designer's approval.

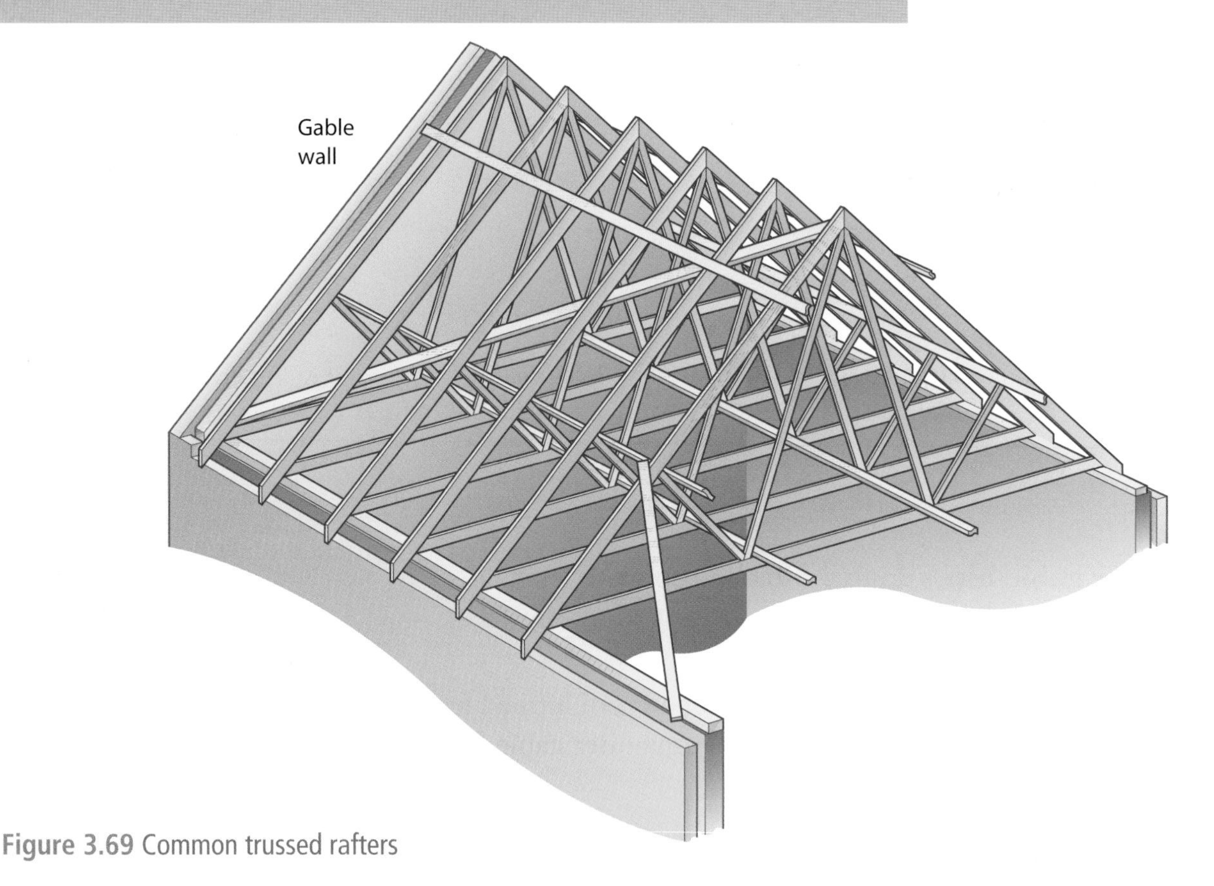

Figure 3.69 Common trussed rafters

Hipped roofs

In a fully hipped roof there are no gables and the eaves run around the perimeter, so there is no roof ladder or bargeboard.

Marking out for a hipped roof

All bevels or angles cut on a hipped roof are based on the right-angled triangle, and the roof members can be set out using the following methods:

- **Roofing ready reckoner** – a book that lists in table form all the angles and lengths of the various rafters for any span or rise of roof.
- **Geometry** – working with scale drawings and basic mathematic principles to give you the lengths and angles of all rafters.

In this unit, we will concentrate on geometry.

Pythagoras' theorem

When setting out a hipped roof, you need to know Pythagoras' theorem. Pythagoras states that 'the square on the hypotenuse of a right-angled triangle is equal to the sum of the squares on the other two sides'. For the carpenter, the 'hypotenuse' is the rafter length, while the 'other two sides' are the run and the rise.

From Pythagoras' theorem, we get this calculation:

$A = \sqrt{B^2 + C^2}$ ($\sqrt{}$ means the square root and ^{2} means squared)

If we again look at our right-angled triangle, we can break it down to:

A = the rafter length – the distance we want to know

B = the rise

C = the run

Therefore, we have all we need to find the length of our rafter (A):

$A = \sqrt{4^2 + 3^2}$

$A = \sqrt{(4 \times 4) + (3 \times 3)}$

$A = \sqrt{16 + 9}$

$A = \sqrt{25}$

$A = 5$

So, our rafter is 5 m long.

Did you know?

The three angles in a triangle always add up to 180°.

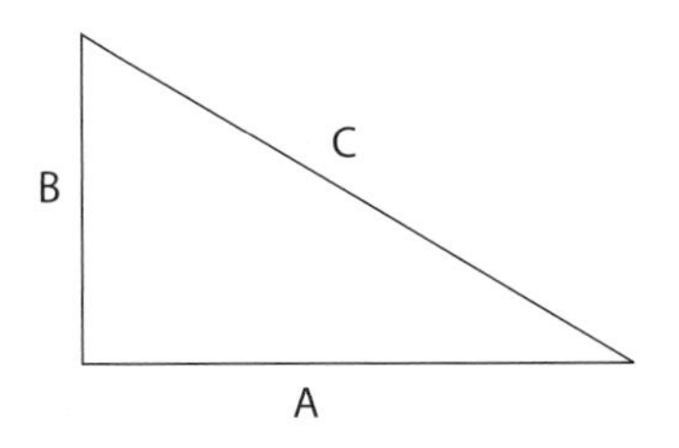

Functional skills

While working through this section, you will be covering **FM 2.2.1** and **FM 2.2.2**, which relate to applying a range of mathematics skills to find solutions, using appropriate checking procedures and evaluating their effectiveness at each stage.

Finishing a roof at the gable and eaves

The two types of finish for a gable end are:

- **a flush finish** – the bargeboard is fixed directly onto the gable wall
- **a roof ladder** – a frame built to give an overhang and to which the bargeboard and soffit are fixed (see Figure 3.70).

The most common way is to use a roof ladder which, when creating an overhang, stops rainwater running down the face of the gable wall.

The continuation of the fascia board around the verge of the roof is called the bargeboard. Usually the bargeboard is fixed to the roof ladder and has a built-up section at the bottom to encase the wall plate.

The simplest way of marking out the bargeboard is to temporarily fix it in place and use a level to mark the plumb and seat cut.

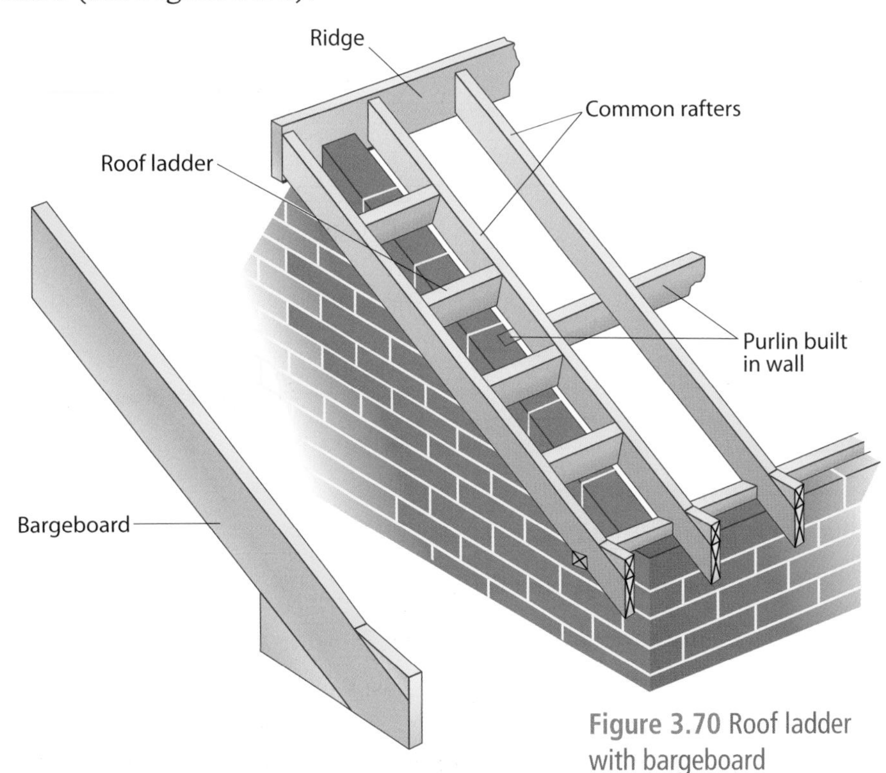

Figure 3.70 Roof ladder with bargeboard

Eaves

Did you know?

Without soffit ventilation, air cannot flow through the roof space, which can cause problems such as dry rot.

The eaves are where the lower part of the roof is finished where it meets the wall, and they incorporate fascia and soffit. The fascia is the vertical board fixed to the ends of the rafters. It is used to close the eaves and allow fixing for rainwater pipes. The soffit is the horizontal board fixed to the bottom of the rafters and the wall. It is used to close the roof space to prevent birds or insects from nesting there, and usually incorporates ventilation to help prevent rot.

There are various ways of finishing a roof at the eaves; we will look at the four most common ones.

Flush eaves

Here the eaves are finished as close to the wall as possible. There is no soffit, but a small gap is left for ventilation.

Open eaves

Open eaves are where the bottom of the rafter feet are planed as they are exposed. The rafter feet project beyond the outer wall and eaves boards are fitted to the top of the rafters to hide the underside of the roof cladding. The rainwater pipes are fitted via brackets fixed to the rafter ends.

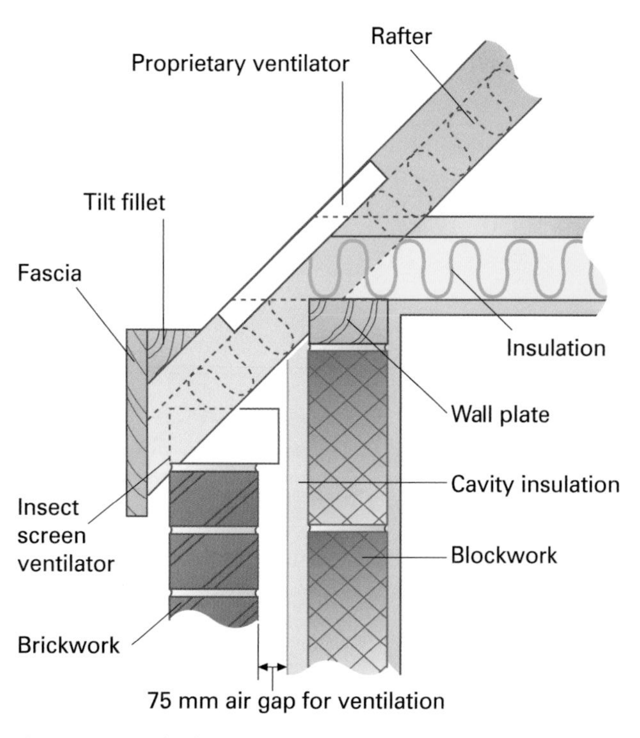

Figure 3.71 Flush eaves

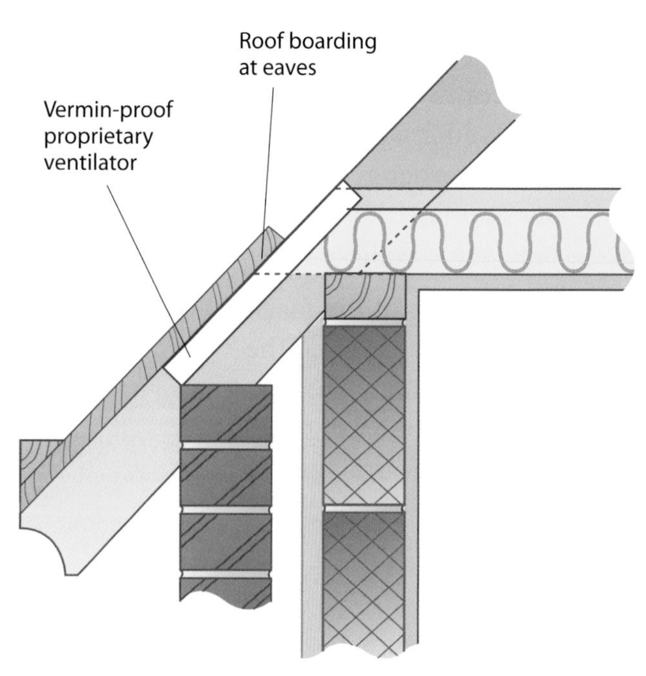

Figure 3.72 Open eaves

Closed eaves

Closed eaves are completely closed or boxed in. The ends of the rafters are cut to allow the fascia and soffit to be fitted. The roof is ventilated either by ventilation strips incorporated into the soffit, or by holes drilled into the soffit with insect-proof netting over them. If closed eaves are to be re-clad due to rot, you must ensure that the ventilation areas are not covered up.

Key term

Sprocket – a piece of timber bolted to the side of the rafter to reduce the pitch at the eaves.

Sprocketed eaves

Sprocketed eaves are used where the roof has a sharp pitch. The **sprocket** reduces the pitch at the eaves, slowing down the flow of rainwater and stopping it overshooting the guttering. Sprockets can either be fixed to the top edge of the rafter or bolted onto the side.

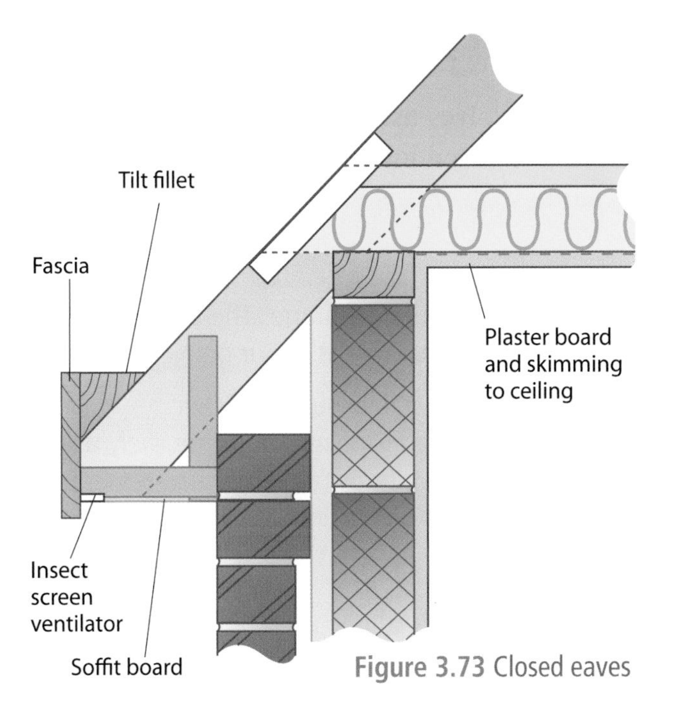

Figure 3.73 Closed eaves

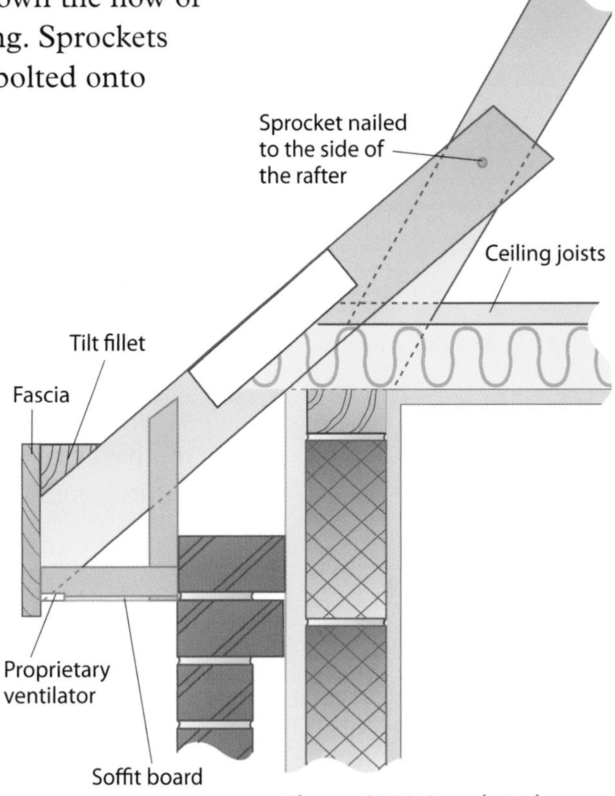

Figure 3.74 Sprocketed eaves

Roof coverings

Once all the rafters are on the roof, the final thing is to cover it. There are two main methods of covering a roof, each using different components. Factors affecting the choice of roof covering include what the local weather is like and what load the roof will have to take.

Did you know?

Where you live may have an effect on your choice of roof: in areas more prone to bad weather, the roof will need to be stronger.

Method 1

This method (outlined on the next page) is usually used in the north of the UK where the roof may be expected to take additional weight from snow:

1. Clad the roof surface with a man-made board such as OSB (oriented strand board) or exterior grade plywood.
2. Cover the roof with roofing felt, starting at the bottom and ensuring the felt is overlapped to stop water getting in.
3. Fit the felt battens (battens fixed vertically and placed to keep the felt down while allowing ventilation) and the tile battens (battens fixed horizontally and accurately spaced to allow the tiles to be fitted with the correct overlap).
4. Finally, fit the tiles and cement on the ridge.

Method 2

This is the most common way of covering a roof:

1. Fit felt directly onto the rafters.
2. Fit the tile battens at the correct spacing.
3. Fit the tiles and cement on the ridge.

Another way to cover a roof involves using slate instead of tiles. Slate-covered roofing is a specialised job as the slates often have to be cut to fit, so roofers usually carry this out.

Working life

You have been asked to quote for building a garage. The client is unsure about what type of roof to have. They ask your opinion.

What type of roof would you select? Just like with the other components of the garage, you will need to think about the likely use of the roof, the shape and design that will fit the client's requirements and the conditions it will probably be experiencing. Why would you select that type of roof? What could influence your selection?

Key term

U-value – a measure of thermal transmittance through a building component, usually a roof, wall, window, door or floor.

Energy conservation in roofing

A roof serves several functions: it protects the building from the elements, directs water into gutters and makes the building look aesthetically pleasing. A roof must also meet the requirements of Part L of the Building Regulations, which details the levels of insulation that must be included to achieve the required **U-value**.

Waterproofing, ventilation and insulation are crucial in making a roof efficient and long lasting, and contribute to the energy efficiency and sustainability of the whole structure.

Remember

A pitched roof is a roof laid to a fall of at least 10°.

Pitched roofs

A pitched roof is made from the following:

- Softwood rafters span from the ridge to the eaves and are fixed to a wall plate, which is strapped to the internal skin of the wall.

- The rafters are then covered with a felt or a breathable membrane, which stops moisture entering the building if the roof tiles are weakened.
- The felt or membrane is held in place with tile battens manufactured from treated softwood, which are fixed to the rafters using galvanised nails.

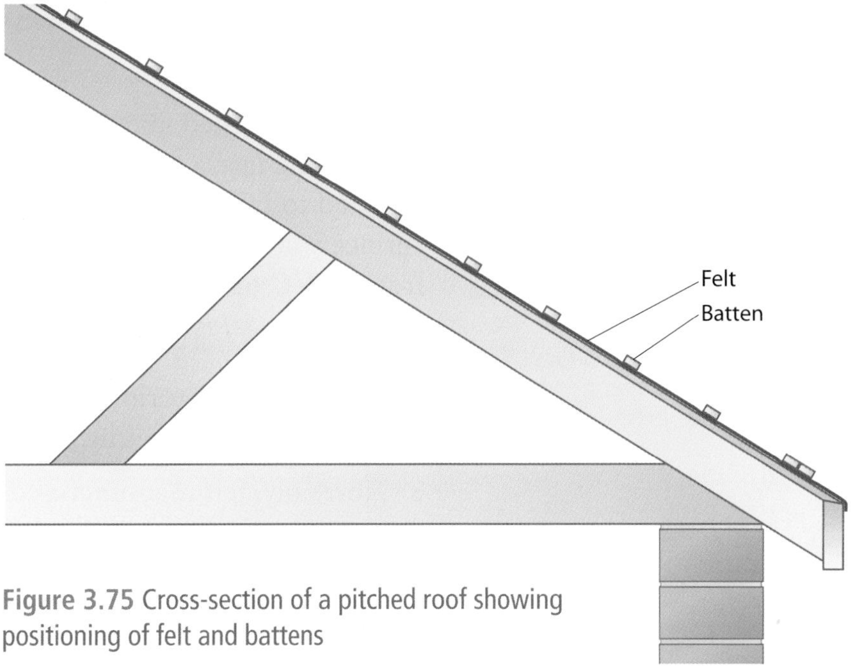

Figure 3.75 Cross-section of a pitched roof showing positioning of felt and battens

Insulation

The insulation can be placed in two locations:

- Between and over ceiling joists – up to 300 mm of mineral wool insulation (manufactured from glass or rock) is laid, 150 mm between the joists and 150 mm at 90° to this, to form the full thickness.
- Between or over rafters – form insulation boards are cut and fitted between roof timbers or, with a loft conversion, are laid over the joists before the tile battens and tiling are fitted.

Find out

Find an insulation manufacturer's website and research the methods of incorporating insulation into a pitched roof.

Waterproofing, damp proofing and ventilation

The ridge is where the two rafters meet at the top of the roof. A fascia and soffit complete the eaves detail of the roof and make it waterproof. Ventilation is provided to prevent the air stagnating, which could lead to the roof timbers rotting. Ventilation may be provided to the roof by placing vents into the soffit or on top of the fascia. In this case, insulation is placed in two layers: one between the ceiling joists, and the other perpendicular to the first layer.

Remember

You will need to make sure that the roof insulation does not block the movement of the ventilating air at the eaves.

Valleys are formed where two roofs meet within an internal corner. Here a valley board is formed using plywood, which is covered with a flashing to stop water entering. All roof abutments, such as chimneys, must have flashings placed around them to make sure that water cannot get in and cause problems with damp and rot.

Functional skills

This task will allow you to practise **FE 2.2.1–2.2.3**, which relate to selecting and using texts, summarising information and ideas, and identifying the purpose of texts.

Roof tiles

Roof tiles tend to be chosen to fit in with the local environment, and are often made of local materials, for example, slate in Wales. Most modern roof tiles are manufactured from concrete and have an interlocking system to stop water getting into the roof space.

Guttering

Did you know?

In the past, guttering used to be manufactured from timber or cast iron.

With both pitched and flat roofs (see below), guttering must be fixed to the fascia board laid to falls, so that rainwater can be directed to fall pipes. The fall pipes connect the guttering to the surface water drainage system. Modern guttering is manufactured from uPVC or aluminum.

Flat roofs

Flat roofs have notoriously short lifespans due to the low pitch and the continual heating and cooling cycle of the environment.

- Normally, felt manufactured with bitumen is used to cover a plywood decking and is laid in three layers, each bonded with bitumen.
- Falls are produced by nailing timber firrings to the roof joists.
- Finally, the roof is covered with a layer of white spar chippings, which reflect the sunlight and prevent solar heat gain from damaging the roof.

With flat roofs, the insulation is set between the joists within the ceiling void or as cut-to-falls insulation, which is laid over the plywood decking and covered with roofing felt. Both methods need a vapour barrier to be fitted, to resist the passage of moisture.

Asphalt is a naturally occurring product. When it is heated up to melting point it can be laid in a single layer onto the prepared roof, forming a single, impenetrable barrier.

Materials used in external walling

External walls come in a variety of types, but the most common is cavity walling. Cavity walling is simply two masonry walls built parallel to each other, with a gap between acting as the cavity. The cavity wall acts as a barrier to weather, with the outer leaf preventing rain and wind penetrating the inner leaf. The cavity is usually filled with insulation to prevent heat loss.

Cavity walls

Cavity walls mainly consist of a brick outer skin and a blockwork inner skin. There are instances where the outer skin may be made of block and then rendered or covered by tile hanging. The minimum cavity size allowed is 75 mm, but the cavity size is normally governed by the type and thickness of insulation to be used and whether the cavity is to be fully filled or partially filled with insulation.

The thickness of blocks used will also govern the overall size of the cavity wall. On older properties, the internal blocks were always of 100 mm thickness. Nowadays, due to the emphasis on energy conservation and efficiency, blocks are more likely to be 125 mm or more.

In all cases, the cavity size will be set out to the drawing, with overall measurements specified by the architect and to local authority requirements.

Once the **foundations** have been concreted the **sub-structural walling** can be constructed, usually by using blocks for both walls (see Figure 3.76).

In some situations trench blocks may be used below ground level and then traditional cavity work constructed up to the damp proof course (DPC). A horizontal DPC must be inserted at a minimum height of 150 mm above ground level to both walls. This is to prevent damp rising, below ground, up through the block and brickwork to penetrate to the inside. The cavity must also be filled with weak concrete to ground level to help the sub-structural walling resist lateral pressure.

Cavity walls above DPC

The older traditional way to build a cavity wall is to build the brickwork first and then the blockwork. Now, due to the introduction of insulation into the cavity, the blockwork is generally built first, especially when the cavity is partially filled with insulation. This is because the insulation requires holding in place against the internal block wall, by means of special clips that are attached to the **tie wires**. In most cases the clips are made of plastic as they do not rust or rot. The reason for clipping the insulation is to stop it from moving away from the blocks, which would cause the loss of warmth to the interior of the building, as well as causing a possible **bridge** of the cavity, which could cause a damp problem.

The brick courses should be gauged at 75 mm per course, but sometimes course sizes may change slightly to accommodate window or door heights. In most instances, these positions and measurements are designed to work to the standard gauge size. This will also allow the blockwork to run level at every third course of brick, although the main reason will be explained in the tie wires section on the next page.

Remember

The correct size must be used for the internal wall, with the cavity size to suit.

Key terms

Foundations – concrete bases supporting walls.

Sub-structural walling – brickwork between the foundation concrete and the horizontal damp-proof-course (DPC).

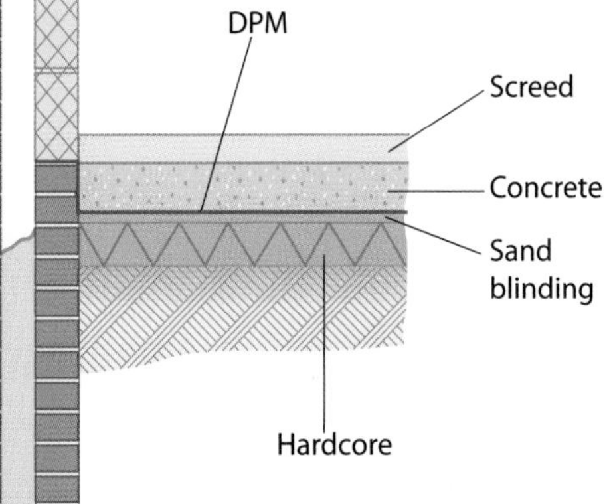

Figure 3.76 Section of sub-structural walling

Key terms

Tie wires – stainless steel or plastic fixings to tie cavity walls together.

Bridge – where moisture can be transferred from the outer wall to the inner leaf by material touching both walls.

Key term

Window head – top of a window.

On most large sites, patent types of corner profile are used rather than building traditional corners (see Figure 3.77). These allow the brickwork to be built faster and, if set up correctly, more accurately. But they must also be marked for the gauge accurately and it makes sense to mark window sill heights or **window heads** and door heights, so they do not get missed, which would result in brickwork being taken down.

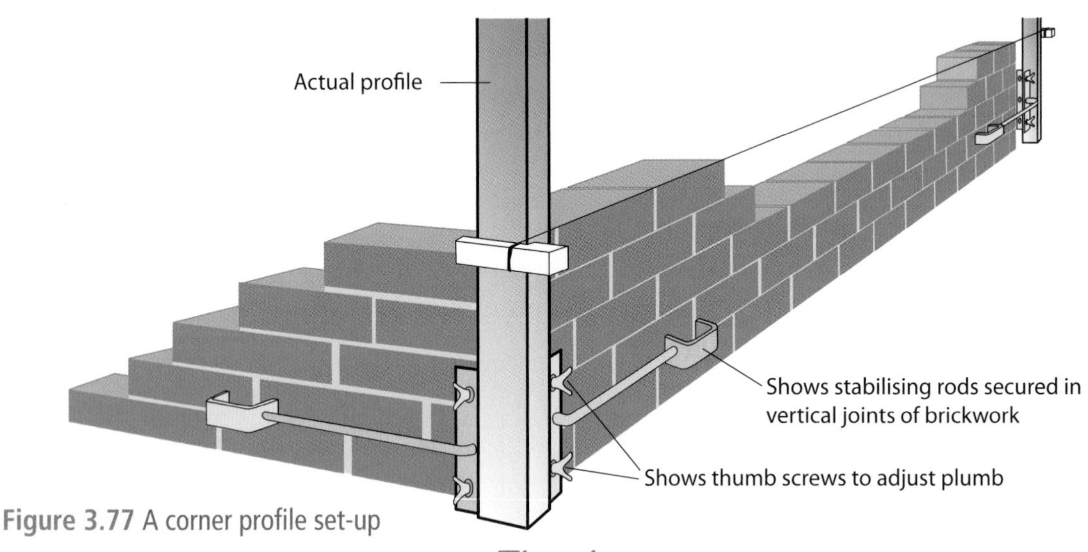

Figure 3.77 A corner profile set-up

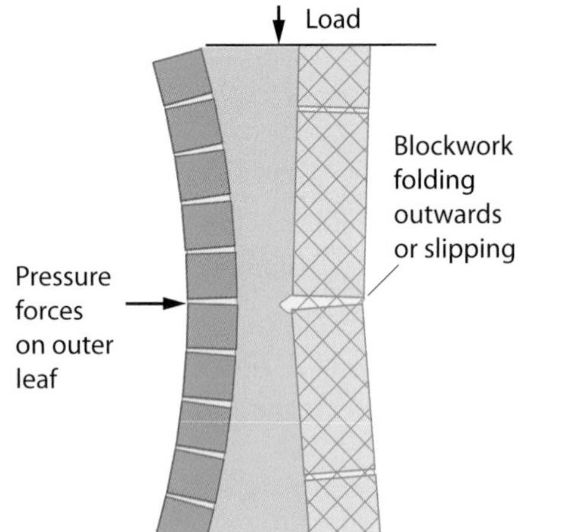

Figure 3.78 Section of wall without tie wire

Tie wires

Tie wires are a very important part of a cavity wall as they tie the internal and external walls together, resulting in a stronger job. If we built cavity walls to any great height without connecting them together, the walls would be very unstable and could possibly collapse (see Figure 3.78).

A tie wire should be:

- rust-proof
- rot-proof
- of sufficient strength
- able to resist moisture.

There are many designs of tie wire currently on the market, with a wide selection suitable for all types of construction methods. One of the most common types used when tying together brick and block leaves is the masonry general purpose tie. These ties are made from very strong stainless steel, and incorporate a twist in the steel at the mid point of the length. This twist forms a drip system, which prevents the passage of water from the outer to the inner leaf of the structure.

Figure 3.79 General purpose tie wire

You must take care to keep tie wires clean when they are placed in the wall: if bridging occurs, it may result in moisture penetrating the internal wall.

The positioning and density of tie wires

In cavity walling where both the outer and inner leaves are 90 mm or thicker, you should use ties at not less than 2.5 per square metre, with 900 mm maximum horizontal distance by a maximum 450 mm vertical distance and staggered.

At positions such as vertical edges of an opening, unreturned or unbonded edges and vertical expansion joints, you need to use additional ties at a maximum of 300 mm in height (usually 225 mm to suit block course height) and located not more than 225 mm from the edge. Tie wires should be bedded into each skin of the cavity wall to a minimum distance of 50 mm.

Key term

Cavity batten – a timber piece in a cavity to prevent mortar droppings falling down the cavity.

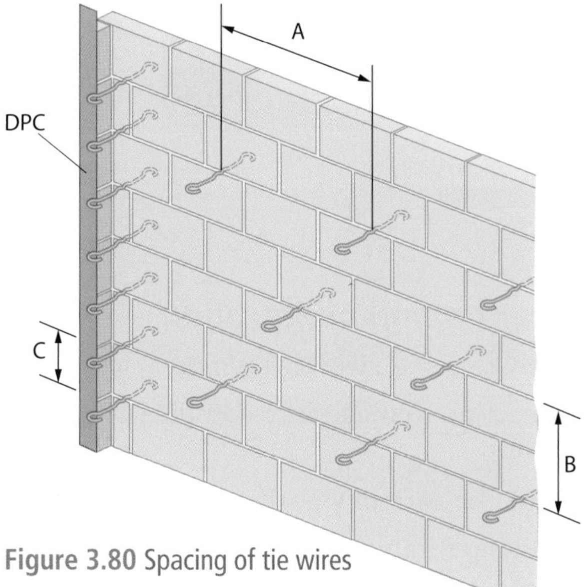

A 900 mm maximum horizontal distance

B 450 mm maximum vertical distance

C Additional ties, 300 mm maximum vertical distance

Figure 3.80 Spacing of tie wires

Did you know?

Any batten can be used as long as the width is the same as the cavity space.

Keeping a cavity wall clean

It is important to keep the cavity clean to prevent dampness. If mortar is allowed to fall to the bottom of the cavity it can build up and allow the damp to cross and enter the building. Mortar can also become lodged on the tie wires and create a bridge for moisture to cross. We can prevent this by the use of **cavity battens** – pieces of timber the thickness of the cavity laid on to the tie wires and attached by wires or string (to prevent dropping down the cavity) to the wall and lifted alternately as the wall progresses.

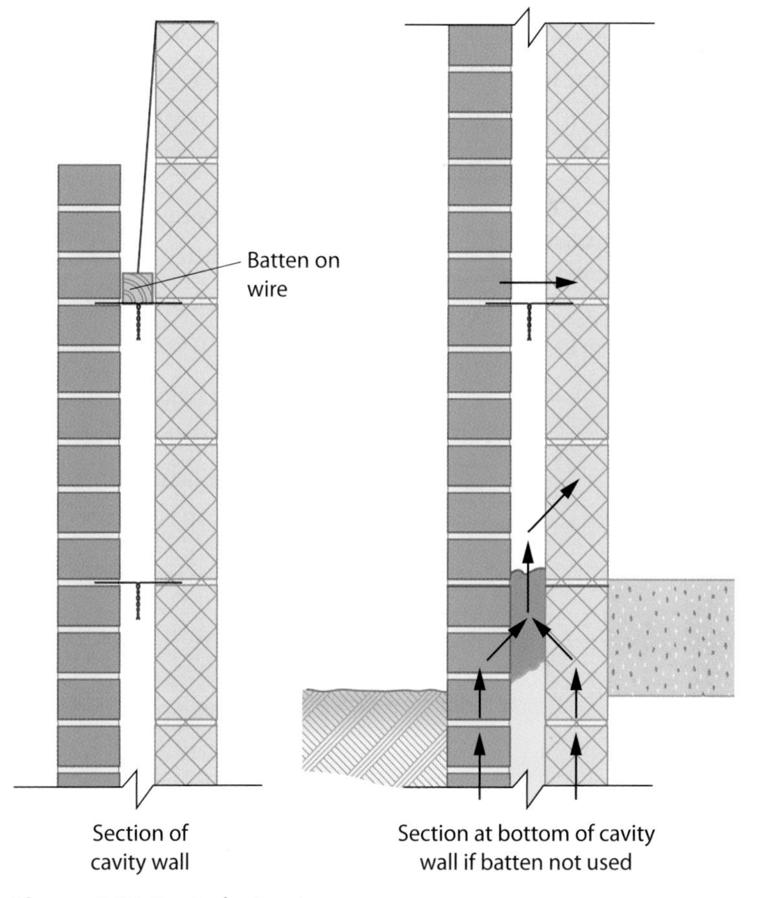

Figure 3.81 Cavity batten in use

The bottom of the wall can be kept clean by either leaving bricks out or bedding bricks with sand so they can be taken out to clean the cavity. These are called core holes and are situated every fourth brick along the wall to make it easy to clean out each day. Once the wall is completed the bricks are bedded into place.

Remember

Clean at the end of each day, as the mortar will go hard. Good practice is to lay hessian across the tie wires and remove it at the end of each day to clean any mortar that has dropped down the cavity.

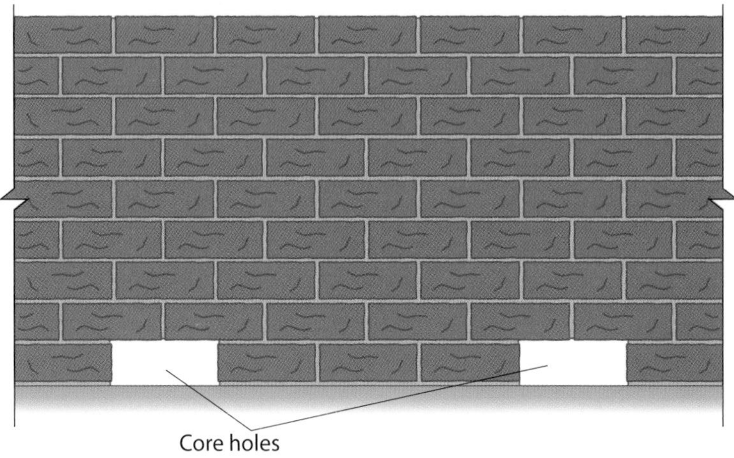

Figure 3.82 Core holes

Key term

Weep hole – small opening left in the outer wall of masonry to allow water out of the building.

Remember

You must read your drawings and specifications carefully to see what is required and always fix insulation to manufacturers' instructions.

Steps to take to prevent damp penetration

- Set out openings carefully to avoid awkward bonds.
- Care is needed in construction to make sure dampness or water does not enter the building.
- DPCs and tie wires should be carefully positioned.
- Steel cavity lintels should have minimum 150 mm bearings solidly bedded in the correct position.
- **Weep holes** should be put in at 450 mm centres immediately above the lintel in the outer leaf.

No insulation has been shown in the drawings because they only show one situation. In most cavity wall construction, insulation of one kind or another will have to be incorporated to satisfy the current Building Regulations.

Fire spread

In addition to the prevention of damp penetration and cold bridging, there is a requirement under the Building Regulations that cavities and concealed spaces in a structure or fabric of a building are sealed by using cavity barriers or fire stopping. This cuts down the hidden spread of smoke or flames in the event of fire breaking out in a building.

Closing at eaves level

The cavity walls have to be 'closed off' at roof level for two main reasons:

- to prevent heat loss and the spread of fire
- to prevent birds or vermin entering and nesting.

This area of the wall is where the roof is connected by means of a timber wall plate bedded on to the inner leaf. The plate is then secured by means of restraint straps that are galvanised 'L' shaped straps screwed to the top of the wall plate and down the blockwork. This holds the roof structure firmly in place and also prevents the roof from spreading under the weight of the tiles, etc. The minimum distance that the straps should be apart is 1.2 m. In some instances they may be connected directly from the roof truss to the wall.

If a gable wall is required, restraint straps should be used to secure the roof to the end wall (see Figure 3.83).

The external wall can be built to the height of the top of the truss so as not to leave gaps, or 'closed off' by building blocks laid flat to cover the cavity above the external soffit line from inside, avoiding damp penetration. In some instances the cavity may be left open, with the cavity insulation used as the seal.

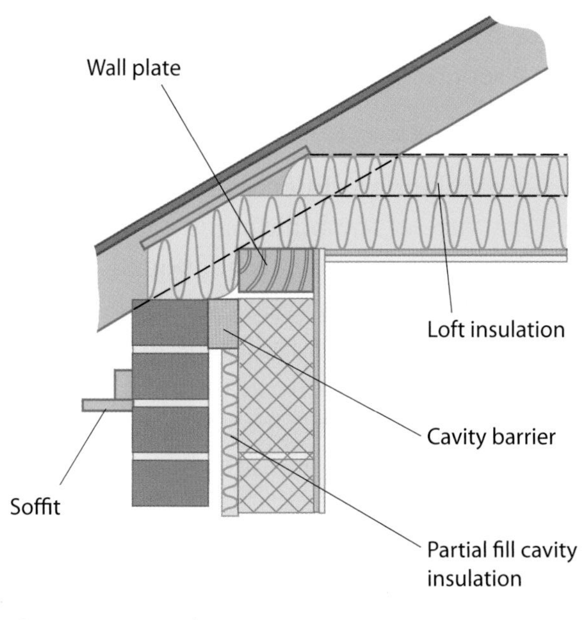

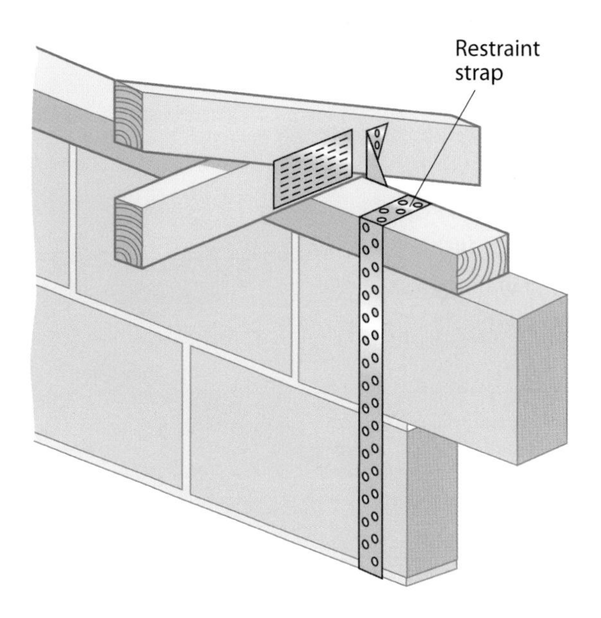

Figure 3.83 Roof section

Remember

When injecting insulation, great care must be taken not to drill the bricks as they will be difficult and costly to replace.

Insulation properties of cavity walls

Cavity walls are insulated mainly to prevent heat loss and therefore save energy. The Building Regulations tell us how much insulation is required in various situations, and in most situations this is stipulated in the specification for the relevant project to obtain planning permission from the local council.

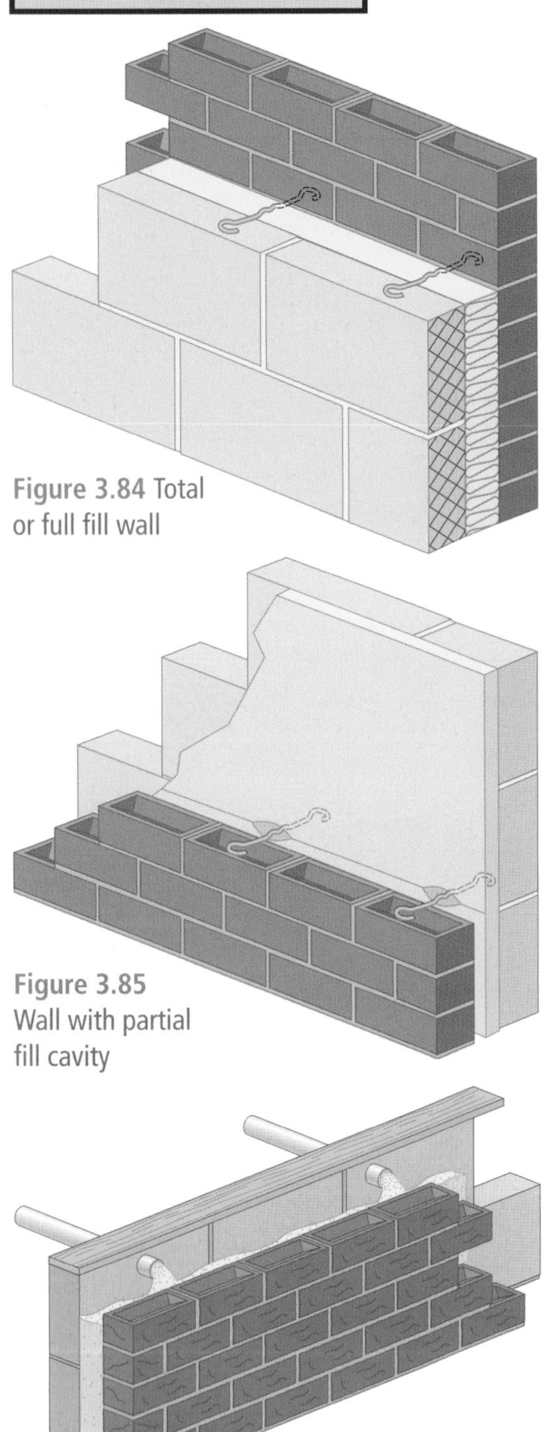

Figure 3.84 Total or full fill wall

Figure 3.85 Wall with partial fill cavity

Figure 3.86 Wall being injected

Cavity insulation can be either Rockwool or polystyrene beads.

There are three main ways to insulate the cavity:

- total or full fill
- partial fill
- injection (after construction).

Total or full fill

Figure 3.84 shows a section of a total fill cavity wall. The cavity is completely filled with insulation 'batts' as the work proceeds. The batts are 450 mm × 1200 mm, are made of mineral fibres, and are placed between the horizontal wall ties.

Partial fill

Figure 3.85 shows a partial fill cavity, where the cavity insulation batts are positioned against the inner leaf and held in place by a plastic clip. More tie wires than usual are used to secure the insulation in place.

Injection

This is where the insulation is injected into the cavity after the main structure of the building is complete. Holes are drilled into the inner walls at about 1 m centres and the insulation is pumped into the cavity (see Figure 3.86). The two main materials used are Rockwool fibreglass or polystyrene granules. The holes are then filled with mortar. If an older property was injected, the holes would be drilled into the external mortar joints.

There are three key points regarding insulation of cavity walls:

- Handle and store insulation material carefully to avoid damage or puncturing.
- Cavities should be clean.

- Read drawing specifications and read manufacturers' instructions carefully.

Timber framed construction

This type of timber-frame construction is a recent method of construction, and is highly efficient at preventing heat loss. In timber-framed construction, the traditional internal skin of blockwork is replaced by timber-framed panels, which support the load of the structure. The breather membrane allows the passage of vapour in one direction only, so the timber can 'breathe' but moisture is not allowed back in – the internal vapour barrier does not allow moisture to pass into the plasterboard.

The insulation within timber-framed construction is kept 'warm' by being placed within the warm side of the construction. It is protected by the vapour barrier and breather membrane.

The outside skin of a timber-framed building can be clad in several different materials, from a traditional brick skin of facing brickwork to rendered blockwork, which is painted.

Timber-kit houses are becoming more and more popular as they can be erected to a windtight and watertight stage within a few days.

Remember

There must be a way for any trapped moisture to be 'breathed out' or the timbers could deteriorate.

Find out

Find a breather membrane supplier and have a close look at the specification for this material.

Find out

Find out the cost of a standard timber-kit panel, then compare the price to building an area the same size in traditional cavity walling.

Constructing timber-kit houses

The building consists of timber sections preformed and then taken to the site and erected. These can be made to specific designs from an architect. Timber-kit house construction starts off in exactly the same way as any house build, with the foundations and the cavity wall built up to DPC level.

The timber panels are then lifted into place (usually by crane) and are bolted together. Once the wall panels are in place, the exterior face brickwork can begin.

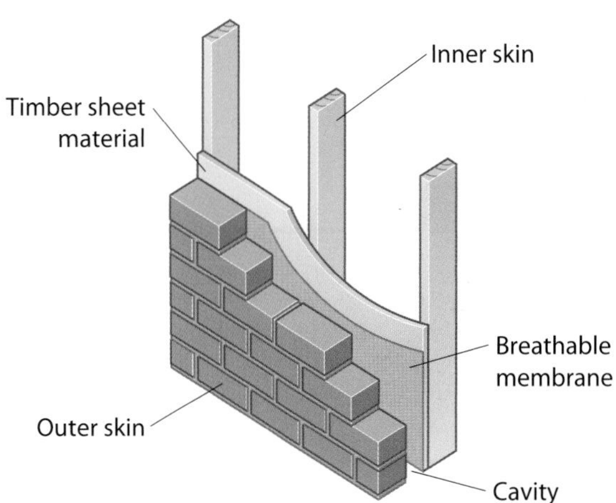

Figure 3.87 Cavity wall on a timber-kit house

There are also other types of exterior walling, such as solid stone or log cabin style. Industrial buildings may have steel walls clad in sheet metal.

Find out

Find out the thermal properties of both timber-kit and traditional cavity wall construction.

Using timber-framed construction

Using timber frame construction delivers a range of advantages. It is quick to erect, saving labour costs, and there is no significant difference in price between block or timber frame. There are also fewer delays due to bad weather.

Figure 3.88 Timber-framed construction

Timber frame systems are suited to brownfield sites with poor soil conditions as construction is lighter than traditional brick and block builds and the modular components are easy to transport to site. Timber frame buildings are environmentally friendly because timber can be responsibly sourced, there is less wastage of materials and timber frame buildings are usually well insulated. Wood is a carbon neutral material and timber frame has the lowest CO_2 'cost' of any commercially available building material. Timber is an organic, non-toxic and naturally renewable building material.

Solid walls

Occasionally, during refurbishment work, you may come across an existing wall that is of solid construction. This tends to be in older houses from the 1980s, where solid brick walls were laid in English bond, with alternate layers of headers and stretchers. This method of construction did not contain a cavity or any insulation, and as such is not very thermally efficient.

Walls for modern-day uninhabited rooms – for example, an outbuilding – can be solid-block walls that are externally rendered to prevent water getting in.

Remember

Solid walls have to have high-quality brickwork and bonding with full mortar joints in order to prevent moisture getting in.

Working life

You are working on the first phase of a new housing complex. The construction type will be timber-framed, with a brick outer skin. You have started the first course above DPC and the general foreman has told you to place the insulation between the breather membrane and the brick skin. You are not happy about this.

What might happen if you do what the foreman says? You will need to think about all the implications of what could happen during the construction, as well as the qualities of the materials that are to be used. What should you do?

Functional skills

This task will allow you to practise **FE 2.2.1–2.2.3**, which relate to selecting and using texts, summarising information and ideas, and identifying the purpose of texts.

Table 3.02 Reasons for using different materials in external walling

Material	Reasons for use
Bricks	Bricks are hardwearing, with a low porosity. Available in a wide range of attractive colours and textures, which enhances aesthetics. Can be used with coloured mortars to good effect. Excellent lifespan.
Blocks	Thermally efficient. Lightweight. Pre-textured ready to receive wet finishes. High dimensional quality.
Insulation	Essential to meet Part L of the Building Regulations. Sits within cavity – does not need clips if fully filled. Thermally efficient component. Relative low cost.
Tie wires	Constructed of stainless steel so they do not corrode over time. Resist lateral forces on the wall. Increase the width of the cavity wall making it more stable.
Timber frames	Lightweight construction method. Very fast to construct. Thermally efficient. Sustainable construction material.
Sheathing plywood	Used to provide strength to the timber frames. External quality to resist moisture. Strong material in shear. Suitable to fix brick ties to.
Breather membrane	Enables the passage of moisture. Acts as a weather protector during construction. Can be stapled to plywood easily.
Vapour barrier	Prevents the passage of moisture, keeping the internal plasterboard dry. Can be stapled to timber studs easily. Retains the insulation in place.
Plasterboard	Protects the DPM underneath. Final finish to timber stud construction. Fire-resistant material.
Fire stop	Piece of insulation wrapped in plastic. Placed in timber frame to prevent the spread of fire.

Comparison of insulation properties between cavity and timber-framed constructions

Existing cavity wall

A house that is over 50 years old will have been built with the level of insulation that was required at that time under legislation – which may mean none at all within the cavity wall. The insulation properties of the existing wall can be upgraded in three ways:

1. The first method is to use a foaming insulation, which is pumped in from the outside. You must take care with any gases that the insulation may give off.
2. The second method is to use a blown glass fibre insulation, which is 'blown' under pressure through holes formed in the external brickwork to fill the cavity.
3. The final method is to clad the internal walls with either insulation-backed plasterboard attached by dot and dab or timber battens clad in plasterboard. This method does reduce the internal dimensions of the room.

Remember

Cost is just one of the factors that will affect the choice of material.

Remember

Any insulation that is retrofitted is normally put in by a specialist contractor after a building survey.

New cavity construction

This construction is much more thermally efficient, as you can look at reducing heat loss through the structure as a whole. Highly efficient internal blocks are used to form the internal skin, with a 100 mm cavity, which is fully filled using a mineral wool insulation material. The traditional brickwork skin is the same as for the older type of construction – this has not changed over the years.

Modern sheet insulation materials can also be used within cavity construction. Manufactured from high-performance rigid urethane, these materials use space age technology to resist the passage of heat, but may require clipping within a cavity to the tie wires. Therefore, the cavity is said to be partially filled, rather than fully filled.

Find out

Research the visual differences between cavity and timber-framed construction by using the Internet to research sketch drawings of a cross-section through a wall, looking for sketches that clearly show the positioning of the insulation within the construction, for either rigid or quilt insulation materials.

Timber-framed construction

This is the most thermally efficient system. It uses insulation (usually quilted) fixed within the timber studs.

Again, rigid insulation products can be used to fill between the timber studs. You can increase the level of insulation by making the timber studs deeper in a mineral wall, or doubling up on rigid insulation boards.

Rigid insulation is normally covered with a foil surface that resists the passage of vapour and slows down the resistance of heat, reflecting it back towards the warm side of the construction.

Find out

Find a rigid insulation manufacturer's details for a wall insulation board, and have a look at the fixing details.

Internal walling

Internal walls are either load bearing – meaning they support any upper floors or roof – or are non-load bearing, used to divide the floor space into rooms.

Internal walls also come in a variety of styles. Here is a list of the most common types:

- **Solid block walls** – simple blockwork, either covered with plasterboard or plastered over to give a smooth finish, to which wallpaper or paint is applied (see Figure 3.89). Solid block walls offer low thermal and sound insulation qualities, but advances in technology and materials mean that blocks manufactured from a lightweight aggregate can give better sound and heat insulation.
- **Solid brick walling** – usually made with face brickwork as a decorative finish (see Figure 3.90). It is unusual for all walls within a house to be made from brickwork.

Safety tip

Load-bearing walls must not be altered without first providing temporary supports to carry the load until the work has been completed.

- **Timber stud walling** – more common in timber-kit houses and newer buildings (see Figure 3.91). Timber stud walling is also preferred when dividing an existing room, as it is quicker to erect. Clad in plasterboard and plastered to a smooth finish, timber stud partitions can be made more fire resistant and sound/thermal qualities can be improved with the addition of insulation or different types of plasterboard. Another benefit of timber stud walling is that timber noggins can be placed within the stud to give additional fixings for components such as radiators or wall units. Timber stud walling can also be load bearing, in which case thicker timbers are used.
- **Metal stud walling** – similar to timber stud, except metal studs are used and the plasterboard is screwed to the studding (see Figure 3.92).
- **Grounds lats** – timber battens that are fixed to a concrete or stone wall to provide a flat surface, to which plasterboard is attached and a plaster finish applied.

Find out

Using the Internet and other sources of information, find out the thermal properties of both a timber stud wall and an internal solid block wall.

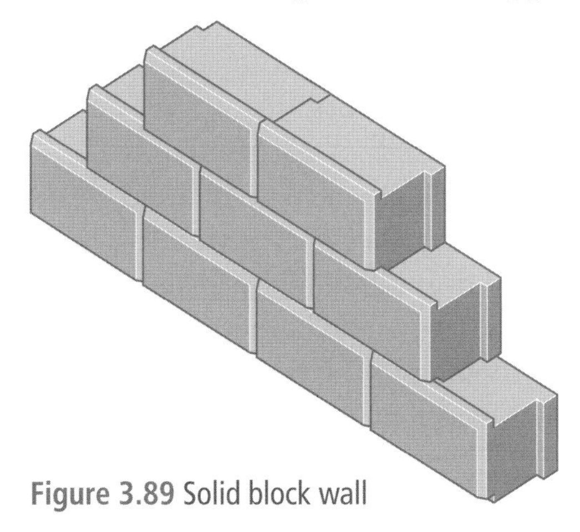

Figure 3.89 Solid block wall

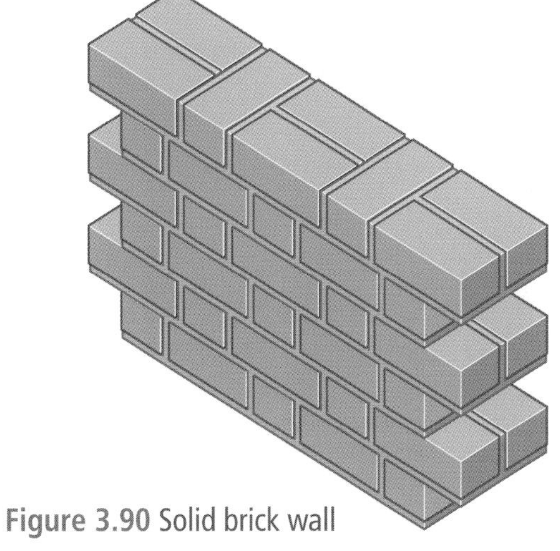

Figure 3.90 Solid brick wall

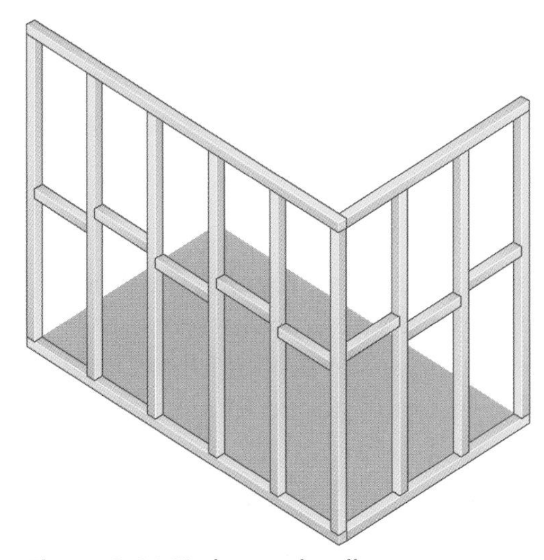

Figure 3.91 Timber stud wall

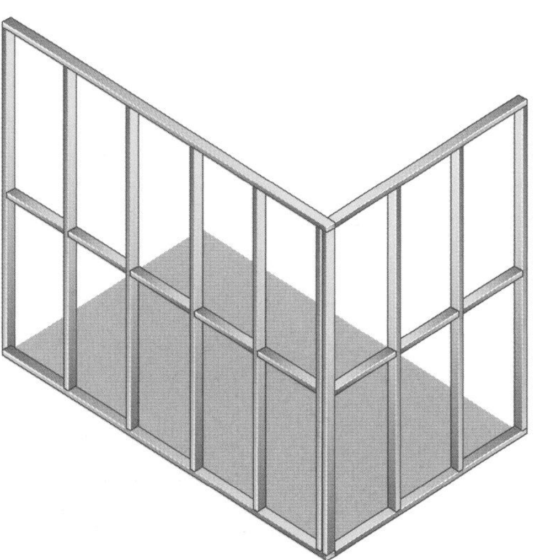

Figure 3.92 Metal stud wall

Working life

You have been tasked with creating an opening in an internal blockwork wall.

What can you do to check if the wall is load bearing? What should be done if the wall is load bearing? What could the consequences be if the wall is load bearing and you create the opening without shoring?

Key term

Load bearing – walls that carry a load from floors, walls and roofs (dead load) and occupancy (live load).

Energy-saving materials in internal walls

As we have already stated, internal walls can either be **load bearing** or non-load bearing. If they are load bearing, they must be made of materials that can resist the load. In this section, you will investigate the sustainability and energy efficiency of the internal walls and linings most often used in modern construction.

Fair-faced blockwork

This is economical in terms of the energy used to produce the desired finish: the blockwork produces the completed finish itself, and no secondary wet or dry finishes (plastering or plasterboard) are applied. If the wall is not to be painted, a good quality block must be chosen; if it is to be painted, paint-grade blocks are needed. You must consider the joints between blocks carefully when looking at the type of pointing required. This must be neat and applied to both sides of the wall.

Fair-faced concrete blockwork is an ideal solution for heavily trafficked areas, sports halls and other walls that need a hardwearing surface. However, it may take several coats of paint to produce an acceptable finish on the blockwork surfaces.

Figure 3.93 Traditional timber stud wall construction

Timber stud walls

These are constructed using regulated timber studs of equal widths. These are attached to head and sole plates, at the top and floor level respectively. Noggins are added to aid stability and fixing of finishes.

Timber studs are a sustainable timber product. Putting up a timber stud wall is an efficient process, as the walls do not have to dry out and no scaffolding is needed. The studs are normally clad with a plasterboard skin on each side, which can be dry-lined or skim-plastered. With this type of wall, it is also quick and easy to distribute services around a building.

Metal stud walls

In this system, timber studs are replaced by pre-formed metal channels. These are fixed together by crimping (using a special tool) or screwing. Metal studs are made with recycled steel, but do have to go through a galvanising process. However, their strength-to-weight ratio is better than timber. Plasterboard finishes are normally screwed into the studs using dry wall screws and a hand-held drill.

Both timber studs and steel channels can be recycled after final use.

Example

The architect is unsure which of the three systems to incorporate into the ground floor internal layout design. You have been asked to discuss the fixing side of the three methods.

Prepare a few notes so that you can discuss these issues with your tutor, who will act as the architect.

Dry lining

This is the process that lines a surface – normally blockwork – with a plasterboard finish.

'Dot and dab' is a dry-lining process. Plasterboard adhesive is applied in dabs to the surface of the wall, and plasterboard is pressed onto the adhesive dabs, then knocked until vertical and aligned with the wall, using a straight edge. A temporary fixing is often used to hold the board in place, and then removed once set. The joints of the plasterboard are bevelled so that a jointing finish tape and filler can be applied to complete the wall.

This process leaves an air gap behind the board, which makes the wall more energy efficient in terms of heat loss. This dry-lining process is useful in refurbishment work, where untidy walls can be easily covered up and upgraded to a quality finish, saving the need to replace whole walls.

Find out

British Gypsum and Knauf are the UK's leading plaster suppliers. Have a look at their websites and investigate the different types of plaster they manufacture. Can you find any information about sustainability on their websites?

Plastered blockwork

This is the least energy-efficient process because the finish has to dry out, using dehumidifiers and a heat source if necessary (for example, in winter). Wet plaster cannot be painted. A plaster coat is normally applied in two layers, the second being a trowelled and polished final finish that produces a smooth wall. There are various types of plasters to suit the surface the plaster is being applied to and the level of wear the wall will be subjected to.

Table 3.03 Reasons for using energy-saving materials in internal wall construction

Material	Reasons for use
Fair-faced blockwork	• Economical single process • Can be left as a self-finish • High-quality block available • Saves on secondary resources and energy • Greater sound resistance
Timber stud walls	• No wet construction • No drying-out time • Renewable resource • Lightweight construction • Traps air within void that can be insulated • Easy hiding of services
Metal stud walls	• Can incorporate recycled steel into manufacture • Good strength-to-weight ratio • Screwed physical fixing of linings • Internal void can be insulated
Dry lining	• Covers up untidy backgrounds • Forms air void behind • Can be used with insulation bonded to plasterboard • Quick, easy method of providing a smooth finish • Does not use wet trades • No drying out period
Plastered blockwork	• Traditional finish • Hardwearing • Easy to repair

Remember

Many factors will affect the choice of a material; fashion is just one.

K3. Know about sustainable methods and materials in construction work

As you work in the construction industry, you will come across a vast range of materials. But which of these are the most sustainable? This section gives an overview of the most common building materials, looking at how they are made and at any issues to do with their use and disposal.

Remember

Sustainability means meeting the needs of future generations, without depleting **finite** resources.

Key term

Finite – a resource that can never be replaced once used.

Sustainable materials used in construction

Table 3.04 Sustainability of common materials

Material	Sustainability
Polystyrene	• Petroleum-based product • Can be recycled • Pollution caused through manufacture • Often used as packaging for equipment • High insulation properties
Polyurethane	• Made from volatile organic compounds that damage the environment • Often turned into foams • Gives off harmful chemicals if burned • Disposal issues: if the foam catches fire, it causes pollution and produces carcinogenic gases
Softwood	• Can be recycled into other products • A managed resource that can be grown over and over again • Provides a pollution filter: takes in carbon dioxide, gives out oxygen • A natural 'green' product • Requires treatment to prevent rot
Hardwood	• Forms part of the tropical rainforest • Intensively felled • Takes a longer time to re-grow than softwood • Expensive resource
Concrete	• Cement production causes CO_2 emissions • Disposal issues – can only be crushed and used as hardcore • Hardwearing with a long lifespan • Relies on a lot of formwork and falsework
Common bricks	• Manufacture involves clay extraction, which has environmental issues • Waste areas form ponds • Uses finite gas energy to fire brick
Facing bricks	• Secondary process to produce the patterns and colours • Useful life of up to 100 years • Uses natural products
Engineering bricks	• Hardwearing, with a long lifespan • Water resistant • Heavy to transport to site, so fuel costs increase
Aggregates	• Extracted from gravel pits, causing environmental damage • After extraction, water fills pits to produce another habitat

Material	Sustainability
Glass fibre quilt	• Can use recycled glass in manufacture • High insulation value and long lifespan • High thermal insulation-to-weight ratio • Economical transport from factory to site
Mineral wool	• Manufactured from glass or rock • Provides a lightweight, high insulation product • Can be used in walls and roof spaces • Can use recycled glass
Plasterboard and plaster	• Disposal issues, creating landfill with high sulphate content • High wastage with cutting to size • Manufacture uses gypsum, a waste by-product • Recycling skip agreements in place with manufacturers
Concrete block	• Uses cement • Can incorporate waste products from power stations • Sustainable manufacturing process • Take-back recycling schemes for waste blocks
Thermal block	• High thermal efficiency • Low weight-to-strength ratio • Uses cement in manufacture additives to produce a block full of air bubbles
Metals	• Can use recycled scrap steel in manufacture • Requires secondary treatment to prevent rust • Non-ferrous metals have a long lifespan
Glass	• Can be completely recycled • Waste from cutting can be recycled • Issues with recycling: toughened glass has been heat treated as a secondary process, so needs special recycling techniques

Find out

Use the Internet to find out more about these types of material and their common uses in constructing domestic and commercial buildings.

Using sustainable materials

The following materials are the most sustainable of those listed in Table 3.4. You will now take a closer look at their use in domestic and commercial dwellings.

Softwood

This naturally occurring timber product is grown more in Scandinavian countries, where growth is slower, producing a more structured wood grain. It must be processed then transported to the UK for further manufacture into timber products. Softwood should be purchased locally to save on transport costs. Timber can be used in the following areas of a building: the roof, floor joists and flooring, second fix joinery products, stud walls, windows and doors. Timber can be completely recycled into a further product, such as chipboard.

Concrete

This is only sustainable in that the finished product has a very long life. Cement production is costly to the environment in terms of energy used and emissions released to the atmosphere. Concrete can be recycled into a crushed hardcore, but no recycled materials can be used in its manufacture. Concrete is used in foundations, lintels, floors and some roof structures.

Figure 3.94 Using softwood to clad the outside of this building is an attractive and sustainable method that requires no treatment

Common bricks

Common bricks are of lower quality and are used in areas where strength is required but facing bricks are not needed. They would be used as coursing bricks in internal skins of external walls and as levelling courses under floor joists.

Find out

Find a local ready-mixed concrete supplier and establish what concrete products they can supply.

Facing bricks

These are used where you can see the external brickwork as they look more attractive, but they are more expensive to make. They are mainly used for the external skin of a cavity wall, or for garden walls where these have to match the house.

Engineering bricks

These are hardwearing high-strength bricks that are water-resistant. They are used below DPC level as they are not affected by rising damp. They are an excellent brick to use within drainage manhole construction.

Find out

Visit a brick manufacturer's website and find out about their sustainable practices.

Aggregates

These are used for several purposes. Primarily they are constituents of concrete mixes, where a blend of coarse and fine aggregate plus cement and water is mixed to form concrete. A more modern application of aggregates is in external landscaping, where they are used to provide attractive areas that resist water run-off into the main drainage system, thereby adding to sustainable drainage for surface water.

Glass fibre quilt and mineral wool

Did you know?

You can use sheep's wool as a modern insulation product. It is a fantastic insulator and can 'breathe out' moisture.

These are used to insulate houses and commercial properties. They are used within ceiling voids above and between the joists. External walls use pre-cut 'batts' of mineral wool, which are built at full thickness into the cavity. Internal walls can also use mineral wool, but it must be of a heavier density, so as to resist the passage of sound. This type of insulation is very thermally efficient and will reduce heat loss from a building.

Safety tip

Always use a dust mask and gloves when handling fibre insulation products.

Figure 3.95 Example of glass fibre quilt

Metals

Using stainless steel in many building materials (such as tie wires) helps overcome corrosion problems. Mild steel products require a secondary process, which could be galvanising, powder coating or just a layer of paint. Metal is used in other products such as lintels, joist hangers and roof truss fixings, and in aluminum windows. Non-ferrous metals are used in water pipes and fittings.

Glass

Glass is a versatile material, used mainly in windows. In certain areas, only safety glass can be used. Glass can be manufactured from up to 80 per cent recycled product. It is transported in standard sheet from the factory to the supplier, who then cuts it to size locally. Any waste can be fully recycled, either locally or via the manufacturer.

Using different sustainable materials

Modern construction methods have been developed in response to the new legislation on sustainability.

Cavity walling

Thirty years ago, it was quite acceptable to build a cavity 75 to 85 mm wide that had very little insulation included in its construction. Modern methods of cavity wall construction now allow for a cavity that is 100 mm wide and is fully filled with glass fibre cavity wall batt insulation, with an engineered lightweight internal skin of blockwork. On page 109 you saw how spaced stainless steel tie wires are used to provide structural integrity to the completed wall, and how the insulation is not clipped.

In commercial buildings, cavity walls tend to be used in the bottom storey, while portal-framed construction with insulated cladding is usually used for the upper half of the unit.

Figure 3.96 Solar panels on a domestic dwelling

Timber-framed construction

Timber-framed construction is a fast, easy and efficient method of producing domestic buildings, and is sustainable in its approach as it uses timber from managed forests.

With timber-framed construction, the insulation is placed internally between the studs and covered with a vapour barrier and plasterboard. A traditional skin of brickwork clads the exterior, and a cavity can be formed by covering the plywood panels with a breathable membrane. Stainless steel ties are screwed to the plywood to provide support for the brickwork.

The floors and roof are traditionally constructed, but the timber-framed panels can be pre-fabricated off site, which can have its own implications for energy efficiency.

> **Find out**
>
> Locate a local timber frame manufacturer and research their methods of construction, to see what external cladding materials can be used to cover the frame.

Alternative construction methods

A number of other energy-efficient construction methods are growing in popularity.

Insulated concrete formwork

This is the 'Lego brick' construction method where hollow, moulded polystyrene forms are literally snapped together to form a wall, with the help of locating **castellations** on each form. Reinforcing rods are added where required and concrete is poured into the moulds and allowed to set. A solid wall is formed that can be rendered externally and plastered internally. Traditional brick cladding can be used externally to clad the structure.

> **Did you know?**
>
> Castellation comes from the word for castle – castellations have the staggered profile you often see at the top of castle walls.

> **Key term**
>
> **Castellation** – having turrets like a castle.

Remember

Pouring concrete requires the use of appropriate PPE.

Insulated concrete formwork is faster and more energy efficient than more traditional construction methods, as it uses less concrete, saves on site resources (for example, no bricklayers are needed), and the insulation is included as part of the formwork. Once constructed, the formwork can simply be rendered and plastered.

Figure 3.97 Energy-efficient insulated framework

Functional skills

This task will allow you to practise **FE 2.2.1–2.2.3**, which relate to selecting and using texts, summarising information and ideas, and identifying the purpose of texts.

You will also be practising **FE 2.1.1–2.1.4** Make a range of contributions to discussions and make effective presentations in a wide range of contexts.

Insulated panel construction

This is a system that uses insulation bonded to the plywood panels that form a structural wall. Traditional cladding – both external and internal – can then be used to cover the insulated panel frames, using the same process as with the timber-framed construction.

This construction method uses pre-formed factory panels that can be readily assembled on site, saving time and resources, and finishing trades can come in earlier than usual, while the outer skin is being completed. The insulated panels are also highly thermally efficient.

Find out

There are several major block manufacturers which produce the thin joint masonry system. Find one of their websites and produce a short presentation that includes:

- the advantages and disadvantages of the system
- why this system is sustainable
- a photograph showing how the blocks are joined.

Thin joint masonry

This is a system that uses high-quality dimensioned blocks that are up to three times the size of normal blocks. A cement-based adhesive is used to bond the locks together using 1 mm tight joints. The starting base course of brickwork has to be very accurate in its setting out and level. Traditional cavities are formed using helix type tie wires that are just driven into the thin joint masonry using a hammer.

This uses lightweight, thermally efficient blocks with no mortar, so it is faster and cheaper to construct, there is less waste, and the end product is very energy efficient.

The impact of sustainability on the different elements of a structure

The sub-structure

Foundations need to be accurately set out to prevent wastage of materials: for example, the concrete in the foundations.

Cement manufacture is one of the processes that uses a high amount of energy and produces high carbon emissions. For example, by making a foundation larger than you require – perhaps because of an overdig by the excavator – you are using more material than you need to. As well as costing the client more, this will waste energy.

Sustainable excavation involves not removing the excavated material from the site. Removing excavated material incurs tip charges, which are part of the landfill tax. Dumping of waste into the land is now not a cheap option. By forming landscaped mounds on site, which can be planted and made into attractive green areas, you can increase the environmentally friendly impact of the excavation and save valuable resources in several ways, including reduced fuel charges, transport costs, air pollution and taxation.

Find out

Find out more about the landfill tax on the Internet or at your local library.

The depth to which foundations have to be taken is governed by Building Control inspection. However, optical and levelling equipment such as profile boards and a traveller must be used to make sure the foundation depth is set out accurately, which will reduce wastage and hence energy use.

Generally, the deeper the foundation, the more expensive and more energy consuming it is to produce. Deep strip and wide strip are expensive, and it could be better to use piled/ground beam foundations and raft foundations. These foundation types do not require deep, wasteful excavations on site. The soil report and site conditions will need careful consideration at the design stage to produce an energy-efficient design that supports the building's loads safely.

Figure 3.98 The depth to which foundations have to be taken is governed by Building Control inspection

Working life

Your client is adamant that all excavated materials are to be removed from site. There is a large, open space at the rear of the development.

How would you convince the client of the benefits of leaving and landscaping the excavated material on site? You will need to think about the possible implications to the land and the project.

Figure 3.99 Blockwork, brick and insulation used in cavity wall construction

Cavity walls

Cavity walls have been used for the past 60 years. They rely on the external brick skin to keep the impact of the weather elements from crossing the cavity. The inner leaf of blockwork is now thermally engineered to trap as much air into its structure as possible, making it lightweight and easy to handle. Openings in cavity walls must be fitted into the construction properly to maximise their energy efficiency.

Insulating timber partitions/floors/roofs

Insulating between timbers – whether joists or studs – is simply a matter of placing either glass/Rockwool or polystyrene between the timbers, with thicker insulation placed between joist roof spaces.

Figure 3.100 Insulation being fitted between timber

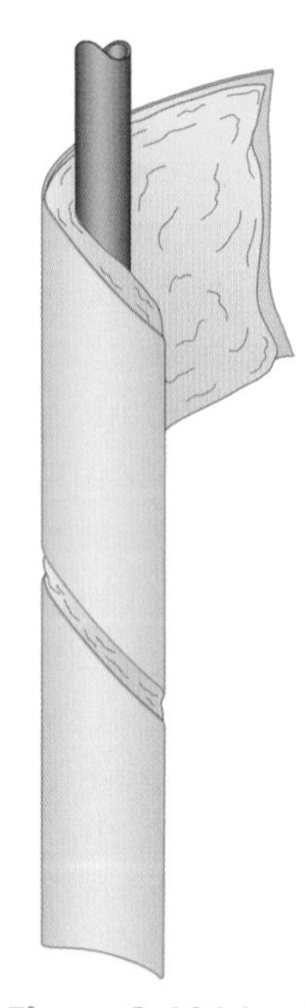

Figure 3.101 Insulating pipework using mineral wool

Insulating water pipes

All water pipes in a loft space or on exterior walls must be insulated to protect them from freezing, which may cause them to crack. There are two methods used when insulating pipes – mineral wool matting or pre-formed moulded insulation:

- **Mineral wool matting** – a small mat is wrapped around the pipes with a bandage and secured with tape or string. The pipes must be completely covered with no gaps, and taps and stopcocks must also be covered.
- **Pre-formed moulded insulation** – this is available to suit different sizes of pipe, and specially formed sections are available for taps and stopcocks. The mouldings can be cut at any angle to fit around bends, and when installed the sections of insulation should be taped together to ensure that they are fully enclosed around the pipes and are tightly butted up to one another.

Insulating water tanks

Any water tanks in the loft space should be insulated around the sides and on the top. The insulation around the sides must extend down to the insulation on the floor of the loft. Insulation jackets are available to fit most sizes and styles of tank.

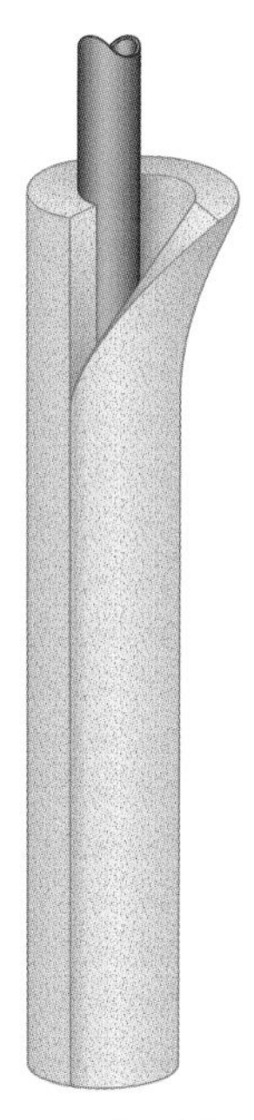

Figure 3.102 Use of pre-formed mouldings to insulate pipework

Figure 3.103 Insulation jacket used to protect water tanks against freezing

Key term

FENSA – Fenestration Self-Assessment scheme, under which window installers agree to install to certain standards.

Find out

Use the Internet and other sources to find out the thermal value of a same-sized piece of polystyrene and mineral wool, then compare the values.

Preventing heat loss in windows and doors

A specialist **FENSA** installer should be used to supply and fit new uPVC windows and doors. These need to meet the agreed insulation standard set by the Building Regulations. The cavity should be closed using a thermal bridging product that insulates across the cavity and acts as a DPM, if required.

Selecting appropriate materials for sustainability

When choosing a material, you should consider a number of factors that impact on energy efficiency and sustainability. You should ask the following questions:

- **Is it recyclable?** Ideally, the material selected should be recyclable, at least at the end of its life, as should any wastage generated during its use or installation.
- **Is it local?** If the material is available locally, this reduces the cost of transporting it from the supplier to the construction site.
- **What are the transport implications?** The weight of the product will have an effect on the cost of its transport and the amount of fuels expended in getting it to the site.
- **How much waste is there?** A product that produces no wastage is valuable in terms of not having to throw away valuable resources and the energy that has been used in manufacture.
- **Is it a natural product?** A product that is produced naturally, for example growing timber, can be replenished time and time again.
- **Is it environmentally friendly?** A material that does not damage the environment in its manufacture and use is a sustainable product that is often referred to as a 'green product'.
- **What is its lifespan?** A longer lifespan, with as little maintenance as possible, means that less energy and resources will have to be put into the product in the future, which saves valuable future resources.
- **Is it worth spending more now?** A material that does not need, for example, painting every five years will save on energy time and costs. Spending a bit more money now will save future expenditure.

Effects of water, frost, chemicals and heat on materials

Many of the materials used in a building project can have their lifespan reduced by the effects of damp, water, frost and chemicals. Treating problems like this early – or, better still, preventing them from happening in the first place – is part and parcel of good environmentally aware building practice.

Remember

All of these defects can be remedied through good design, quality workmanship and materials, and effective maintenance.

Effect of water

Masonry

Efflorescence is one cause of the effects of moisture migrating within a new brick structure. Water moves through the mortar into the brickwork causing salts to move to the outside of the newly constructed wall. This process is known as primary efflorescence. The water evaporates leaving a white deposit on the surface of the brick. It can be brushed off and the action of weathering will eventually remove the effect.

Find out

Have a look at some of the newly constructed houses in your location and see if you can spot some efflorescence appearing on new brickwork.

Concrete

The effect of water on concrete is to discolour its initial fresh new look from the airborne pollution that is present in the atmosphere and which is picked up within rainfall. Water staining on concrete, where the designer has not detailed for the run-off from the building, is another obvious effect. Water action within less porous concrete can result in the leaching of alkaline ('lime leaching') through the concrete that form deposits very much like those within cave structures.

Figure 3.104 An example of efflorescence on brickwork

Timber – wet and dry rot

This is caused by two different elements. Wet rot is caused by excessive exposure to moisture, which eventually causes the breakdown of the timber cells so that they rot and deteriorate. Dry rot is caused by a fungus that sends out long threads, which attack and eat away at the cells of the timber, causing structural damage.

Metal

Rust is caused by the oxidation of a ferrous metal. For the reaction to occur, three things must be present: metal, water and oxygen. Rust forms as an orange deposit on the surface of the metal. If it goes untreated, it will continue to eat away at the metal, eventually compromising its strength.

Find out

The client is not happy with the efflorescence that is appearing on the first phase handover of homes and has asked for some remedial action. Research this effect and make recommendations to the client, justifying your remedy.

Find out

What do the terms 'Grade A' and 'Grade B' mean on engineering bricks?

Find out

The Building Regulations specify the inclusion of DPCs and DPMs into the sub-structure construction. Find out which section of the regulations cover this aspect and establish exactly what is required.

Damp

Damp can cause many problems in a building, and the damage it causes can be costly and difficult to repair. A range of methods and materials can be used to help stop damp getting into a structure:

- **DPCs** – placed within the skins of the external walls; always 150 mm above finished ground level.
- **Slate** – used traditionally as a barrier to resist the passage of moisture; still found in older external cavity walls; can still be used as a DPC, but commonly replaced by plastic DPC.
- **Engineering brick** – Grade A quality brick, so dense it naturally resists the passage of water; can be used below DPC to act as an additional barrier to rising damp.
- **Pitch polymer** – a bitumen-type DPC sometimes seen in external walls; can squeeze out of the joint due to pressure and heat; seldom used today.
- **PVC** – the most common type of DPC in use today; economical and easy to bed into the joint; often textured to bond to the mortar above and below the joint.
- **Injected** – chemical DPC is injected into the wall to prevent the passage of moisture; used as a cheap refurbishment technique as you do not have to remove brickwork; internal plaster often removed up to 1 m high and wall replastered with a renovating plaster after injection work is complete.
- **DPMs** – placed under floors to prevent rising damp from entering warm floor; tucked into DPC within the external wall, thus 'tanking' the construction against damp.
- **Visqueen** – common trade name for a 1200-gauge plastic product tough enough to withstand tearing but which will bend around corners. Note: hardcore beneath DPM should always be sand-blinded to prevent rupture of DPM.
- **Bituthene** – a sheet material manufactured from bitumen that can be laid in sheets, lapped, and used both horizontally and vertically; mainly used for waterproofing basement walls; must be used in accordance with the manufacturer's instructions.

Effects of frost

Spalling

Spalling is the action of freezing water on the surface of a porous brick. The freezing action causes water to expand, pushing flakes of brickwork from the brick face. The photograph in Figure 3.105 clearly illustrates this action, which leaves the stronger mortar joints proud of the original brickwork face.

Figure 3.105 The effects of spalling on brickwork

Thermal expansion

The Building Regulations state clearly the specification for expansion joints within brickwork external walls. These allow the brickwork to expand in the summer months and contract in the winter months without showing signs of cracking. Thermal movement may appear as complete cracks across bricks vertically.

Timber and concrete are good at resisting the effects of frost or freezing water. Exposed timber will maintain its structural integrity for many years. Concrete, once damaged, allows water to enter; if this meets any exposed reinforcing bars, then problems will occur.

Effects of chemicals

Concrete can be attacked by certain chemicals.

Carbon dioxide from the air can react with certain chemicals in the concrete. This decreases the protective alkaline which ensures that the steel reinforcement does not rust. Once this bond is broken, then rusting can occur as the concrete expands and spalls away from the reinforcement bars.

Sulphates that come into contact with the cement in concrete can attack the chemical bond of the cement, affecting its strength. Sulphate-resistant cement (SRC) and not ordinary Portland cement (OPC) must be used where this is a risk, for example, where groundwater is contaminated.

Sulphates can also attack the brickwork cement mortar, especially with a chimney where coal is being burned. Sulphates eat away the cement mortar joints, the reaction causes the joint to expand and the chimney can start to lean over.

Chlorides have been used in the past to alter the setting times of concrete. Unfortunately, over time these affect the strength of concrete, which can have a disastrous effect if the material is under excessive loading conditions.

Did you know?

Brickwork expansion joints will need taking right through the cavity and the internal block skins.

Find out

Research the properties of SRC cement and when you need to use it instead of OPC.

Safety tip

With any chemical process, always refer to the supplier's safety data sheet for instructions.

Alkali silicate is present in some concrete aggregates. This can react with the cement and water and expand, causing a pop-out of the concrete surrounding the reaction. This can have an adverse effect on structural concrete.

Acid rain – a product of burning fossil fuels – also has an effect on certain building materials. Natural stone is mostly affected, as acid rain eats away at the surface of the stone, causing long-term damage.

Effects of heat and fire

Masonry

Brickwork and blockwork cope well in a fire within a dwelling or commercial property. Extreme heat can cause masonry to expand and crack, but this would be after a substantial time from evacuation of a building. Fire is used in a kiln to harden bricks so they are more than capable of staying stable in a fire. Smoke does, however, cause blackening to the surfaces of bricks and blocks. This can to some degree be washed off, but may leave a permanent stain that requires decoration internally and power washing externally.

Water, as we have seen previously, when combined with a frost, or freezing conditions, causes spalling to the surface of the brickwork or the breakdown of the mortar joint due to weathering.

Concrete

Concrete can spall under the influence of fire. If the reinforcement lies near to the surface of the concrete and this heats up then it will expand at a different rate to the surrounding concrete. This expansion can cause cracking to the concrete structure. Concrete is fairly resilient to water damage and will dry out after wetting; but if the water contains contaminants then this can lead to the chemical attack of the cement within the concrete.

Find out

Consult your local newspaper's website and look up a building fire that has occurred in your area. Look for photographs to see what damage was caused.

Timber

Heat and the presence of oxygen cause the combustion of timber by fire. The surface chars and eventually breaks down the structural integrity of the timber until it is burnt right through. Smoke damage can discolour timber which will then require decoration if it has not caught fire and charred. Water can damage timber by wetting which expands the hygroscopic material and causes dimensional change to the timber, which will eventually rot if this wetting persists.

Metal

Metal does not react well in a fire. As it heats up, the molecular structure weakens and it loses up to half its strength at over 500°. This can cause the collapse of structures as the steel melts slightly and warps under extreme heat. However, this does take some time and may not affect the evacuation of a structure. Water reacts with exposed metal, as we have seen, to form rust. Surface rust is not harmful, but continual exposure to the elements of unprotected metal will result in severe corrosion.

Remember

The Building Regulations will make provisions for the building structure in the event of a fire.

Treating building materials to prevent the effects of deterioration

Many building materials can be treated with chemical products in order to extend their lifespan and to reduce the effects of deterioration over time.

Figure 3.106 Treated timber decking

Intumescent paint

This is a chemical-based paint that reacts with high temperatures. It foams and expands around the steelwork it is painted upon to protect it from the heat – giving the occupants time to escape.

Paints

These are used to protect timber, brickwork and metal work from the effects of water. They form a microscopic seal across the surface, preventing water penetration past this layer.

Water repellent

This can be painted onto the surface of exposed brickwork to allow water to run off the brick face more easily. This prevents it standing and absorbing into the brickwork, because, as we have seen, if it freezes it can become a problem.

Vac Vac and tanalising

These are chemical pressure impregnation systems to chemically treat timber from fungi and insect attack. They fill the pore holes within the timber and kill off any wood-boring insect that attacks its structure. Fungi includes the dry rot spores and other cellulose-eating organisms.

This timber treatment alters the colour of the natural timber often to a green shade.

Sulphate-resistant cement (SRC)

As we have already mentioned, this is a type of cement containing chemicals and mineral aggregates, which resists the attack of sulphates within concrete and is used where this need is detected by testing.

Injected damp-proof courses

These use chemicals of various kinds and mixtures to form a chemical barrier within the brickwork after they have cured and filled the pores in the bricks. These have to be injected through holes drilled into the outer wall and sometimes in inner walls.

Methods used to rectify and prevent deterioration

There are methods that can be used to protect and repair damaged materials once they have been put in place.

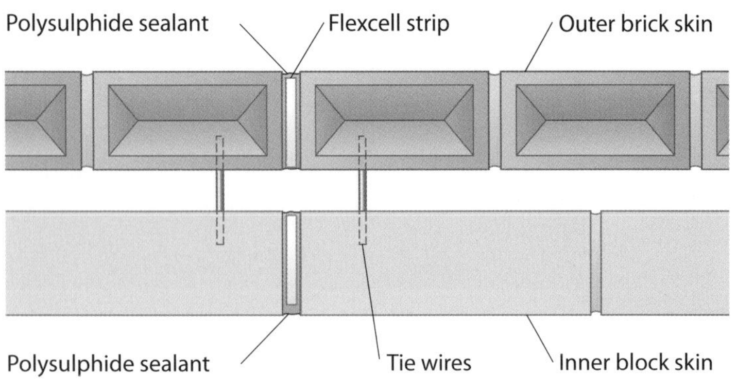

Figure 3.107 Flexcell strip

Masonry and concrete

Masonry and concrete make a movement joint that allows the brick and block panels to expand and contract under thermal movement. A compressible board is used within the joint which is sealed with a polysulphide joint to prevent water ingress. Joints within concrete slabs are very similar, but the Flexcell is replaced with a material that is more resilient.

Find out

Have a look around at the built environment where you live and work. Analyse the brick and concrete buildings and see where the movement joints have been formed.

Timber

Cutting and splicing in new timber is an age-old prevention method that saves having to remove the whole length of timber, manufacture a new piece and re-fit. Timber can be initially protected by pressure impregnation, or it can be simply painted or stained using a flexible coating as the timber will expand and contract. This method of coating must be maintained to extend the lifespan of the timber and any damaged areas repaired quickly.

Insecticides or fungicidal washes will remove mould growth on timber, but ventilation and quick drying of wet surfaces is the key to preventing this damage.

Methods used on metals to give galvanic protection and protective coatings

Galvanising using a hot dip process provides a good layer of protection to ordinary mild steel. The steel is fed through a bath of molten zinc, which coats the metal and gives it long lasting protection against water damage.

High-quality paints that are developed for bonding to steel work should be used instead of galvanising where colour is required. A primer base coat is applied, followed by several other coats that build up the film of paint. Powder coating steelwork is another process that can be used to coat the steel with a colour protective layer.

Find out

Find out the largest size of steelwork that can be powder coated, and gather some pictures of examples.

Effects of adverse weather on building materials used in domestic and commercial buildings

Water can be driven into a structure through:

- the roof tiles and onto the underlay
- open windows or poor seals
- badly maintained brickwork mortar joints.

This can lead to damage. The recent spate of floods within the UK have shown just what damage can occur to a property – even if the water did not enter the ground floor. Excessive rising damp can occur, which causes damage to internal finishes, timber wall plates and floor timbers if the water contains sewage particles. High volumes of floodwaters can cause buildings to move on their foundations from the water pressure or can, in fact, push over walls.

High levels of snowfall can produce heavy loads on structures that could lead to cracking or collapse of the roof structure. High winds each autumn also account for some damage to chimneys and walls as brickwork is blown over.

FAQ

How do I know if the materials I am using are strong enough to carry the load?

The specification will give you the details of the sizes and types of material that are to be used. You should use this document when you need to know which materials to use.

Do I have to fix battens to the wall before I plasterboard it?

No, the method is called dot and dab and can be used where plaster is dabbed onto the back of the plasterboard and then pushed onto the wall.

How can I check my carbon footprint?

There are various websites that allow you to calculate your own footprint, such as http://carboncalculator.direct.gov.uk/index.html. Use one of these to see how environmentally friendly you are being.

How do I know what the properties of materials are?

The supplier of the materials should be able to provide you with a data sheet that lists all the properties.

Check it out

1. Explain the reasons behind good environmental design and describe the key examples of this. Use the Internet and other sources of information to produce examples of each of these in practice.
2. In an area designated by your trainer, carry out a site survey. Make a detailed record of your results and complete a method statement explaining the process you followed.
3. Sketch the different ways of supporting trench excavations and explain the benefits of each.
4. Explain the factors that influence the design of a foundation and describe the benefits and features of each type of foundation.
5. Describe the difference between a single and a double floor on a suspended timber floor. Use sketches to illustrate your explanation.
6. With the aid of a sketch, explain the purpose of a joist hanger and how it functions.
7. Use sketches to show the different types of fall on a flat roof.
8. Describe why a cavity needs to be kept clean and outline the three steps to take to prevent damp penetration in a cavity wall.
9. Sketch sections through a suspended timber floor, showing how the flooring is supported.
10. Describe an I-type joist and the purpose of a roof ladder.
11. Explain how the effectiveness of insulation is measured and describe the qualities of the main types of insulation, explaining why insulation material is used.
12. A client is considering moving away from traditional brick cavity walls to produce domestic dwellings. Describe the benefits of two alternative methods that could be used and make a recommendation of which is best, explaining your reasons.
13. Identify key sustainable materials and explain their qualities, using sketches to illustrate their use.
14. Explain how to identify if a material is sustainable, and some of the methods used to protect timber and steel from decay.
15. You are working for a client who builds doctors' surgery centres. The current contract you are working on is a large medical centre. You have been asked by the client to recommend the finishes for the front of the centre due to your experience in using current sustainable materials. Assist the client by describing the advantages of cedar boarding over rendered blockwork.

Getting ready for assessment

The information contained in this unit, as well as continued practical assignments that you will carry out in your college or training centre, will help you with preparing for both your end of unit test and the diploma multiple-choice test. It will also aid you in preparing for the work that is required for the synoptic practical assignments.

The information in this unit will build on the information that you may have acquired during Level 2, Unit 2003, and will help you understand the basics of your own trade as well as basic information on several other trade areas.

You will need to be familiar with:

- new technologies and methods in construction
- energy efficiency in new construction buildings
- sustainable methods and materials in construction work.

It is important to understand what other trades do in relation to you and how their work affects you and your work. It is also good to know how the different components of a building are constructed and how these tie in with the tasks that you carry out. You must always remember that there are a number of tasks being carried out on a building site at all times, and many of these will not be connected to the work you are doing. It is useful to remember the communication skills you learnt in Unit 3002, as these will be important for working with other trades on site. You will also need to be familiar with specifications and contract documents to know the type of construction work other crafts will be doing around you on site. It is important that working drawings are precise in order to complete the structure accurately. You will need to be able to sketch a section through building elements and components.

A big focus of this unit has been on energy efficiency and sustainability in building. It is important to understand both the methods, and the benefits of these methods, when working on new constructions. For learning outcome three, you will need to be able to select from and use a range of sustainable materials, choosing the correct equipment for the project. When working with these materials you will need to be able to identify how these can be affected by deterioration over long and short periods of time. As well as the impact of this on how the materials are used, you will also need to be able to organise the safe storage and protection of these building materials. Relating this information back into planning an environmentally friendly and sustainable project will be an important skill – both for your course and professionally.

This unit has explained a variety of different construction methods, including modern, sustainable and energy efficient, and the elements that are used in the construction. Although you will not be working on all these elements, you need to be familiar with the work undertaken on them in order to plan when to carry out your own work. You will need to be knowledgeable about different materials and their properties.

Good luck!

CHECK YOUR KNOWLEDGE

1. Possible effects from global warming can be:
   - a rising sea levels
   - b increased temperatures
   - c higher humidity levels
   - d all of the above
2. What will investigation of the site prior to construction look at?
   - a Proximity of nearest bench mark
   - b Position of trees
   - c The type of materials likely to be needed
   - d All of the above
3. Which type of foundation should be used on sloping ground?
   - a Stepped
   - b Piled
   - c Pad
   - d Wide strip
4. What is a tamper board used for?
   - a Fixing joists
   - b Compacting concrete
   - c Levelling foundations
   - d Covering windows
5. In cavity walling where both the outer and inner leaves are 90 mm or thicker, you should use tie wires at what maximum horizontal distance?
   - a 450 mm
   - b 600 mm
   - c 750 mm
   - d 900 mm
6. Which value is used to describe insulation quality?
   - a R-value
   - b B-value
   - c I-value
   - d K-value
7. What is described as 'a lightweight construction method, very fast to construct, thermally efficient, sustainable construction material'?
   - a Bricks
   - b Blocks
   - c Timber frames
   - d Sheathing plywood
8. What is the timber component laid onto walls to support joists and trusses?
   - a Purlin
   - b Wall plate
   - c Sleeper plate
   - d Truss clip
9. What is strutting and bridging used in?
   - a Cavity walls
   - b Timber floors
   - c Metal studwork
   - d Rafters
10. What is spalling caused by?
    - a Freezing water
    - b Extreme heat
    - c Chemicals
    - d Dry rot

Functional skills

In answering the Check it out and Check your knowledge questions, you will be practising **FE 2.2.1** Select and use different types of texts to obtain relevant information, **FE 2.2.2** Read and summarise succinctly information/ideas from different sources and **FE 2.2.3** Identify the purposes of texts and comment on how effectively meaning is conveyed.

You will also cover **FM 2.3.1** Interpret and communicate solutions to multistage practical problems.

Unit 2008

Know how to carry out first fixing operations

Carpenters and joiners will undertake many different types of work on site. One of the main areas, and probably the most important, is called first fixing. First fixing is the name given to the work carried out before plastering takes place, and includes studwork, ground lats, stairs, windows and doors. This unit may be familiar to you from Level 2 and will provide a brief recap of the key information and skills you need in order to carry out first fixing operations.

This unit supports TAP Unit 2 Install first fixing components, and delivery of the five generic units.

This unit contains material that supports the following NVQ unit:

- VR 09 Install first fixing components

This unit will cover the following learning outcomes:

- Know how to fix frames and linings
- Know how to fit and fix floor coverings and flat roof decking
- Know how to erect timber stud partitions
- Know how to assemble, erect and fix straight flights of stairs, including handrails

K1. Know how to fix frames and linings

A key aspect of first fixing is knowing how to fix frames and linings. In this section, you will learn the skills needed to work with door frames and linings and window frames and linings. Knowing how to identify and select the correct timber is also covered in this section.

Door linings and frames

Door linings or casings

Door linings or casings are lightweight internal frames that can be fixed into position by nailing through the jambs into fixing blocks, pads, plugs, or directly into a timber stud wall. Blocks or pads are usually built into the door opening during construction by the bricklayer. Three or four will be needed on each jamb.

Where fixing blocks or pads have not been used, it may be necessary to plug the opening, using a plugging chisel to chase out four mortar joints on each side of the frame. Wooden plugs can be cut from scrap timber, driven into these slots and sawn off plumb. The sawn ends must be kept exactly in line and square (see Figure 28.01).

The door opening should have been built with a clearance of about 20 mm, so that the lining fits loosely and can easily be plumbed and aligned with the walls. Packing will be needed between lining and brickwork so that it is held securely in place once fitted.

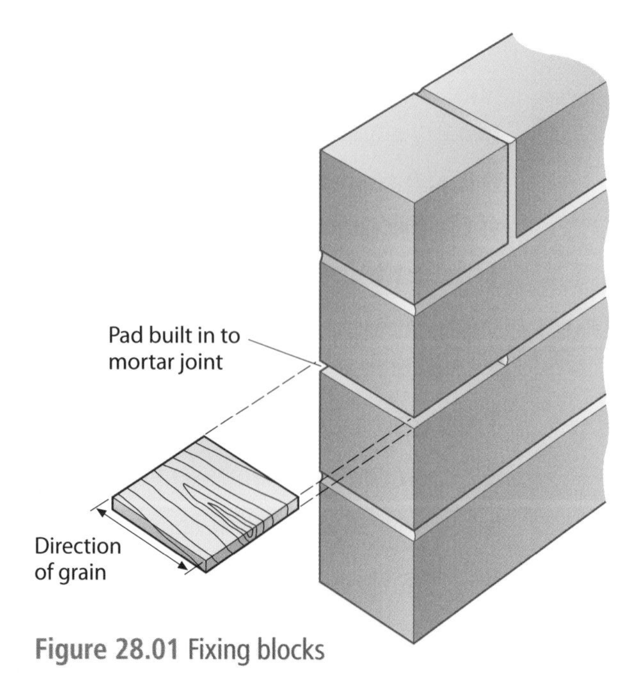

Figure 28.01 Fixing blocks

The lining can now be offered into the opening and checked that it is plumb and level. It is normal practice to fix one jamb first, making sure that it is plumb and straight and will suit the line of the finished plasterwork. Once this is done, the second jamb can be fixed and sighted through to make certain that it is exactly parallel.

It is good practice to make a temporary fixing at the top and bottom of each jamb, leaving the fixing protruding. The lining can then be checked finally for accuracy and position, and adjustment made before it is finally fixed (see Figures 28.02 and 28.03).

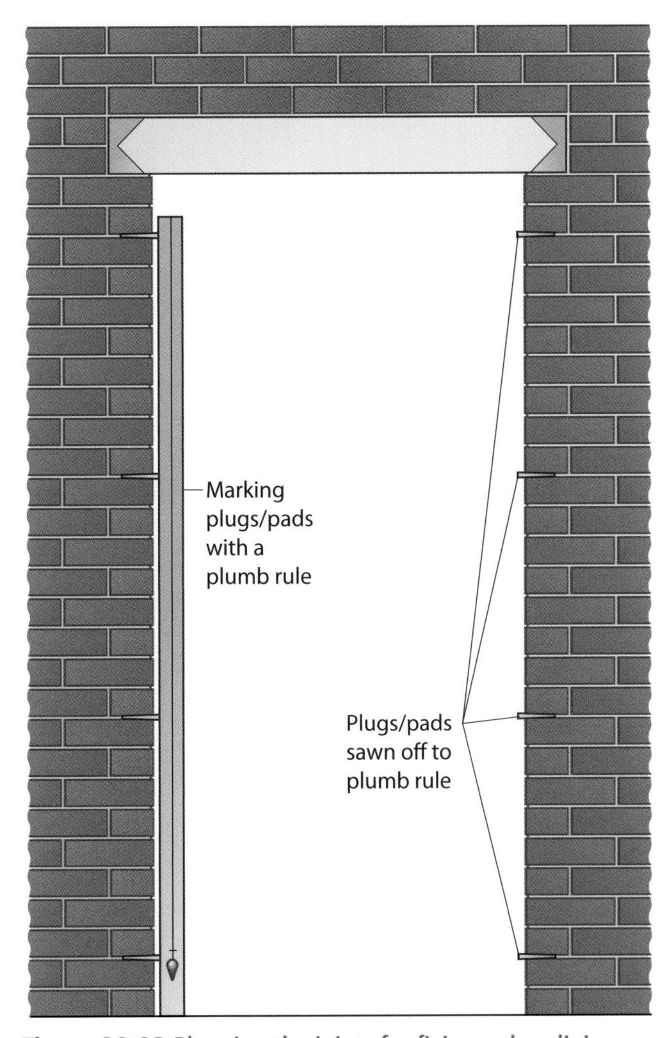

Figure 28.02 Plugging the joints for fixing a door lining

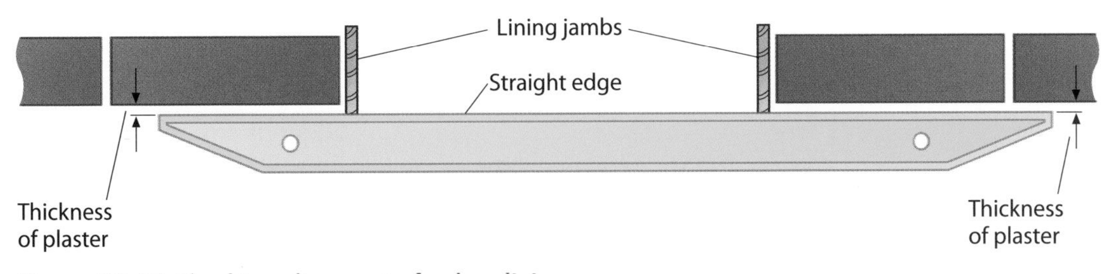

Figure 28.03 Checking alignment of a door lining

Step-by-step: Door lining

1

Measure down

Step 1 Remove the door stops, which should have been nailed lightly. Check the finished floor level (FFL) in relation to the bottom of the lining. The most accurate method is to measure down from a datum line (usually 1 m or 900 mm above FFL) and pack lining to suit.

2

Wedge the lining

Step 2 Wedge the lining roughly in position by putting a wedge above each leg in the gap between the lintel and the head of the lining. The head should be checked with a spirit level to see that it is level. The lining can now be packed down the sides with hardboard or plywood, at the top fixing positions. The top of the lining should be moved to ensure equal projection either side of the wall.

3

Fix the lining

Step 3 Fix the lining through the packing at the top with two nails at each fixing point. An alternative is to use screws, which has become common practice.

Step 4 Plumb the lining on the face side and edge using a level. Pack the bottom position as required and fix through one set of packing. Check that the fixed leg is square. Then complete the fixings on the same leg. The amount of fixing points should be around five, but not less than four.

5

Mark the inside width

Step 5 Remove the stretcher from the bottom of the lining, hold it at the head and mark the inside width to make a pinch rod.

6

Fit the pinch rod

Step 6 The pinch rod can now be fitted in the bottom position of the lining and the lining packed to suit.

Step 7 Check the leg to see that it is plumb and check the alignment of the frame. The bottom can be fixed; then – moving the pinch rod to ensure the frame is parallel – fix the intermediate positions. The head will only require fixing if the opening exceeds normal width. All nails should be punched to about 3 mm below the surface. The door stops can now be replaced and protection strips fixed if required.

Door frames

Door frames are usually made from heavier material than door linings, with solid rebates. They are generally used for external doors. Frames can be fixed to timber stud partitions using 75 mm oval nails. Alternatively, they can be screwed to solid walls using plastic plugs.

Built-in frames

The majority of frames are 'built-in' by the bricklayer as the brickwork proceeds. Prior to this the frame has to be accurately positioned, plumbed and levelled, and struts temporarily inserted by the joiner.

Functional skills

While working through this section, you will be practising **FE 2.2.1–2.2.3**, which relate to reading and understanding information. You will also cover **FM 2.2.1** Apply a range of mathematics skills to find solutions.

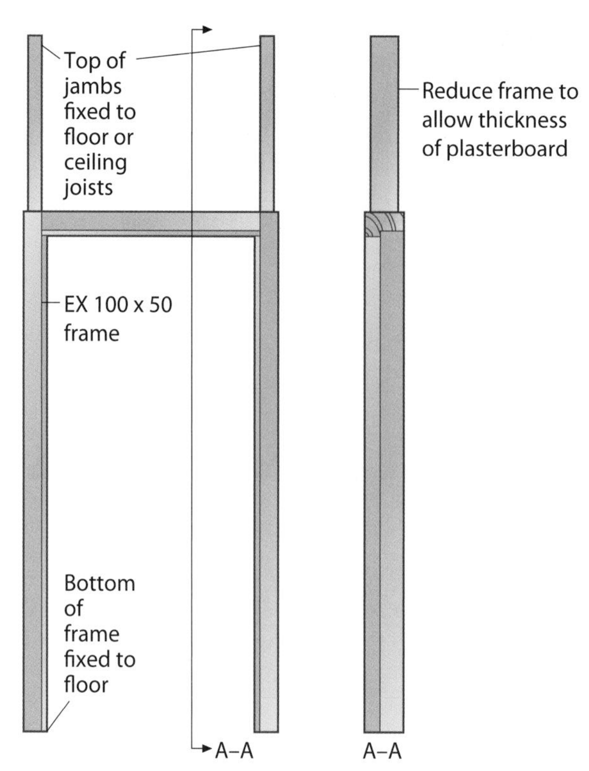

Figure 28.04 Plain storey frame

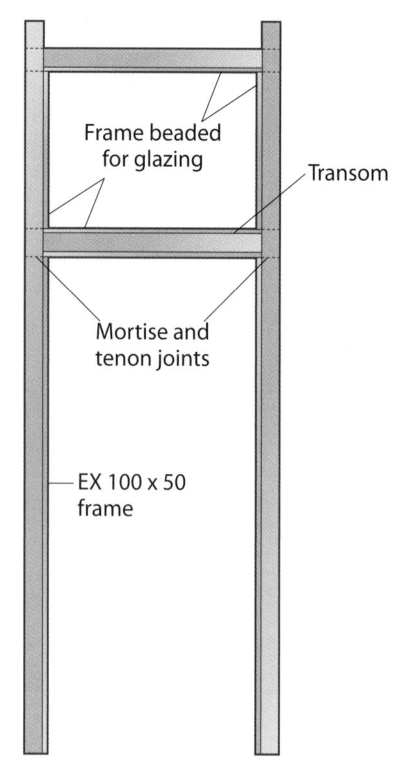

Figure 28.05 Fanlight storey frame

Storey frames

Storey frames are designed to give stability when doorways or fanlights are placed in thin non-loading walls. They are known as storey frames because their height is from floor to floor (one storey). They are used to borrow light from one room to another.

Fire-resistant frames

Fire-resistant frames and frames for fire doors are different to other frames in that they have an intumescent strip which is rebated into the frame. This can help prevent the spread of fire, as it expands when heated and seals the gap between frame and door.

Window frames and linings

Different types of windows have different parts, but the majority will have:

- **a sill** – the bottom horizontal member of the frame
- **a head** – the top horizontal member of the frame
- **jambs** – the outside vertical members of the frame
- **a mullion** – intermediate vertical member between the head and sill
- **a transom** – intermediate horizontal member between the jambs.

If a window has a sash, the sash parts will consist of:

- **a top rail** – the top horizontal part of the sash frame
- **a bottom rail** – the bottom horizontal part of the sash frame
- **stiles** – the outer vertical parts of the sash frame
- **glazing bars** – intermediate horizontal and vertical members of the sash.

Main window types

Traditional casement windows

Casement windows comprise a solid outer frame with one or more smaller and lighter frames within it, called sashes or casements.

These are the main components of a casement window:

- **Frame** – consists of a head, sill and two vertical jambs. Intermediate members are incorporated to form openings for sashes or to alter the design. The vertical sections are known as mullions and horizontal members are transoms.
- **Opening casement** – consists of top and bottom rails with two stiles. If a casement is divided up, these members are known as glazing bars. When glass is fixed into the main frame itself, this is known as direct glazing.

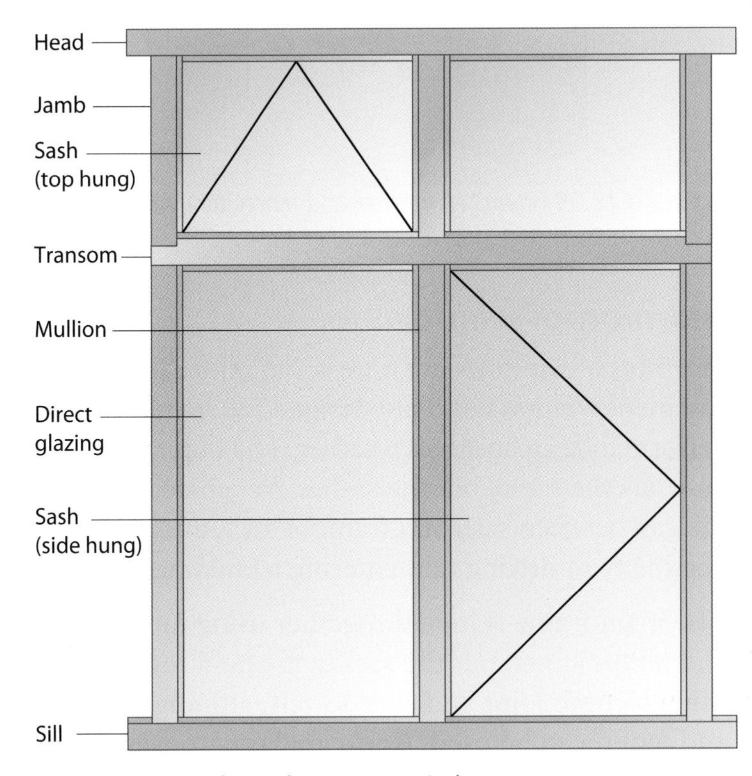

Figure 28.06 Traditional casement window

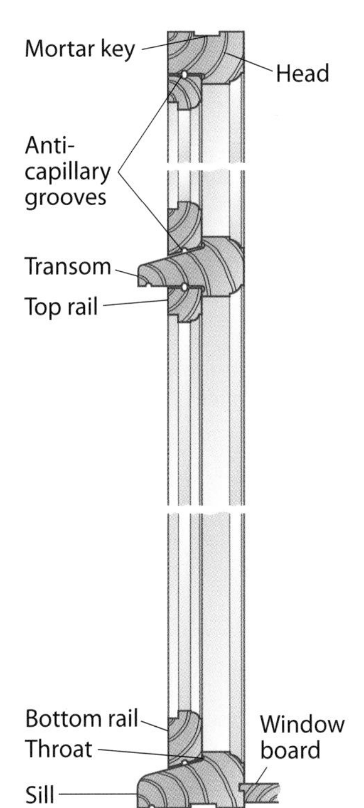

Figure 28.07 Section through traditional window, incorporating top and side hung sashes

Key term

Horns – these are extensions to the frame that protect it during storage or transport, and are cut off before installation.

- **Joints** – all joints used are mortise and tenon: mortises for the head and sill, tenons for the jambs. Casements are haunched mortise and tenons and held together using wedges or draw pins and star dowels, draw pins being more suitable when the **horns** are to be cut off later.
- **Weather proofing** – a groove is formed around the outside edges of the head, sill and jambs to provide a mortar key. Grooves, known as anti-capillary grooves, are also incorporated along the inside of the rebates to prevent water passing into the building; the transom and sill have a 'throat' to stop water penetrating and are sloped to allow rainwater to run off.

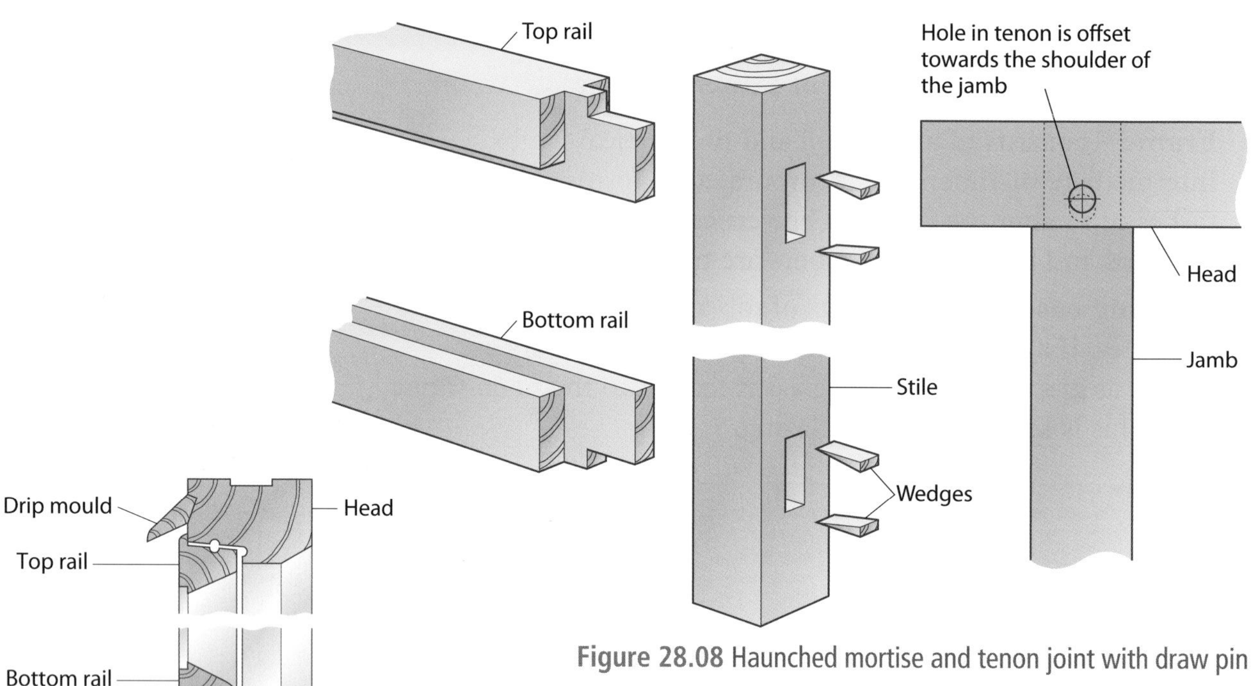

Figure 28.08 Haunched mortise and tenon joint with draw pin

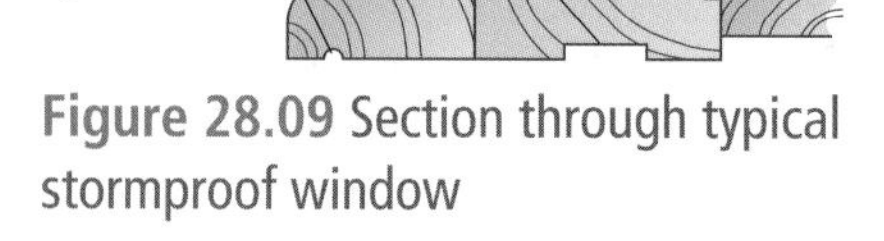

Figure 28.09 Section through typical stormproof window

Stormproof windows

Stormproof windows are a variation on traditional casement windows and are designed to increase their performance against bad weather. The outer frame is basically the same, but the sashes are rebated to cover the gap between sash and frame. This reduces the possibility of driving rain entering a building.

The main frame is joined together using mortise and tenon joints, but the sashes are held together using a comb joint and star dowels, although mortise and tenon can be used. Anti-capillary grooves are incorporated along the inside of the rebates to prevent water passing into a building.

Assembly of frames and sashes

Windows may be delivered fully constructed. However, if they have to be assembled, the procedure below shows how this should be done.

Key term

Squaring rod – any piece of straight timber long enough to lie across the diagonal.

Step-by-step: Assembly of frames and sashes

1

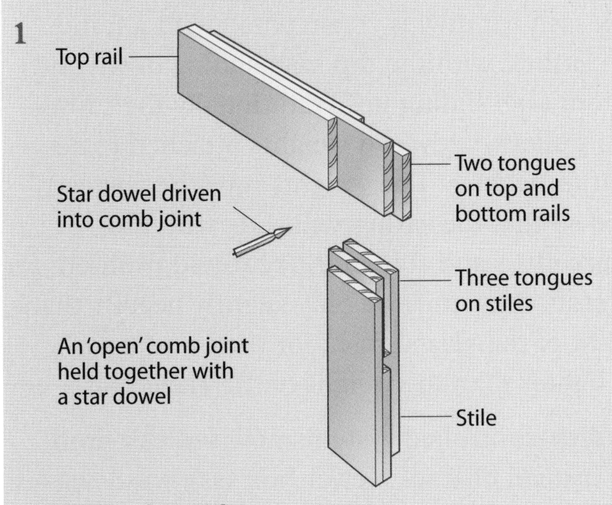

Joining of main frame

Step 1 Prior to gluing, frames and sashes should be assembled dry to check the joints, sizes and that they are square.

2

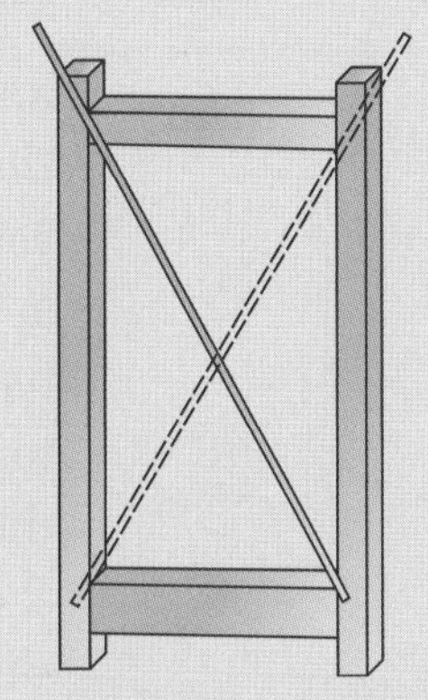

Check diagonals with a squaring rod

Step 2 Lay a **squaring rod** across the diagonals. Mark the length of one diagonal on the rod, then do the same for the other diagonal. If the pencil mark is in the same place, the frame is square. New joints may have to be made if the frame is distorted.

3

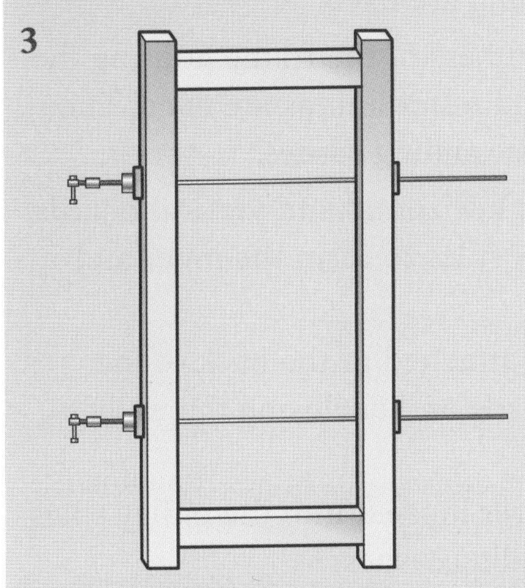

Glue together and cramp

Step 3 Once a sash or frame has been assembled dry and checked, a waterproof adhesive can be applied to the faces of the tenons and shoulders, and the items glued and cramped together to dry.

4

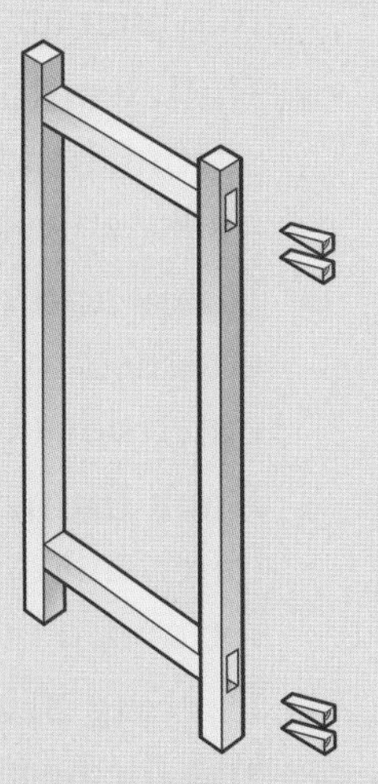

Secure with wedges

Step 4 Check again that the sash is square and not in wind, as well as ensuring the overall size is correct. Then wedges should be lightly tapped into each tenon joint and, if everything is correct, they can be driven in fully – outside wedges first – and star dowels inserted if required.

Box frame sliding sash window

Often referred to as a box sash window, this is rarely used nowadays – the casement window is preferred as it is easier to manufacture and maintain. Box sash windows are mainly fitted in listed buildings or in buildings where like-for-like replacements are to be used. Sometimes box sash windows are fitted because the client prefers them.

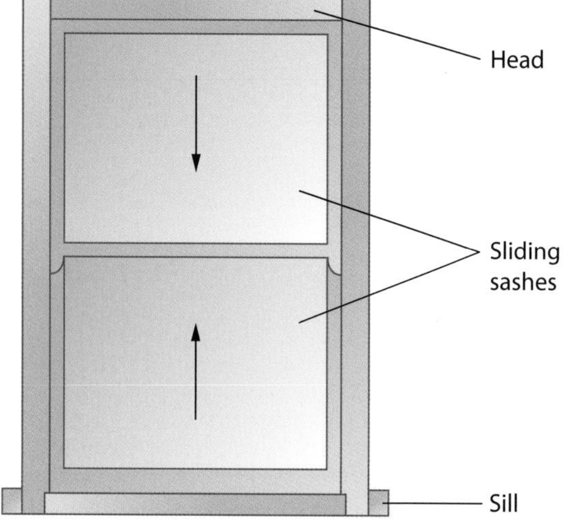

Figure 28.10 Box frame window

The box sash window is constructed with a frame and two sashes, with the top sash sliding down and the bottom sash sliding up. Traditionally the sashes work via pulleys, with lead weights attached to act as a counterbalance. The weights must be correctly balanced so that the sashes will slide with the minimum effort and stay in the required position: for the top sash, they must be only slightly heavier than the weight of the glazed sash; for the bottom sash, slightly lighter than the weight of the glazed sash.

Most pulley-controlled systems now use cast-iron weights instead of lead. Newer box sash windows use helical springs instead of the pulley system.

The parts of a box sash window frame are as follows:

- **Head** – the top horizontal member housing the parting bead, to which an inner and outer lining are fixed.
- **Pulley stiles** – the vertical members of the frame housing the parting bead, to which inner and outer linings are fixed. The pulley stiles act as a guide for the sliding sashes.
- **Inner and outer linings** – the horizontal and vertical boards attached to the head and pulley stiles to form the inner and outer sides of the boxed frame.
- **Back lining** – a vertical board attached to the back of the inner and outer linings on the side and enclosing the weights, forming a box.
- **Parting bead** – a piece of timber housed into the pulley stiles and the head, separating the sliding sashes.
- **Sash weights** – cylindrical cast iron weights used to counterbalance the sashes. These are attached to the sashes via cords or chains running over pulleys housed into the top of the pulley stiles.
- **Sill** – the bottom horizontal part of the frame housing the pulley stiles and vertical inner and outer linings.

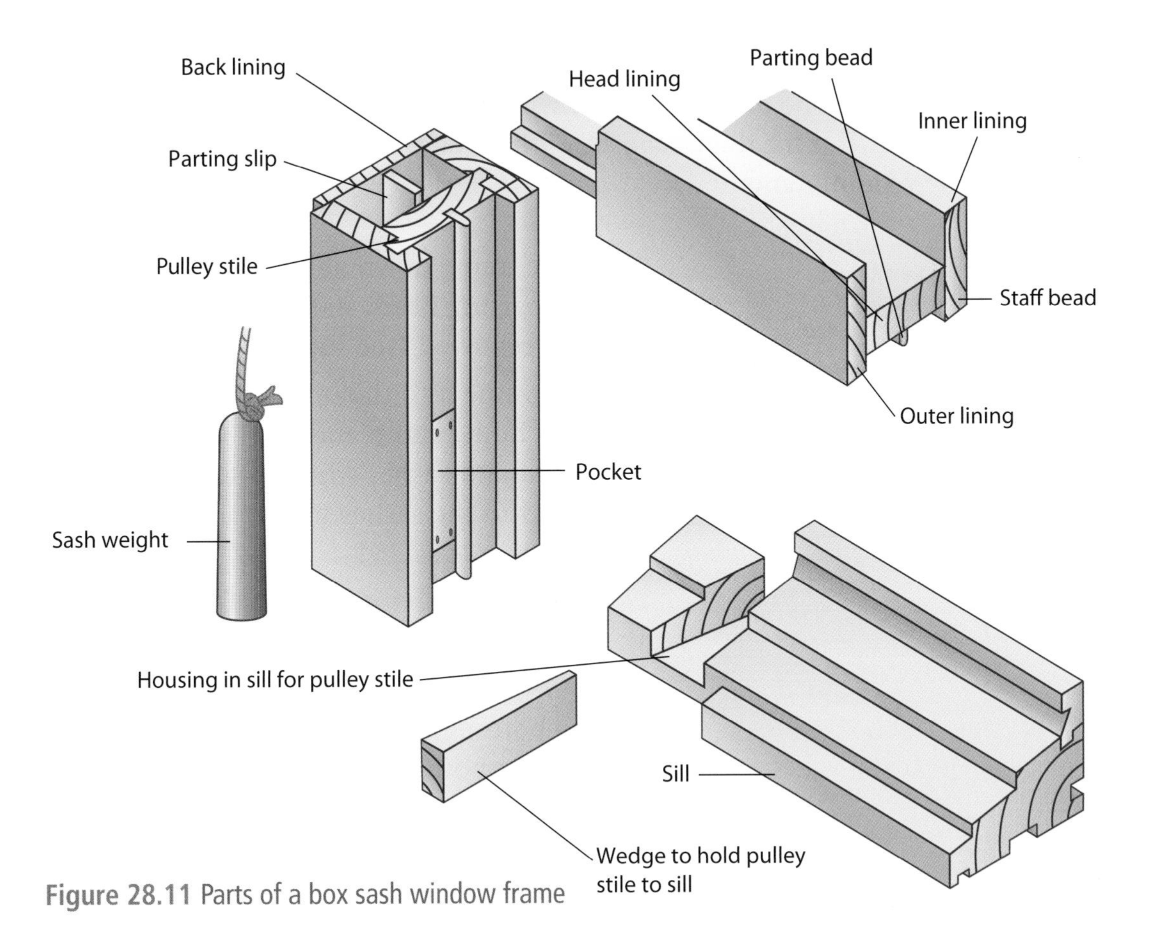

Figure 28.11 Parts of a box sash window frame

- **Staff bead** – a piece of timber fixed to the inner lining, keeping the lower sash in place.
- **Parting slip** – a piece of timber fixed into the back of the pulley stiles, used to keep the sash weights apart.
- **Pockets** – openings in the pulley stile that give access to the sash cord and weights, allowing the window to be fitted and maintained.

The sash in a box frame is made slightly differently to normal sashes, and has the following parts:

- **Bottom rail** – the bottom horizontal part of the bottom sash.
- **Meeting rails** – the top horizontal part of the bottom sash and the bottom horizontal part of the top sash.
- **Top rail** – the top horizontal part of the top sash.
- **Stiles** – the outer vertical members of the sashes.

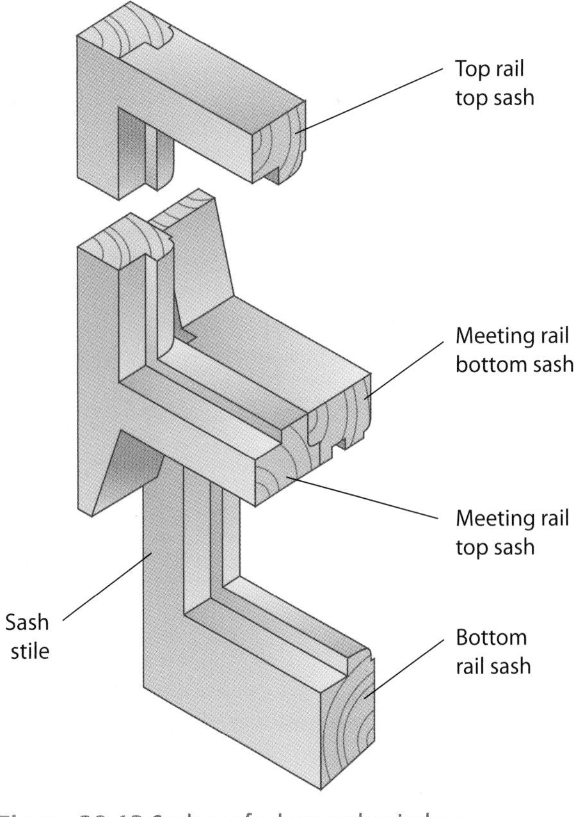

Figure 28.12 Sashes of a box sash window

Installing a box sash window

This can be done in one go with the sashes and weights already fixed, but for the purposes of this book we will show a more traditional method.

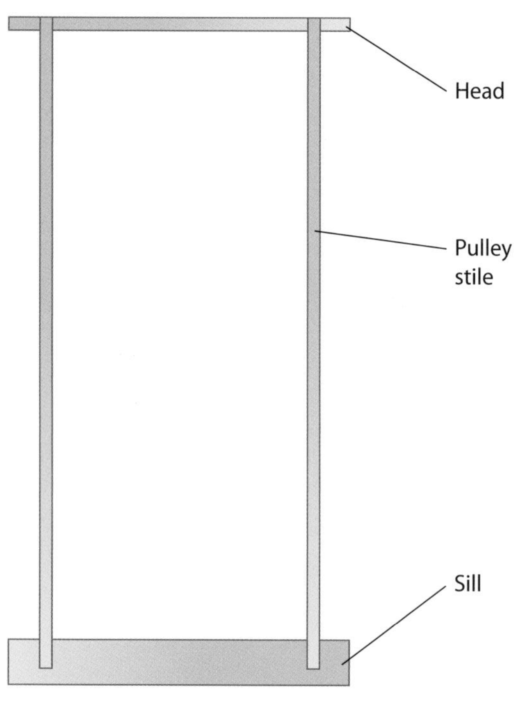

Figure 28.13 Stripped-down frame

1. Strip the frame down and have the sashes, parting bead, staff bead and pockets removed (see Figure 28.13).
2. Fix the frame into the opening, ensuring that it is both plumb and level, and fit the sash cord and sashes (the fitting of the sash cord to the sashes will be covered in Unit 3009, pages 251–53).
3. Slide the top sash into place and fit the parting bead, to keep the top sash in place.
4. Fit the bottom sash, then the staff bead, to keep the bottom sash in place.
5. Seal the outside with a suitable sealer, fit the ironmongery and make good to the inside of the window.

Figure 28.14 Top sash in place

Figure 28.15 Top and bottom sashes in place

Functional skills

When following construction processes and choosing materials, you will be practising **FM 2.2.1** Apply a range of mathematics to find solutions and **FM 2.2.2** Use appropriate checking procedures and evaluate their effectiveness at each stage.

You will also cover **FM 2.3.1** Judge whether findings answer the original problem and **FM 2.3.2** Communicate solutions to answer practical problems.

Pivot windows

Centre hung or pivot windows are used mainly in high rise buildings or similar areas, as they allow the outside of the window to be cleaned easily from the inside. The window consists of a solid frame into which the casement is fitted and hung on pivot hinges in the centre.

Did you know?

The back edge of a window board can be rebated to form a tongue to fit into a groove on the inside sill of a window.

Timber identification and selection

Timber is divided into the categories hardwood and softwood, although these names do not describe the properties of the timber as you will be working with it. Instead these divisions come from botany, and are a scientific classification.

Timber can decay – more information on this can be found in Unit 3009. Defects can also be found in timber. Some present a serious structural weakness in the timber; others do little more than spoil its appearance. Defects can be divided into two groups:

- **seasoning defects** – bowing, springing, winding, cupping, shaking, collapse, case hardening
- **natural defects** – heart shakes, cup shakes, star shakes, knots.

Tables 28.01 and 28.02 on pages 154–56 are a handy reference to the key properties of some of the more common woods you will encounter.

Find out

Find examples of each type of these natural and seasoning defects, either in actual pieces of wood or from pictures and the Internet. Explain what causes these and how they could be prevented.

Table 28.01 Identification of softwoods

Name	Main sources	Identification	Properties/description	Uses
Douglas fir	USA, Canada		Straight grained and resilient: easy to work with by hand or machine. One of the hardest softwoods; can take heavy, continuous wear. High resistance to acids and decay. Good gluing and high insulation qualities.	First-class joinery, light and heavy structural work, glued laminated work and plywood.
Larch	Europe		Very strong and durable, average to work with. Sawing and machine can leave a decent finish, but loose knots can be troublesome.	Fencing, gate posts, garden furniture, railway sleepers.
Pitch pine	North-eastern USA		Similar strength to Douglas fir. Works moderately easily, but resin is often a problem. Can clog saw-teeth and cutters and adhere to machine tables. Good finish is obtainable. Glue can be used, also nails and screws.	Shipbuilding, polished softwood joinery and church furniture.
Redwood (also known as Scots pine)	Europe		Moderately strong for its weight, with average durability. Works easily with both hand and machine tools, but ease of working and quality of finish dependent upon the size, number of knots and amount of resin. Capable of a smooth, clean finish. Can be glued, stained, varnished and painted. Takes nails and screws.	Depending on quality, it can be used for interior or exterior work and can be used for most tasks such as carcassing and finish joinery.
Western red cedar	Canada, USA		Not as strong as redwood, but has naturally occurring oils to prevent insect attack. Non-resinous and straight grained. Good workability with both hand and machine. Does not need treating as the wood will stand up to severe weather and will turn a silvery colour when exposed.	Externally for good quality timber buildings, saunas, etc.
Whitewood (also known as European spruce)	Europe		Similar to redwood in strength and durability. Good to work with by hand and machine, takes glue, nail and screws well and can produce a good finish.	Similar uses to redwood.

Table 28.02 Identification of hardwoods

Name	Main sources	Identification	Properties/description	Uses
Ash	Europe		Straight grained, very tough and flexible. Although tough, ash works and machines quite well, and finishes to a reasonably smooth finish. It can be glued, stained and polished and takes nails and screws well.	Furniture, boat building, sports equipment, tool handles, etc.
Beech	Europe		Hard, close grained and durable, with a fine texture. Works fairly well by both hand and machine, capable of a good smooth surface. Takes glue, stains and polish well and produces an excellent veneer.	Furniture, kitchen utensils, wood block floors, etc.
Mahogany	Africa		Interlocked grain, reasonably strong for its weight and moderately resistant to decay. Fairly easy to work with hand and machine, takes glue finish, nails and screws well.	High-class joinery, furniture, boat building and plywood veneers.
Mahogany	Cuba		Extremely strong for its weight, also has an interlocking grain. Good to work with and takes glue, nails, etc. very well, can be polished to an excellent finish.	As for African mahogany, but considered to be superior.
Maple rock	Canada, north east USA		Fine grained, with excellent finishing qualities and average durability. Can be difficult to work and is hard to nail or screw.	Panelling, flooring, furniture and snooker cues.
Oak	Europe		Very strong, with English oak the strongest. Good resistance to bending and shearing. Workability is quite tough but can be planed to a good finish. Susceptible to fungi attack. Ironwork will disfigure and leave a bluish stain. Glues well but difficult to screw and nail.	High-class joinery, panelling, doors, exposed roofing, etc.

Name	Main sources	Identification	Properties/description	Uses
Sapele	West Africa		Harder than mahogany, with similar strength properties to oak. Works fairly well with hand and machine tools, but interlocked grain is often troublesome in planing and moulding. Takes screws and nails well, glues satisfactorily, stains readily, and takes an excellent polish.	Furniture, veneers, etc.
Teak	India, Java, Thailand		Greasy to touch and resistant to both insect and fire. Strong, very durable, moderately elastic, and hard, but brittle along the grain. Workability can be difficult – tools soon blunt. Capable of a good finish if cutting edges are kept sharpened. Can be glued, stained and polished. It holds screws and nails reasonably well, but inclined to be brittle.	High-class joinery, furniture, boat building, etc.
Walnut	Europe		A relatively tough wood, with good resistance to splitting. Easy to work with by hand and machine and takes glue, nails and screws well. Can be polished to an excellent finish. Staining likely if in contact with iron under damp conditions.	Furniture, veneers, etc.

K2. Know how to fit and fix floor coverings and flat roof decking

Flat roof decking will be covered in depth in Unit 3008. You will need to select the correct timber and material when working with floor coverings and decking – durability is key, as softer woods will get easily damaged from the constant footfall on the floor.

Softwood flooring

Softwood flooring can be used at either ground or upper floor levels. It usually consists of 25 × 150 mm tongued and grooved (T&G) boards. The tongue is slightly off centre to provide extra wear on the surface that will be walked on.

Functional skills

To select the correct materials for a job, you will need to read and use design specifications. This will allow you to practise **FE 2.2.1** Select and use different types of texts to obtain relevant information, **FE 2.2.2** Read and summarise succinctly information/ideas from different sources and **FE 2.2.3** Identify the purposes of texts and comment on how effectively meaning is conveyed.

When boards are joined together, the joints should be staggered evenly throughout the floor to give it strength. They should never be placed next to each other, as this prevents the joists from being tied together properly. The boards are either fixed with floor brads nailed through the surface and punched below flush, or secret nailed with lost head nails through the tongue. The nails used should be 2½ times the thickness of the floorboard.

The first board is nailed down about 10–12 mm from the wall. The remaining boards can be fixed four to six boards at a time, leaving a 10–12 mm gap around the perimeter to allow for expansion. This gap will eventually be covered by the skirting board.

There are two methods of clamping the boards before fixing (shown in Figure 28.18 on the next page).

Hardwoods can also be used as floorboards. They are more decorative and much more expensive.

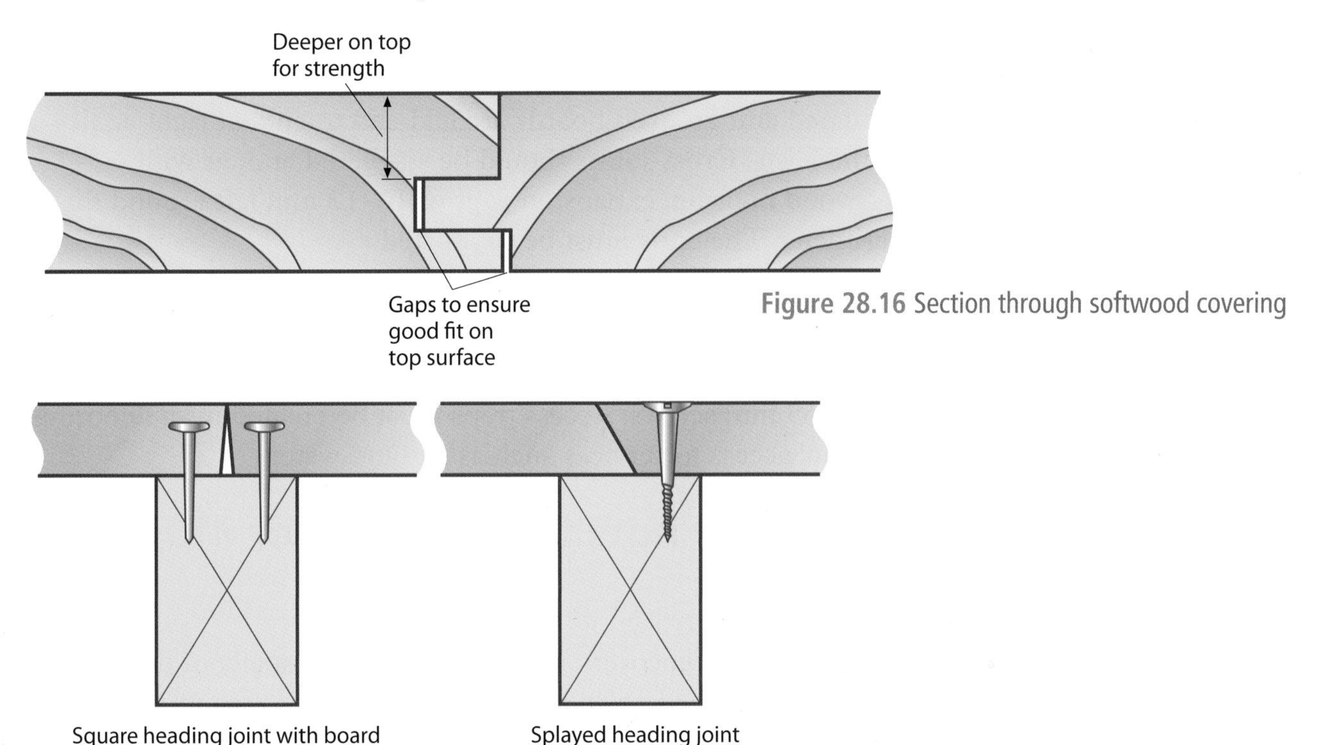

Figure 28.16 Section through softwood covering

Figure 28.17 Square and splayed heading

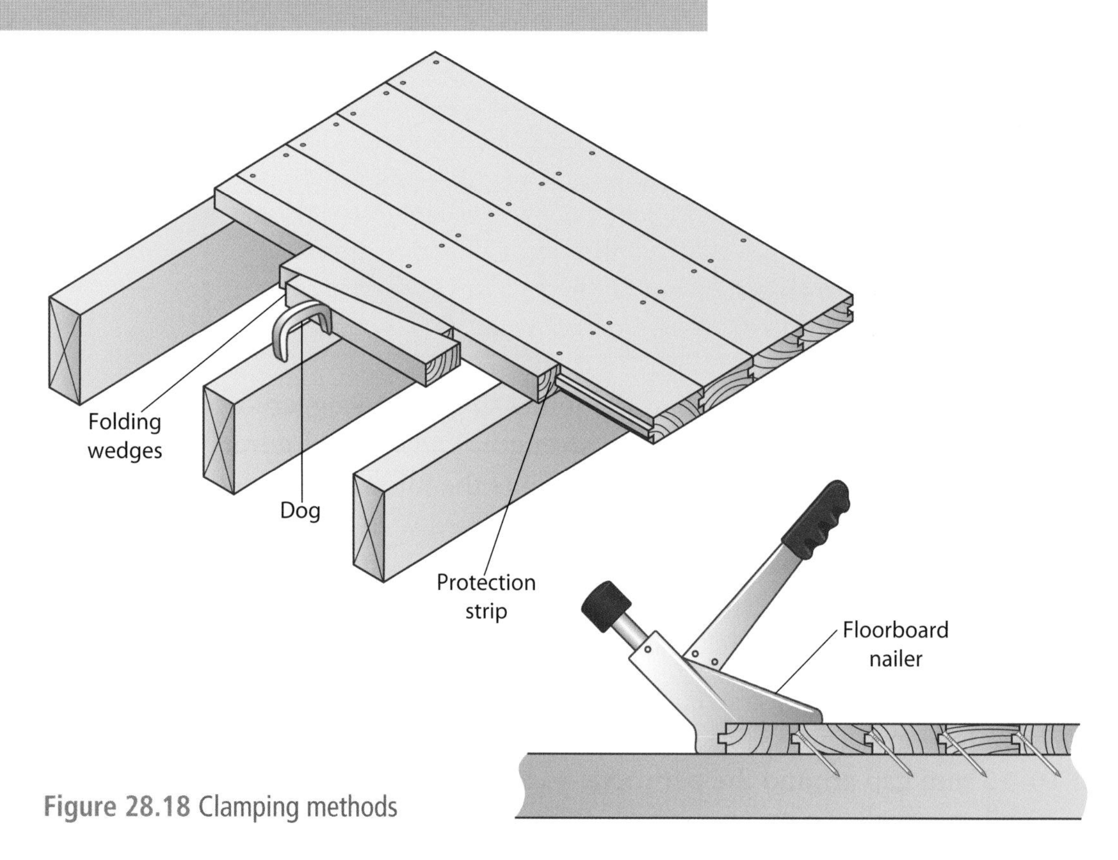

Figure 28.18 Clamping methods

Chipboard flooring

Flooring-grade chipboard is increasingly being used for domestic floors. It is available in sheet sizes of 2440 × 600 × 18 mm and can be square edged or tongued and grooved on all edges, the latter being preferred. If square-edged chipboard is used, it must be supported along every joint.

Tongued and grooved boards are laid end to end, at right angles to the joists. Cross-joints should be staggered and, as with softwood flooring, expansion gaps of 10–12 mm left around the perimeter. The ends must be supported.

When setting out the floor joists, spacing should be set to avoid any unnecessary wastage. The boards should be glued along all joints and fixed using either 50–65 mm annular ring shank nails or 50–65 mm screws. Access traps must be created in the flooring to allow access to services such as gas and water.

Remember

Access traps should be taken into consideration and the sheets or boards should be cut so that the traps can be formed. Access traps can be formed by finding the nearest joists and cutting between them to create an opening, using a floorboard saw, drilling a hole or using a jigsaw.

K3. Know how to erect timber stud partitions

Timber studwork covers partitions or walls, usually of light construction, used to divide a building or large area into compartments. The NVQ syllabus only covers non-load bearing walls.

Basic construction

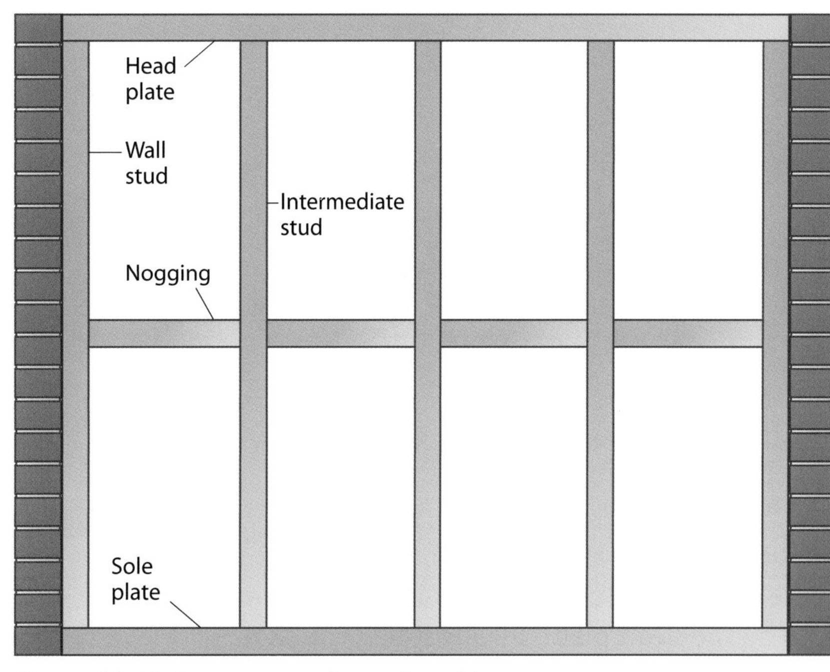

Figure 28.19 Components of a stud partition

Timber partitions are made up of a head and sole plate, with studs at regular centres fixed between them. These are stiffened up with noggings, which also provide extra fixings for sheet material, usually plasterboard, and fixing points for heavy components such as wash basins, toilets, etc.

The timber used is normally 75 mm × 50 mm for heights up to around 2.4 m and 100 mm × 50 mm for studwork above this height. Planed all round timber is preferred because its cross-section is uniform and is better to handle.

Partitions are either made in-situ or pre-made. In-situ components are cut, fixed together and fitted on site. Pre-made partitions are assembled on site or in a factory, for erection on site later.

Find out

When were stud partitions introduced to the building trade? What was in use before?

The intermediate studs are measured and fixed to suit the sheet material to be fixed. For 9.5 mm plasterboard, studs are spaced at 400 mm centre to centre and for 12 mm plasterboard at 600 mm. These sheets should cover 2.4 m × 1.2 m. A sheet size of 2.4 m × 0.9 m would have studs spaced at 450 mm.

The majority of stud partitions are fixed together by butt joints and skew-nailed. Another method is to use framing anchors, which are strong and quick to fit.

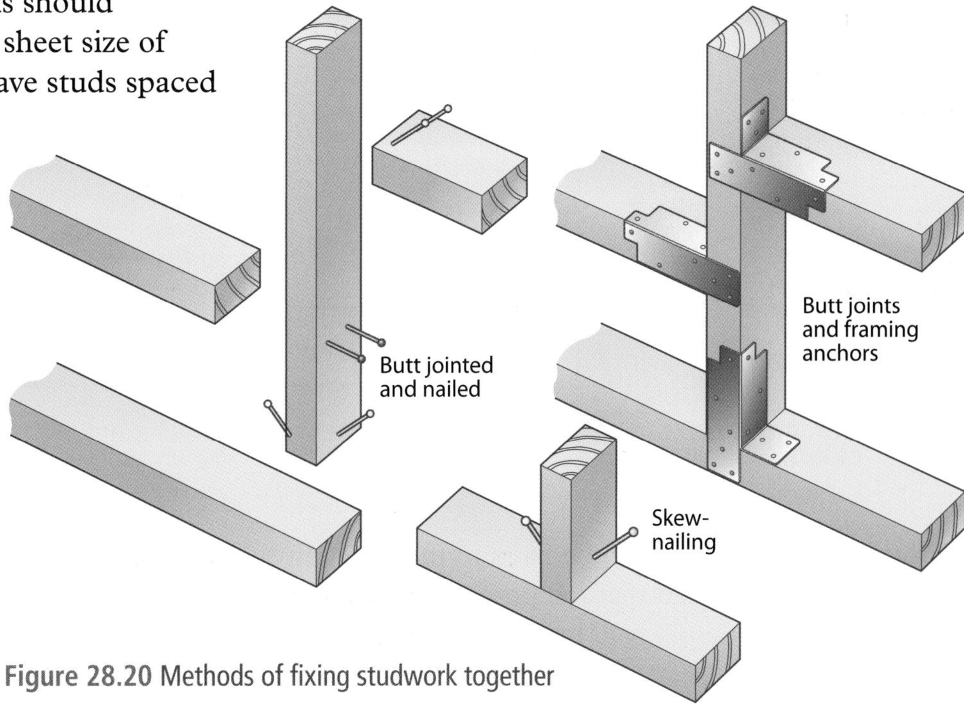

Figure 28.20 Methods of fixing studwork together

Step-by-step: In-situ partitions

1

Mark out positions

Step 1 Mark out positions for partitions on the floor. Then check the ceiling to see how to fix the head plate. Ideally it will run at a right angle to the joists but, if not, further noggings should be fixed between each joist to provide a fixing point. The head plate can now be fixed to the ceiling, usually with 100 mm wire nails or screws.

2

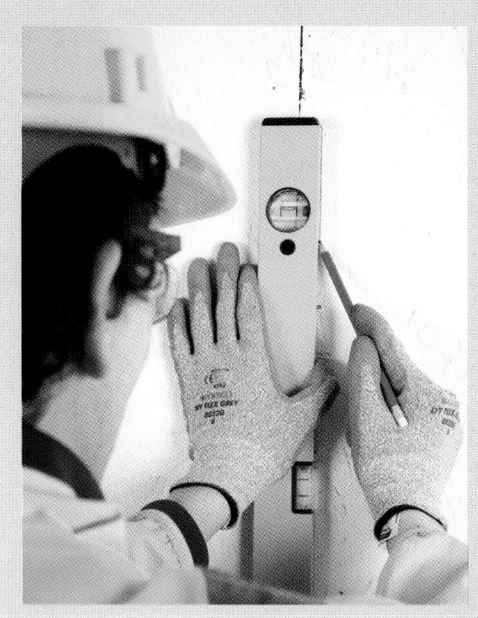

Plumb down

Step 2 Plumb down from the head plate with a plumb bob, or a level and straight edge, and mark the position on the floor.

3

Fix the sole plate

Step 3 The sole plate can be fixed in the same way as the head plate. If on a concrete floor, it should be plugged and screwed.

4

Fix studs

Step 4 Cut the wall studs and fix to the wall by plugs and screws. They should be skew-nailed to the head and sole plate.

Position intermediate studs

Step 5 Mark the position of intermediate studs on the side of the head and sole plates at the appropriate centres, that is 400, 450 or 600 mm depending on sheet material size and thickness.

Skew-nail the studs

Step 6 Measure each stud individually, as they may vary in length between head and sole plate, then skew-nail with 100 mm wire nails. Studs should be a tight fit and, with the bottom in place, the tip should be forced over into position.

Through-nail central noggings

Step 7 If overall partition height is greater than 2.4 m, position noggings centrally at the top edge of the boards. Through-nail one end and skew-nail the other.

Fix additional noggings

Step 8 Fix additional noggings at various heights from 600 mm, 1200 mm to 1800 mm to give extra strength and rigidity to the frame, and provide extra fixing points for the sheet material and any components that may have to be fixed to the walls.

Forming openings

Where an opening is to be formed in a studwork partition, studs and noggings should provide fixing points for items such as door linings, windows, hatches, etc. They also provide fixings for the edges of any sheet material to be used. Details are shown in Figures 28.21–28.24.

Figure 28.21 Noggings provide fixings for sheet materials

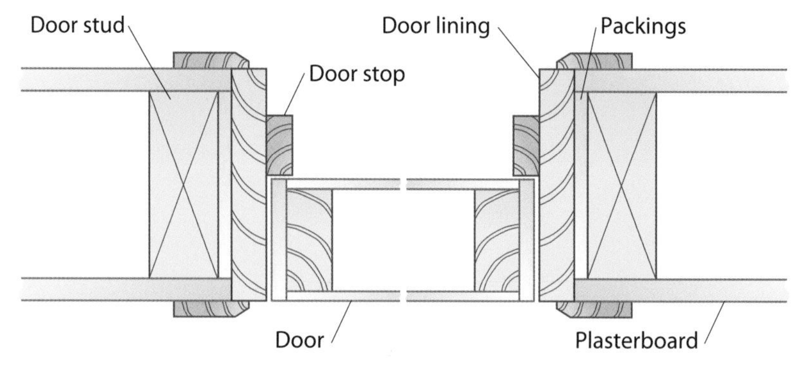

Figure 28.22 Section through a door opening

Did you know?

Stud partitions can be clad with plasterboard, plywood, or almost any sheet material. They can be insulated with rock wool, glass wool, mineral wool or polystyrene sheets.

Remember

When a partition has to be returned at right angles or when a 'tee' junction is required, extra studs must be fixed to provide support and fixing for the covering material.

Figure 28.23 Opening formed to take a window lining

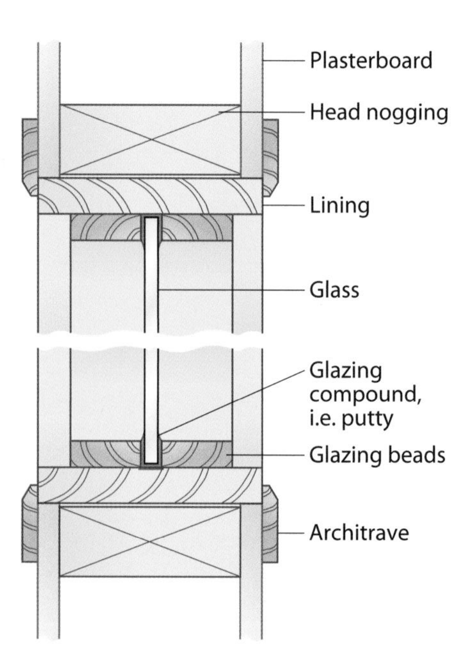

Figure 28.24 Section through a window lining

Pre-made partitions

Pre-made partitions are the same specification as in-situ partitions, but can be made up either in a workshop or on site. They are made slightly under size in height and width to allow for the frame to be offered into position.

Joints are usually butt-jointed and nailed, but can also be housed-in or framing anchors used.

Folding wedges can be used to keep the frame in position while fixing takes place.

Holes and notches

It is common practice for pipes and cables to be concealed within timber partitions, though it is recommended that they are kept to a minimum so as not to affect partition strength.

Positioning is important. Holes for carrying cables, and notches for water pipes, should be kept away from areas where there is a possibility they may be punctured by a nail or screw. This includes areas where, for example, kitchen units, cupboards, skirting boards, dado or picture rails may be fixed.

Figure 28.25 Holes for pipework

Notches can be protected by fixing a metal plate over them.

Figure 28.26 Notches with wire

Joist and stud coverings

Joists are decked with a suitable flooring material, and the underside of the joists is usually covered with plasterboard.

There are two main ways of plasterboarding a room.

Method 1

The plasterboard is fixed to the stud or joist with the back face of the plasterboard showing. The plasterer covers the whole wall or ceiling with a thin skim of plaster, leaving a smooth finish (see Figure 28.27 on the next page).

Method 2

The plasterboard is fixed with the front face showing, and the plasterer uses a special tape to cover any joints; then a ready-mixed filler is applied over the tape and is used to fill in the nail and screw holes (see Figure 28.28). Once the taped area is dry, the plasterer then gives the area a light sand to even it out.

Figure 28.27 Plastered wall

Figure 28.28 Taped wall

Plasterboard comes with a choice of two different edges, and the right edge must be used:

- **Square edge** – for use with Method 1; the whole wall/ceiling is plastered.
- **Tapered edge** – for use with Method 2, so that the plasterer can fix the jointing tape.

As well as having the correct edging, the right type of plasterboard must be used.

Various types of plasterboard are available so a colour-coding system is used, although this may vary from manufacturer to manufacturer:

- **Fire-resistant plasterboard (usually red or pink)** – used to give fire resistance between rooms such as party walls.
- **Moisture-resistant plasterboard (usually green)** – used in areas where it will be subjected to moisture, such as bathrooms and kitchens.
- **Vapour-resistant plasterboard** – comes with a thin layer of metal foil attached to the back face to act as a vapour barrier; usually used on outside walls or the underside of a flat roof.
- **Sound-resistant plasterboard** – used to reduce the transfer of noise between rooms and can also help acoustics; used in places such as cinemas.
- **Thermal check board** – comes with a foam or polystyrene backing; used to prevent heat loss and give good thermal insulation.
- **Tough plasterboard** – stronger than average plasterboard, it will stand up to more impact; used in public places such as schools and hospitals.

Remember

If you double sheet plasterboard, the joints in the two layers must be staggered.

Sometimes plasterboard is doubled up – the walls are 'double sheeted' – to give fire resistance and sound insulation without using special plasterboard.

Plasterboard comes in a variety of sizes, but the most standard size is 2400 × 1200 × 12.5 mm.

Cutting plasterboard

Plasterboard can be cut in two main ways: with a plasterboard saw or with a craft knife. The plasterboard saw can be used when accuracy is needed, but most jobs will be touched up with a bit of plaster so accuracy is not vital. Using a craft knife is certainly quicker.

Step-by-step: Cutting plasterboard

1

Step 1 Mark the line to be cut.

2

Step 2 Score along the line.

3

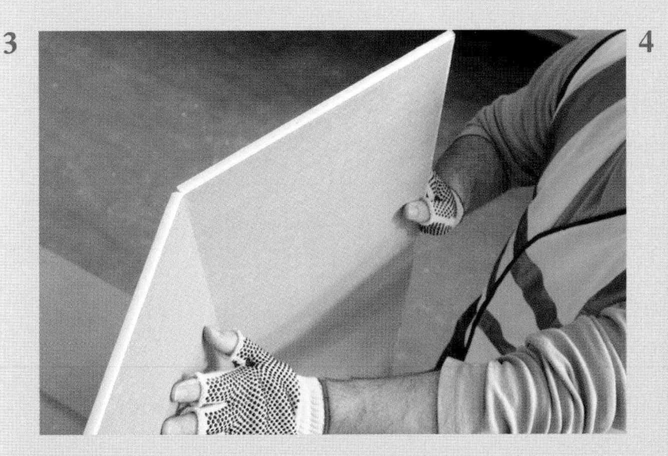

Step 3 Split the board in two.

4

Step 4 Separate the pieces by running the knife along the back.

5

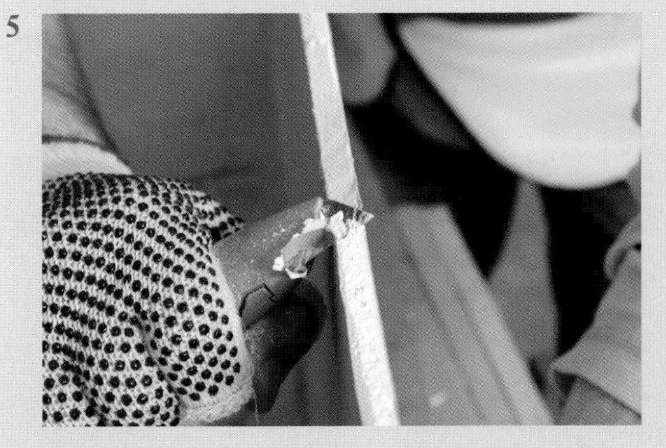

Step 5 Trim the cut with a slight back bevel for a neat finish.

Remember

When fixing plasterboard using nails or screws, be aware of where the services are. Otherwise, you could hit an electricity cable or a water pipe.

Remember

Dot and dab is a skilled technique, as with this method getting the boards plumb and flat can be difficult.

Functional skills

Using the information contained in these tables will give you the opportunity to practise **FM 1.2.1b**, relating to interpreting information from sources such as diagrams, tables, charts and graphs.

Fixing plasterboard

Plasterboard can be fixed by three different methods:

- **Nails** – use clout or special plasterboard nails. Make sure the nail is driven below the surface, leaving a dimple or hollow that can be filled by plaster later. Nails must be galvanised – if not, when the plaster is put on to fill the hollows, the nails will rust and will 'bleed'. This will stain the plaster and can show through several coats of paint.
- **Screws** – these must be galvanised to prevent bleeding.
- **Dot and dab** – used where the plasterboard is fitted to a pre-plastered or flat surface such as blockwork. There is no need for a stud or frame to fix the plasterboard, increasing the room size. The plaster is mixed up and dabbed onto the back of the board, before the board is pushed directly onto the wall.

K4. Know how to assemble, erect and fix straight flights of stairs, including handrails

Stairs are a means of providing access from one floor to another. They are made up of a number of steps, and each continuous set of steps running in one direction is called a flight. Hence, a staircase may have several flights.

The construction and design of a stair is controlled under the Building Regulations 2000: Part K. These specify the requirements for different types of stairs depending on the type of building and what it is to be used for.

Tables 28.03 to 28.05 detail some of the requirements in the regulations.

Table 28.03 Regulations for rise and going

Description of stair (to meet)	Max. rise (mm)	Min. going (mm)	Range (to meet pitch limitation) (mm)
Private stair	220	220	155–220 rise with 245–260 going or 165–200 rise with 220–350 going
Common stair	190	350	155–190 rise with 240–320 going
Stairway in institutional building (except stairs used only by staff)	180	280	n/a
Stairway in assembly area (except areas under 100 m square)	180	250	n/a
Any other stairway	190	250	n/a

Table 28.04 Regulations for minimum balustrade heights

Description of stair	Minimum balustrade height (mm)	
	Flight	Landing
Private stair	840	900
Common stairway	900	1000
Other stairway	900	1100

Table 28.05 Regulations for width of stairs

Description of stair	Minimum width (mm)
Private stair giving access to one room only (except kitchen and living room)	600
Other private stair	800
Common stair	900
Stairway in institutional building (except stairs used only by staff)	1000
Stairway in assembly area (except areas under 100 m square)	1000
Other stairway serving an area that can be used by more than 50 people	1000
Any other stairway	800

Installation of stairs

Installing a staircase is a major part in the construction of the building. The same basic procedures are followed, but there will be differences depending on the type of staircase and the site conditions.

A staircase is a large and fairly expensive item and should be handled with care when delivered to site and offloaded. It is normally delivered in completed form as far as possible, but components such as the handrail, balusters and newel posts will have to be fitted to suit the site requirements.

Did you know?

Spindles, used in balustrades, should be spaced so that a sphere of 100 mm diameter cannot be pushed through.

Step-by-step: Installation of stairs

1

Fix the wall string

Step 1 Fix the wall string, cutting it off at floor level to suit the skirting height.

2

Cut the seat cut

Step 2 Use a hand saw to cut the seat cut on the string at the foot of the stairs.

3

Check level

Step 3 The staircase is now level on the floor.

4

Cut away the top tread

Step 4 Steps 1 and 2 can be repeated at the top with the underside cut out to sit on to the floor trimmer and the top tread cut away so that it sits on the trimmer.

5

Mortise the outer string

Step 5 Mortise the outer string into the newel posts at each end.

6

Fix newel posts

Step 6 Fix newel posts in place. The bottom newel can be held in position using various methods and this will depend on the composition of the floor. For rigidity and maximum strength, the top newel should be notched over the trimmer joist and screwed or bolted to it.

7

Fix the wall string to the wall

Step 7 Fix the wall string to the wall in approximately four places below the steps, usually with 75 mm screws and plugs. Fixing should be below the steps unless access is difficult, in which case the resulting hole should be plugged. The balustrades and handrail can be fitted once the stairs are secure.

8

Protect staircase from damage

Step 8 Once the staircase has been fitted, it should be protected to prevent damage. Strips of hardboard should be pinned to the top of each tread with a lath to ensure the nosing is protected. Use the same method to protect the newel posts.

Working life

Scott and Nick are working in a large house plasterboarding the stud partition walls. When they come to the last wall, they find a note left by the plumber saying 'Don't plasterboard'. They are on price work and Nick says that they should just finish the work and get paid. Scott says that they should ask the plumber first.

- Who is right?
- What could the consequences be if they do plasterboard – and if they don't?

Functional skills

Working with other trades is an important part of working on site. Dealing with problems like this will allow you to practise **FM 2.3.1** Judge whether findings answer the original problem and **FM 2.3.2** Communicate solutions to answer practical problems.

FAQ

I have laid chipboard flooring and the floor is squeaking. What causes this and how can I stop it?

The squeaking is caused by the floorboards rubbing against the nails – something that happens after a while as the nails eventually work themselves loose. The best way to prevent this is to put a few screws into the floor to prevent movement, but be cautious: there may be wires or pipes under the floor.

My maths isn't very good so I often struggle with the geometry and trigonometry skills needed, especially for roofing. What can I do?

Your training will include some basic numeracy skills lessons, and if you struggle with maths you should take advantage of them. Be patient and pay attention to any help and advice you are given regarding these sometimes complicated calculations. You will probably find that the more practical experience you get, the better your maths skills will become.

Check it out

1. Draw a sketch of a window, labelling the component parts.
2. What is a mortar key used for in a window frame?
3. Explain why it is important to assemble windows flat.
4. Write a method statement explaining how you can check whether a window frame is square while you assemble it.
5. Name two ways of fixing a window board into position.
6. How do you level a window board?
7. State and describe three materials used as window boards.
8. Explain what types of weight are used in a box sash window.
9. Name three types of bay window.
10. List five components you will find on a flight of stairs.
11. What is the minimum headroom allowed as stated in the Building Regulations covering stairs?
12. Name the vertical member used to provide support to a handrail and infill on an open balustrade.
13. After stairs have been fitted, what three measures can be taken to prevent damage during building work?
14. Name three ways of fixing grounds.
15. What personal protective equipment (PPE) should be worn when using masonry nails?

Getting ready for assessment

The information contained in this unit, as well as continued practical assignments that you will carry out in your college or training centre, will help you with preparing for both your end of unit test and the diploma multiple-choice test. It will also aid you in preparing for the work that is required for the synoptic practical assignments.

The information contained within this unit will aid you in learning how to identify and calculate the materials and equipment required for first fixing.

You will need to be familiar with:

- preparing materials for carrying out a range of first fixing tasks
- fixing frames and linings
- fitting and fixing floor coverings and flat roof decking
- erecting timber stud partitions
- assembling, erecting and fixing straight flights of stairs.

This unit will have made you familiar with these common first fixing tasks. For learning outcome two, you have seen the types and sizes of timber that can be used for floor coverings and flat roof decking. You will need to use your knowledge of these types of timber to select the correct one for any particular fixing job you may be working on. You will also need to select the correct hand tools and equipment to work with a particular material in a particular situation.

This unit has shown the types and methods of fixing joist coverings and forming openings to services under floors. You will now need to use this knowledge in your practical tasks to install floor coverings and flat roof coverings accurately, and to working drawings and specifications. You will also have to identify the best methods of leaving openings for services, as well as being able to identify any problems with the service openings.

Before you start work on the synoptic practical test, it is important that you have had sufficient practice and that you feel that you are capable of passing. It is best to have a plan of action and a work method that will help you. You will also need a copy of the required standards, any associated drawings and sufficient tools and materials. It is also wise to check your work at regular intervals. This will help you to be sure that you are working correctly and help you to avoid problems developing as you work.

Your speed at carrying out these tasks will also help you to prepare for the time limit that the synoptic practical task has. But remember, don't try to rush the job as speed will come with practice and it is important that you get the quality of workmanship right.

Make sure you are working to all the safety requirements given throughout the test and wear all appropriate personal protective equipment (PPE). When using tools, make sure you are using them correctly and safely.

Good luck!

CHECK YOUR KNOWLEDGE

1. The purpose of a window is:
   - a to allow natural light into a room
   - b to conserve heat
   - c to protect from weather
   - d all of the above
2. When assembling a window frame, you can check that it is square by:
   - a measuring the length and width
   - b using the 3:4:5 method
   - c using a squaring rod
   - d using a spirit level
3. A window that projects from the front of a building is called a:
   - a bay window
   - b stormproof window
   - c casement window
   - d box sash window
4. The vertical part of a step on a staircase is called a:
   - a tread
   - b riser
   - c newel
   - d string
5. The combination of a tread and a riser is called a:
   - a stair
   - b string
   - c step
   - d stairwell
6. On a staircase, a string is:
   - a an opening formed in the floor to accept a stair
   - b something the handrail is fixed to
   - c the height of a step
   - d the board that the treads and risers are fitted to
7. The maximum pitch for a private stair is:
   - a 38°
   - b 40°
   - c 42°
   - d 44°
8. The top of a staircase must be fixed to a:
   - a trimmer joist
   - b trimmed joist
   - c trimming joist
   - d bridging joist
9. The timber part of a stud partition that stiffens the structure and allows a fixing for heavy components is called:
   - a nogging
   - b nugging
   - c nigging
   - d nagging
10. Extra studs are placed at corners because:
    - a they are stronger
    - b they offer a fixing for cladding
    - c they are weaker
    - d they allow for doorways
11. Partitions can be made more fire resistant by:
    - a using insulation
    - b double boarding
    - c using larger studs
    - d using more studs
12. A frame that runs from floor to floor is called a:
    - a door frame
    - b fan light frame
    - c storey frame
    - d door lining
13. The joint used in the construction of a door lining is called a:
    - a mortise and tenon joint
    - b butt joint
    - c housing joint
    - d tongued housing joint
14. The purpose of grounds is:
    - a to provide a flat surface to allow cladding
    - b to allow easy fixing for skirting boards
    - c to provide a line for the plasterer to work to
    - d all of the above

Unit 2009

Know how to carry out second fixing

Once plasterwork has been completed, you will be in a position to carry out second fixing. Second fixing is the name given to all joinery work undertaken after plasterwork has been finished, and includes side hung doors, units and ironmongery. In this unit you will learn the skills needed to confidently complete second fixing. This unit may be familiar to you from Level 2, so this will give you a reminder of some of the key issues and skills you need to know.

This unit supports TAP Unit 3 Install second fixing components, and delivery of the five generic units.

This unit contains material that supports the following NVQ unit:

- VR 10 Install second fixing components

This unit will cover the following learning outcomes:

- Know how to install side hung doors and ironmongery
- Know how to install mouldings
- Know how to install service encasements and cladding
- Know how to install wall and floor units and fitments

K1. Know how to install side hung doors and ironmongery

Internal doors

This section has been designed to provide you with the knowledge and understanding to select and hang internal doors.

> **Functional skills**
>
> While working through this section, you will be practising **FE 2.2.1** Select and use different types of texts to obtain relevant information, **FE 2.2.2** Read and summarise succinctly information/ideas from different sources and **FE 2.2.3** Identify the purposes of texts and comment on how effectively meaning is conveyed.
>
> If there are any words or phrases you do not understand, use a dictionary, look them up on the Internet, or discuss with your tutor.

Types of internal door

Framed doors

Framed doors are doors made from hardwood or softwood and are constructed using mortise and tenon joints (described in Unit 2008), or dowelled joints. The frame is rebated or grooved, into which a board panel can be fitted; alternatively, glass could be used and held in place with beading. Figure 29.01 shows an exploded view of a door, including the types of joint used.

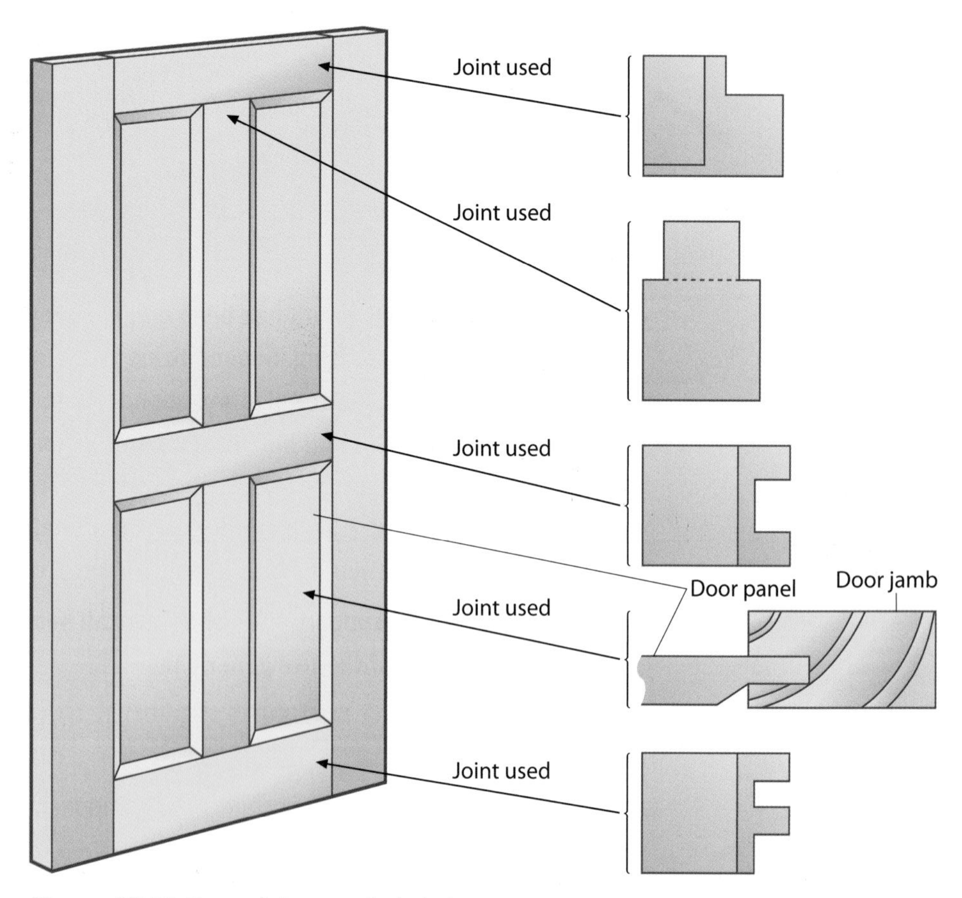

Figure 29.01 Framed door, exploded view

Flush doors

Flush doors are lightweight, cheap and simply made. Most consist of a softwood frame, which is stapled together and houses a hollow-core material (usually cardboard honeycomb that has the appearance of egg boxes) (see Figure 29.02). They are then faced with hardboard or plywood.

Flush doors often come with fitting instructions and these indicate which **stile** contains a lock block (an extra block of wood included in the door frame to take the door lock). Failing this, a symbol is normally printed on the top or bottom edge that indicates where the lock block can be found.

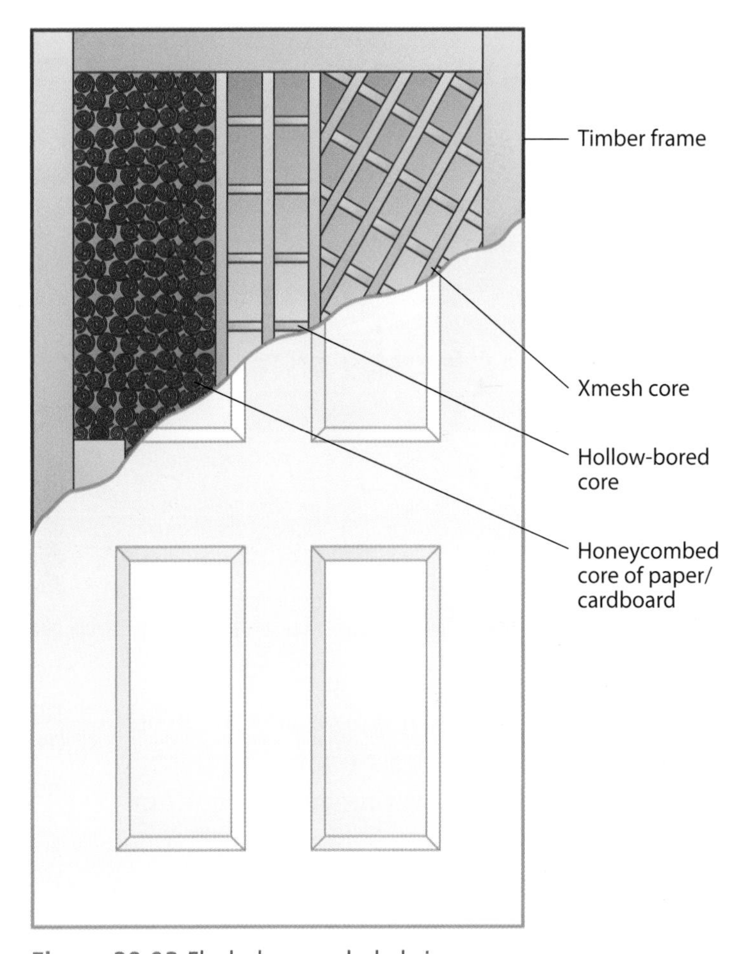

Figure 29.02 Flush door, exploded view

Fire-resisting doors

Fire-resisting doors have a core of solid, fire-retarding material and are frequently flush panelled. Their ratings are designated by their performance (resistance to penetration by flames or smoke through splits or gaps) with the prefix FD. For example, FD30 has a 30-minute rating, and an S suffix indicates the ability to resist smoke. The manufacturers also insert a coloured plastic plug into the door frame to indicate the rating. They are available in standard sizes and in thicknesses of 44 mm for the FD30 and 54 mm for the FD60.

Did you know?

A material that swells when subjected to heat or damp is called intumescent.

Fitting a door

The following actions should be completed before starting to fit a door:

1. Check the building plans and door schedule to determine which type of door is to be used and in which direction it should operate. Failing this, ask the supervisor or client.
2. Check the frame is square and aligned properly.
3. Measure the new door to make sure it will fit within the frame and check it is not twisted.
4. Cut off horns, if fitted.
5. Find where the lock block is located. The other stile is the hanging stile to which hinges will be attached.

Key term

Stile – the longest vertical timber in the frame of a door, etc.

Did you know?

Doors should normally conceal the largest area of the room when open to allow maximum privacy to those inside.

Step-by-step: Fitting a door

1

Make sure there is a gap between the door and the frame

Step 1 There must be a gap of 2 mm between the head of the door and the head of the frame, and at least 6 mm clearance between the floor covering and the bottom of the door. If the door is too tall, always cut excess off the bottom rail.

2

Cut door to width and shape stiles

Step 2 There must be a gap of 2 mm between both door stiles and the frame. If necessary, plane evenly from both sides of the door. The door should now fit in the frame. With the hanging stile against the frame, wedge the door up so there is a 2 mm gap at the top and mark, if necessary, to shape the hanging stile to match the frame. Plane off as necessary.

3

Bevel the leading edge

Step 3 If necessary, mark and shape the opposite stile (containing the lock block); then slightly bevel the edge that will lead into the frame, so that it will not catch when closing.

4

Mark hinges

Step 4 With the door fitting snugly in the frame and wedged up to give 2 mm clearance at the top, mark where the hinges are to go, on both door and frame at the same time. The top hinge normally sits 150 mm from the top of the door and the bottom hinge 225 mm from the bottom of the door.

5

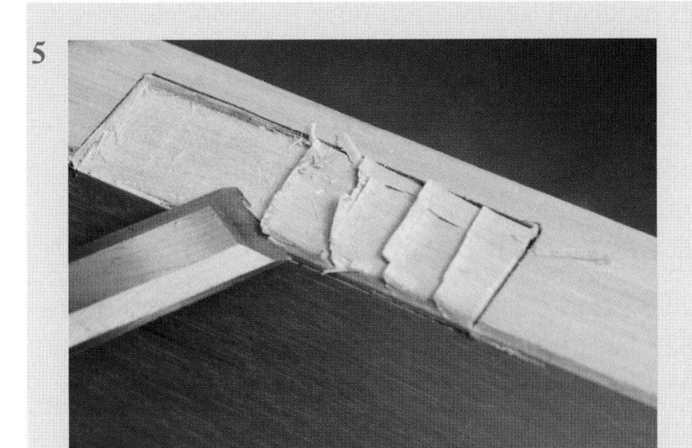

Mark and cut hinge recesses

Step 5 Remove the door from the frame and accurately mark where the hinges are to go, with the aid of a square and marking gauge. Do the same on the frame and then chop out recesses on both door and frame.

6

Fit hinges

Step 6 Fit the hinges into the door recesses, putting in all the screws. Take the door to the frame and fit the hinges into the frame recesses, securing the hinges with one screw. Check the door swings without binding. Adjust until there is an equal gap on both sides and the correct clearance at the top, and no resistance is encountered when closing the door. Fit all remaining screws on the frame hinges.

Lastly, fit all remaining ironmongery, or furniture, as instructed by the schedule or client.

External doors

This section aims to provide you with the knowledge and understanding needed to select, hang and fix the required ironmongery to an external door.

Types of external door

External doors are always solid or framed – a flush or hollow door will not provide the required strength or security.

In some cases external doors are made from uPVC. uPVC doors require little maintenance and the locking system normally locks the door at three or four different locations, making it more secure. High-quality external doors are usually made from hardwearing hardwoods such as mahogany or oak, but can be made from softwoods such as pine.

External doors come in the same dimensions as internal doors, except external doors are thicker (44 mm rather than 40 mm).

Did you know?

Because the hinge knuckle goes on the outside edge of the door, the arc followed by the inside edge is different; so a door requires a bevelled lead-in edge to close properly.

Remember

When fitting a fire-resisting door, all tolerances, as specified in the Building Regulations, must be adhered to, so check with the supervisor.

Did you know?

A framed, ledged and braced door comes with the bracing unattached. The bracing can then be attached to suit the side on which the door is hung.

On older properties external doors often have to be specially made, as they tend not to be of standard size.

Framed, ledged and braced doors

Framed, ledged and braced (FLB) doors consist of an outer frame clad on one side with tongued and grooved boarding, with a bracing on the back to support the door's weight.

FLB doors are usually used for gates and garages, and sometimes for back doors. When hanging an FLB door it is vital that the bracing is fitted in the correct way; if not, the door will start to sag and will not operate properly.

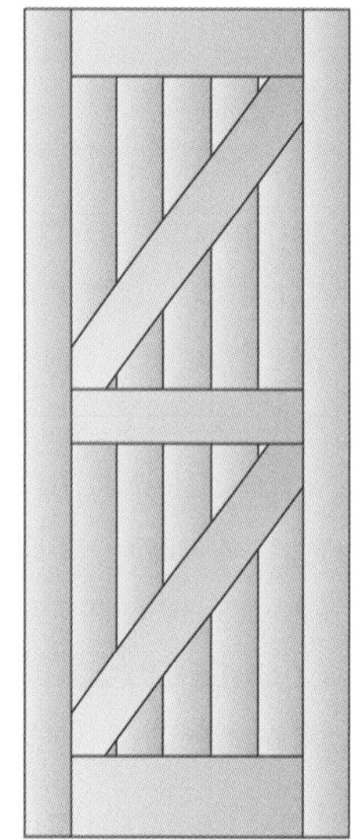

Figure 29.03 Front and back of a framed, ledged and braced door

Panelled doors

Panelled doors consist of a frame made up from stiles, rails, muntins and panels. Some panel doors are solid, but most front and back doors have a glazed section at the top to allow natural light into the room.

Half- and full-glazed doors

Half-glazed doors are panelled doors with the top half of the door glazed. These doors usually have diminished rails to give a larger glass area.

Full-glazed doors come either fully glazed or with glazed top and bottom panels, separated by a middle rail (see Figure 29.05).

Full-glazed doors are used mainly for French doors or back doors, where there is no need for a letterbox.

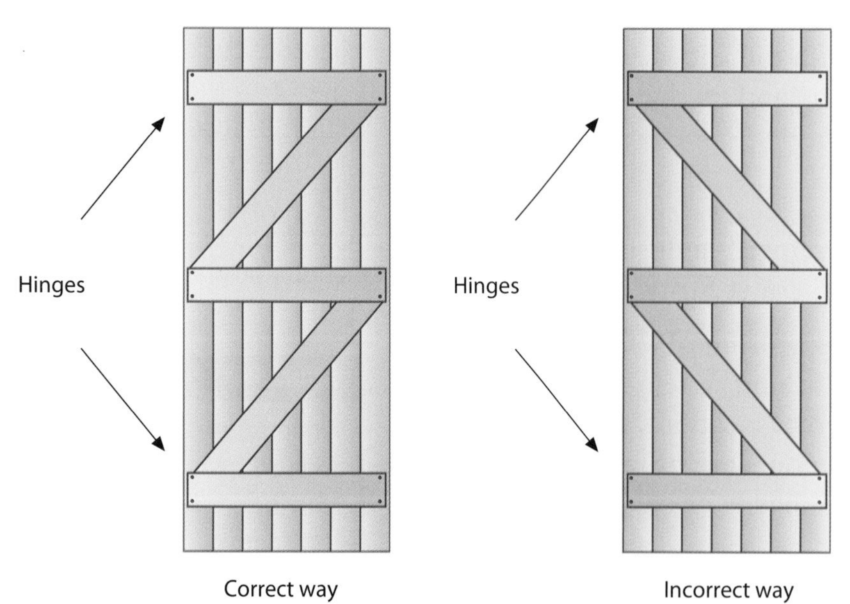

Figure 29.04 Bracing fitted correctly and incorrectly

Stable doors

As their name implies, stable doors are modelled on the doors for horses' stables and they are now most commonly used in country or farm properties. A stable door consists of two doors hung on the same frame, with the top part opening independently of the bottom. The make-up of a stable door is similar to the framed, ledged and braced door, but the middle rail is split and rebated as shown in Figure 29.06.

To hang a stable door, first secure the two leaves together with temporary fixings, then hang just like any other door, remembering that four hinges are used instead of three.

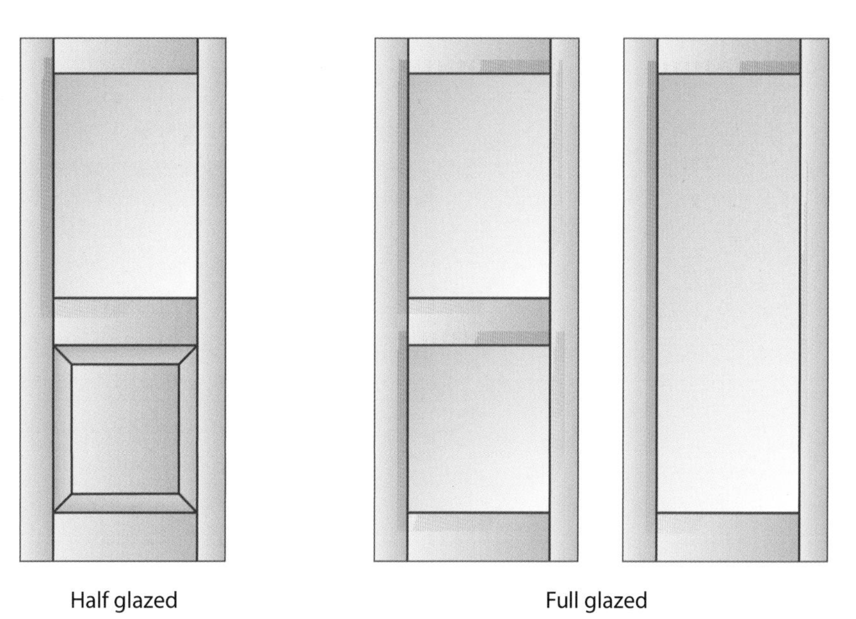

Figure 29.05 Half-glazed door and two types of full-glazed door

Hanging an external door

External doors usually open inwards, into the building. Externally opening front doors are usually only used where there is limited space, or where there is another door nearby which affects the front door's usage.

Hanging an exterior door is largely the same as hanging an interior door, except that the weight of the door requires three hinges sited into a frame rather than a lining. Because of this, there is usually a threshold or sill at the bottom of the door frame.

If a water bar is fitted into the threshold to prevent water entering the dwelling, you will need to rebate the bottom of the door to allow it to open over the water bar. The way the door is then hung will depend on which side of the door is rebated. A weatherboard must also be fitted to the bottom of an external door, to stop driving rain entering the premises.

As an exterior door is exposed to the elements it is important that the opening is draught proofed: a draught-proofing strip can be fitted to the frame, or draught proofing can be fitted to the side of the door.

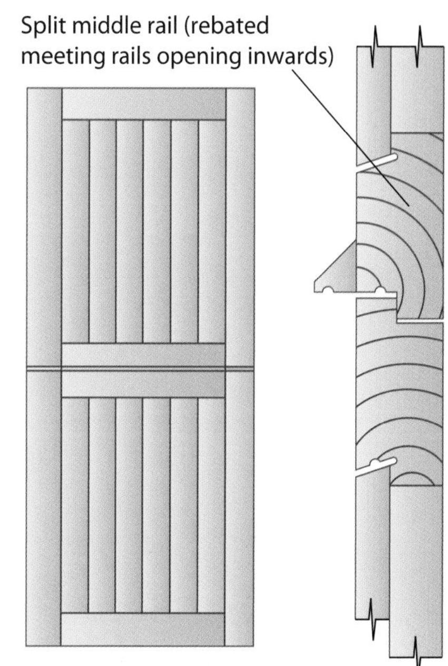

Figure 29.06 Stable door with section showing the rebate on the middle rails

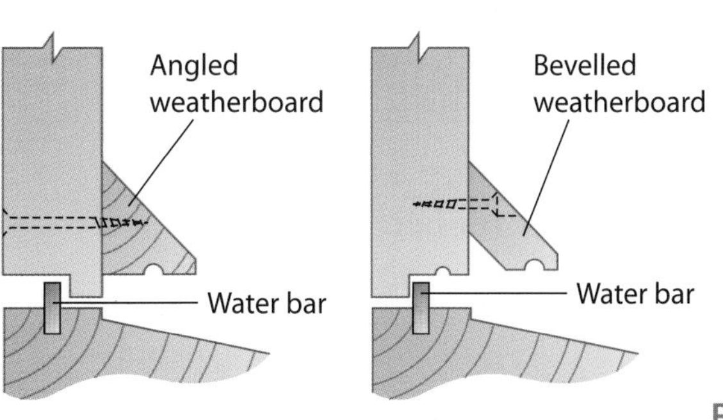

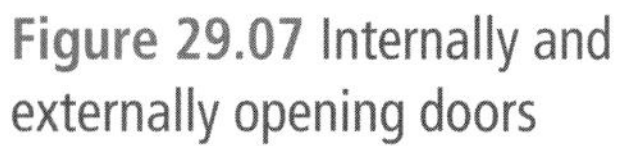
Figure 29.07 Internally and externally opening doors

Ironmongery

Table 29.01 Ironmongery

Item	Use
Hinges	Usually butt hinges. Framed, ledged and braced doors often use T hinges. Three 4-inch butt hinges are usually sufficient, though it is advisable to use security hinges (hinges with a small steel rod fixed to one leaf, with a hole on the other leaf) to prevent the door being forced at the hinge side.
Letter plate	Position depends on type of door. Usually fitted into the middle rail, but could be fitted in the bottom rail (for full-glazed doors) or the stile (with the letter plate fitted vertically). Fitted by either drilling a series of holes, or cutting out the shape with a jigsaw, or using a router with a guide.
Spy hole	Fitted to solid doors, or doors with no glass, as a security measure. Drill a hole the correct size, unscrew the two pieces, place outer part in the hole, then re-screw the inner part to the outer part.
Security chain	Allows the door to be opened without allowing the person outside in. Chain slides into a receiver and, as the door opens the chain tightens, stopping the door from opening.

Step-by-step: Fitting a mortise lock/latch

1

Step 1 Mark out lock position (usually 900 mm from the floor to the centre of the spindle) and mark width and thickness of lock on the stile.

2

Step 2 Drill a series of holes to correct depth.

3

Step 3 Remove excess timbers to leave a neat opening.

4

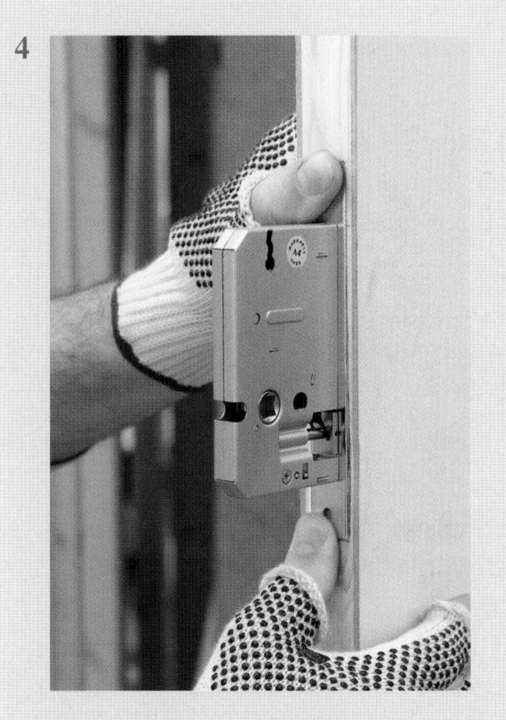

Step 4 Mark around the faceplate and remove the timber so the faceplate sits flush with the stile.

5

Step 5 Drill out spindle and keyhole.

6

Step 6 Fix the lock and handles in place.

7

Step 7 House the striking plate in the frame.

Find out

There are several types of ironmongery available on the market to use on doors. Using the Internet and other resources, such as catalogues, identify ten other types of ironmongery and suggest where these might be used.

Functional skills

This task will allow you to practise **FE 2.2.1** and **2.2.2**, which relate to selecting and using texts and summarising information and ideas.

A mortise lock on its own does not usually provide sufficient security for an exterior door, so most doors will also have one of the following:

- **Mortise deadlock** – fitted like a mortise lock except that it has no latch or handles, so an escutcheon is used to cover the keyhole opening. It is usually fitted three-quarters of the way up the door.
- **Cylinder night latch** – preferred to a mortise deadlock, as it does not weaken the door or frame as much. It is also usually fitted three-quarters of the way up the door. The manufacturer will provide fitting instructions with the lock.

Double doors

Here, two doors are fitted within one single larger frame/lining, with 'meeting stiles', that is two stiles that meet in the middle, rebated so that one fits over the other. They are usually used where a number of people will be walking in the same direction, while double swing doors are used where people will be walking in different directions. Hanging these doors is the same as any other door, though extra care must be taken to ensure that the stiles meet evenly, and that there is a suitable gap around the doors.

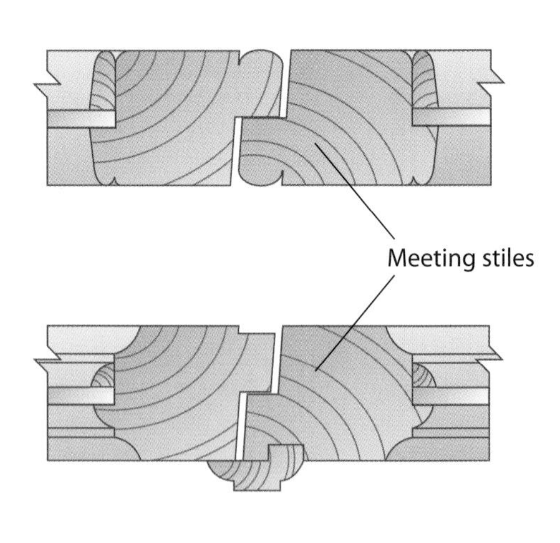

Figure 29.08 Meeting stiles

Extra ironmongery is required on double doors as follows:

- **Parliament hinges** – these project from the face of the door and allow the door to open 180°.

Figure 29.09 Parliament hinge

- **Rebated mortise lock** – similar to a standard mortise lock, but the lock and striking plate are rebated to allow for the rebates in the meeting stiles.
- **Push/kick plates** – fixed to the meeting stiles and the bottom rails to stop the doors getting damaged.
- **Door pull handles** – fixed to the meeting stiles on the opposite side from the push plates, these allow the door to be opened.
- **Barrel bolts** – usually fixed to one of the doors at the top or the bottom to secure the door when not in use.

Figure 29.10 Rebated mortise lock

Most double doors also have door closers to ensure that they will close on their own, to prevent the spread of fire or draughts throughout the building.

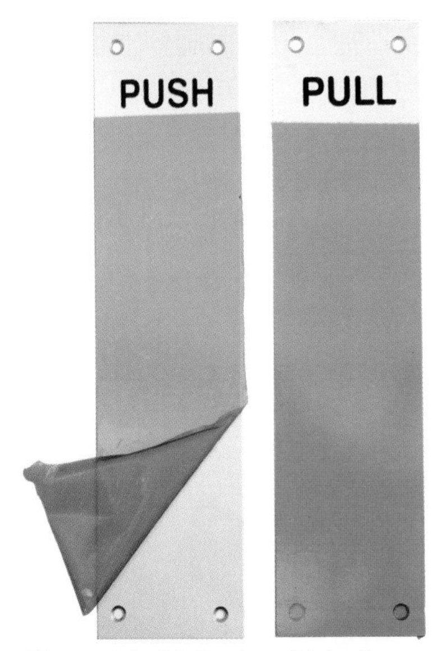

Figure 29.11 Push or kick plates

Figure 29.12 Pull handle

Figure 29.13 Barrel bolt

Concealed door closers

Concealed door closers work through a spring and chain mechanism housed into a tube. To fit a concealed door closer, you must house the tube into the edge of the hinge side of the door, with the tension-retaining plate fitted into the frame.

Overhead door closers

Overhead door closers are fitted to either the top of the door or the frame above. They use a spring or a hydraulic system fitted inside a casing, with an arm to pull the door closed. There are different strengths of overhead door closer to choose from depending on the size and weight of the door. Table 29.02 on the next page shows the strengths available.

Remember

When choosing an overhead door closer, you must take into account any air pressure from the wind. If the pressure is strong, you may require a more powerful closer.

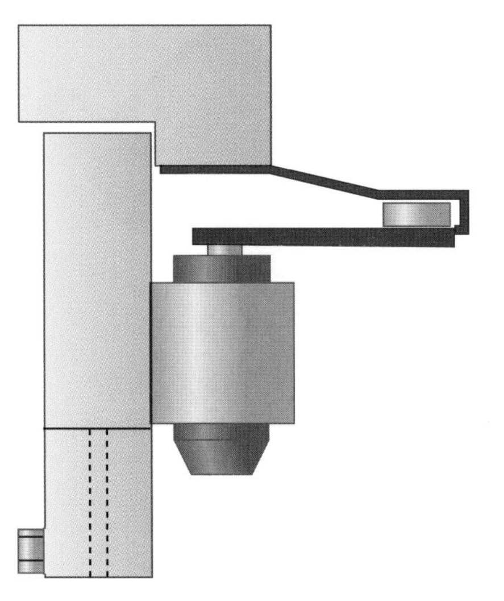
Figure 29.14 Closer fitted to door

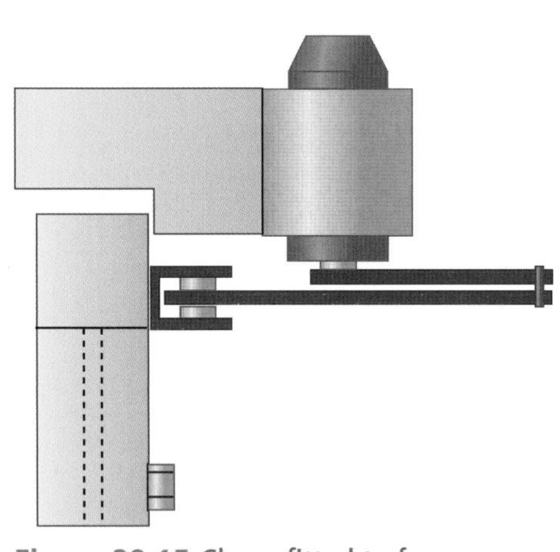
Figure 29.15 Closer fitted to frame

Key terms

Helical – in a spiral shape.

Helical hinge – a hinge with three leaves, which allows a door to be opened through 180°.

Table 29.02 Strengths of overhead door closures

Power no.	Recommended door width	Door weight
1	750 mm	20 kg
2	850 mm	40 kg
3	950 mm	60 kg
4	1100 mm	80 kg
5	1250 mm	100 kg
6	1400 mm	120 kg
7	1600 mm	160 kg

There are a number of different ways to fit an overhead door closer, depending on where the door is situated. The two main ways are:

- fitting the main body of the closer to the door, with the arm attached to the frame (see Figure 29.14)
- inverting the closer so that the main body is fitted to the frame and the arm is fitted to the door (see Figure 29.15).

Double swing doors open both ways. The main difference work-wise is in the ironmongery. Double swing doors need special hinges, and must have some form of door closer fitted. It is usually best to combine these, in one of two ways:

- **Helical hinges** – three-leaf hinges with springs integrated into the barrels. Fitted just like normal hinges, and the tension on the springs are adjusted via a bar inserted into the hinge collar.
- **Floor springs** – a floor spring is placed at the bottom and a pivot plate at the top. The floor spring is housed into the floor, and the bottom of the door is recessed to accept the shoe attached to the floor spring.

Figure 29.16 Helical (or double-action) hinge

The pivot plate at the top is attached to the frame, and a socket is fixed to the top of the door.

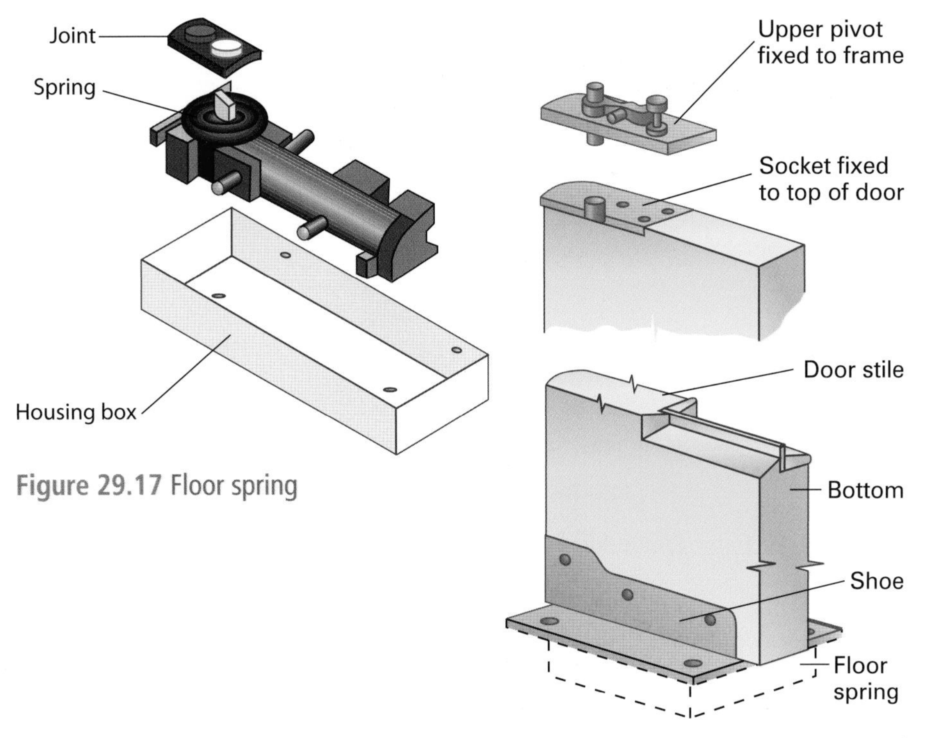

Figure 29.17 Floor spring

Figure 29.18 Floor spring and top of door fixing

K2. Know how to install mouldings

This section will provide you with the knowledge and understanding to enable you to select and fit internal mouldings.

Typical timber mouldings found within a building

Moulding refers to the pattern put on a length of timber. This is normally done by a machine such as a spindle moulder. However, specialist hand planes could be used to match existing patterns. Where you will find these within a room is shown on Figure 29.19 on the next page.

Mouldings are usually produced using premium grade timber or MDF, and should only be fixed at the second fixing stage, once a building is weathertight and plastered.

Architrave

Architraves provide a decorative finish around internal openings, especially doors, and cover the joint between frame and wall finish. They are available in lengths ranging from 2.1 m to 5.1 m, increasing in 300 mm multiples; or as sets consisting of 2 × 2.1 m legs (the sides of the opening) and 1 × 900 mm head (the top of the opening). Architraves are between 50 mm and 75 mm wide and are usually 19 mm to 25 mm thick before they are planed.

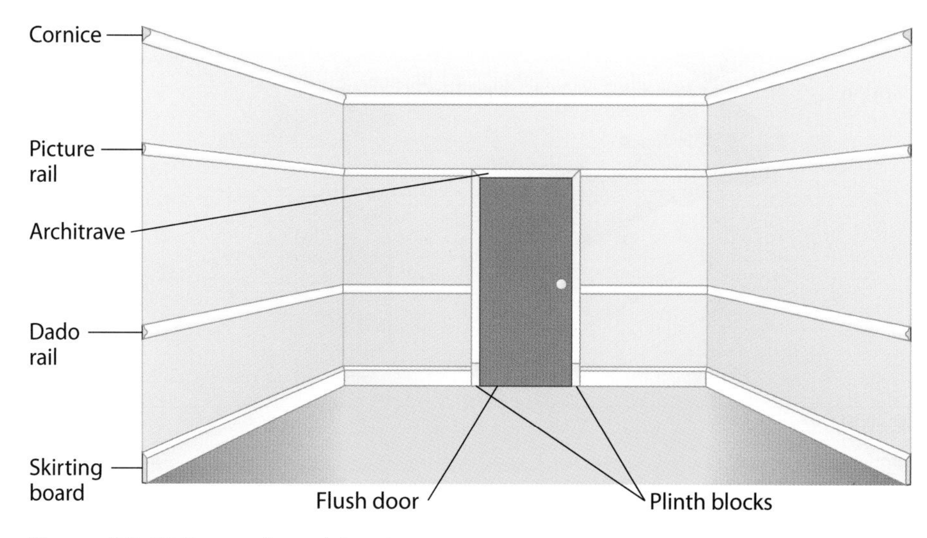

Figure 29.19 Types of moulding in a room

Key terms

Quirk – the hollow in a moulding.

Arris – the sharp edge formed when two flat or curved surfaces meet.

Step-by-step: Fitting an architrave

1

Drawing a working margin

Step 1 Architraves are kept back from the front edge of a frame by 6 mm–10 mm, which is known as the margin. This is the line we work to when fitting architraves. Hence, as Figure 29.33 shows, draw the margin on the front edge of the legs and head of the frame until they meet.

2

Marking the architrave

Step 2 With the heel (narrow point or inner edge) of the architrave to the margin, place the leg on the floor and mark the architrave where the margin marks intersect each other on the frame.

3

Mitring the architrave legs

Step 3 Place the architrave into a mitre box and cut using a tenon saw. Alternatively, use a combination square to mark a 45° angle on the architrave and cut freehand.

4

Fixing the architrave legs

Step 4 Position the cut leg against the margin marks and fix by nails. (When possible try and nail through the **quirks** in the moulding to help hide the nails.) If the surface is uneven you may need to shape the back of the architrave to fit to it before finally fixing, as described in the next section, 'Scribing to walls'.

5

Fixing the architrave head piece

Step 5 Cut a 45° angle on the head piece and offer it to the angle of the leg that has just been fixed. If you have a good fit, repeat Step 2 with the head and then the remaining leg. Should the mitres not go together then the head piece will need planing until it does.

Remember

Use a block plane when planing end grain.

6

Finishing off the architrave

Step 6 Punch all nails below the surface and remove the **arris** from the toe (wide part of the architrave – its outer edge).

Scribing to walls

Scribing is to mark the profile of something onto the surface against which it is to be butted. In the case of architraves (or skirting) it is used so that the back can be shaped to fit against an uneven surface.

Step-by-step: Scribing to walls

1

Temporarily fix forward of the door frame

Step 1 Lightly nail the architrave 20 mm to 25 mm in front of the door frame or lining of the opening, and parallel to it.

2

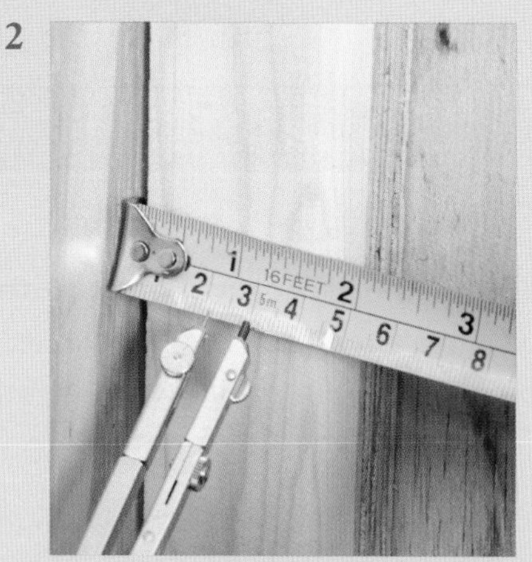

Measure to the margin mark

Step 2 Measure the distance between the front edge of the architrave and the margin mark.

3

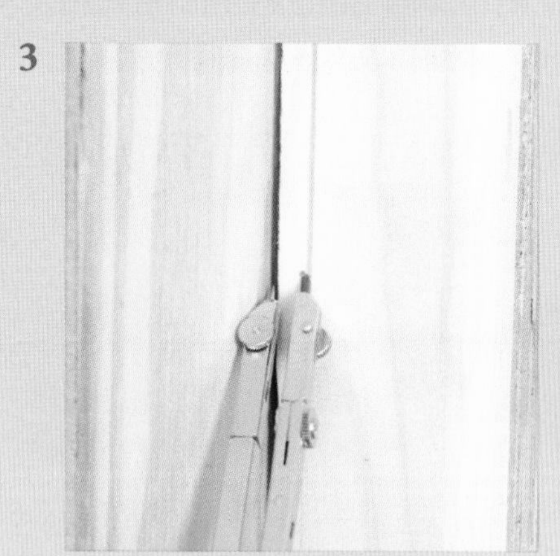

Copy the surface shape onto the architrave

Step 3 Cut a block, or set a pair of compasses, to the measurement you have just worked out. Copy the shape of the uneven surface onto its face by running a compass, or block and pencil, down the uneven surface.

4

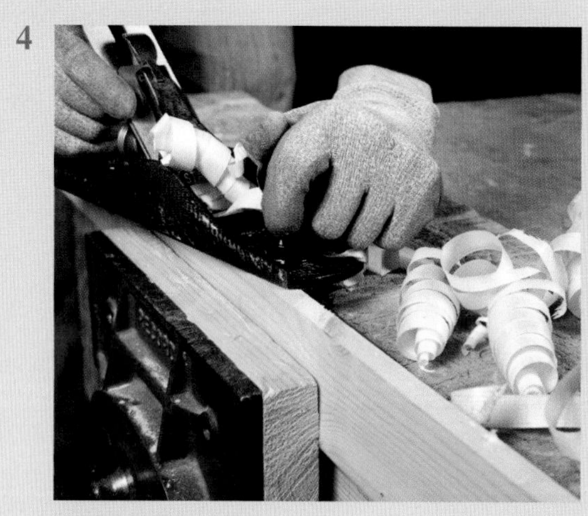

Shape the architrave

Step 4 Take the architrave off the frame. Using a saw or a plane, cut down to the line. Try in position, making sure the front of the architrave is level with the margin mark. Fix with nails.

Skirting board

Skirting boards are used to provide a decorative finish between floor and walls. They also protect the wall finish from damage. Skirting boards are available in lengths ranging from 1.8 m up to 6.3 m, increasing by multiples of 300 mm. They are between 75 mm and 175 mm deep and 19 mm to 25 mm in thickness.

There are only ever three joints to cut when fitting skirting boards:

- internal joints, which should be scribed
- external joints, which should be mitred
- angle-lengthening joints.

How to fit skirting

The boards should be fixed on the top edge and at the bottom, and nails or screws made as inconspicuous as possible:

- Masonry nails should be **dovetailed**.
- Always plug the hole made by the screw or nail, using either a wooden plug or filler.

Remember

Skirting boards may have to be fitted clear of the floor to fit block flooring, etc. so always check plans carefully.

Safety tip

When fixing masonry nails, always use protective glasses.

Key term

Dovetailed nails – pairs of nails angled in towards each other.

Scribed joint

Step-by-step: Method 1

1

Copy moulding shape to be scribed

Step 1 Place the square end of the board to be cut against the fixed board and copy the shape of the moulding onto the face of the board to be cut.

2

Back cut to the line

Step 2 Back cut with a coping saw and/or tenon saw to the line. Keep trying in place until a good joint has been achieved and then fit in place.

Step-by-step: Method 2

Mitred skirting board with end grain to be shaped

Step 1 Using a mitre block, internally mitre across the width of the skirting board to be shaped. This will now show the end grain on the board.

Step 2 Using a coping saw and/or tenon saw, back cut the mitre by letting the saw blade follow the line where the straight grain on the fixed board meets the end grain that is showing on the face of the board being shaped. Try in position and fit when a good joint has been achieved.

Step-by-step: Creating a mitre joint

1

Mark floor lines

Step 1 Lay the skirting board flat against the wall and draw a line on the floor where the board touches it, until the line reaches the adjoining wall. Repeat this with the board flat against the other wall. When the board is removed the lines on the floor should meet at a point near to the angle of the wall.

2

Bisect the angle

Step 2 Place the stock of a sliding bevel right into the angle of the wall and move the blade around until it touches where the lines on the floor meet. You have just bisected the angle!

3

Mark the mitre angle on the skirting board

Step 3 Place the bevel on the top edge of the skirting board and draw the line for the mitre. Cut accurately. Try in position and fit when a good joint has been achieved. Any adjustments can be made using a block plane.

Angle-lengthening joint

Angle-lengthening joints are used to join two pieces of board together in length. This is required if the skirting board available is too short for the wall.

Simply cut a 45° angle across the width of one board and do the opposite cut on the board to which it will be joined. Put them together and skew-nail through the joint to hold it in position. The finished joint is shown in Figure 29.20.

Plinth block

Plinth blocks are also referred to as architrave blocks. They are fixed at the bottom of architraves and allow skirting to run up to them (see Figure 29.21).

They are used because:

- it is then possible to use skirting board that is thicker than the architrave
- skirting can have a moulded back edge
- they provide protection to the moulding on the architrave
- they look elegant.

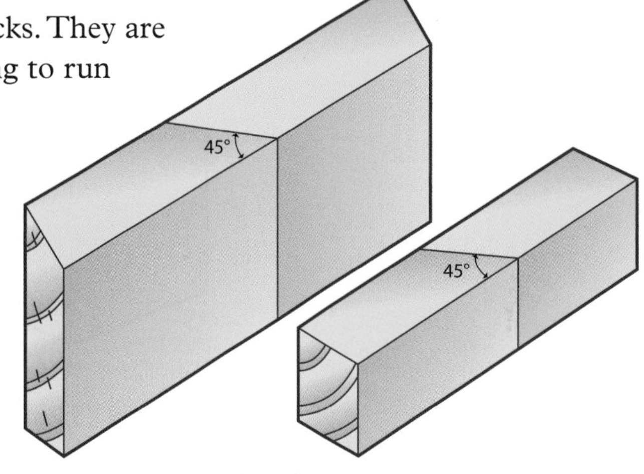

Figure 29.20 Angle-lengthening joint

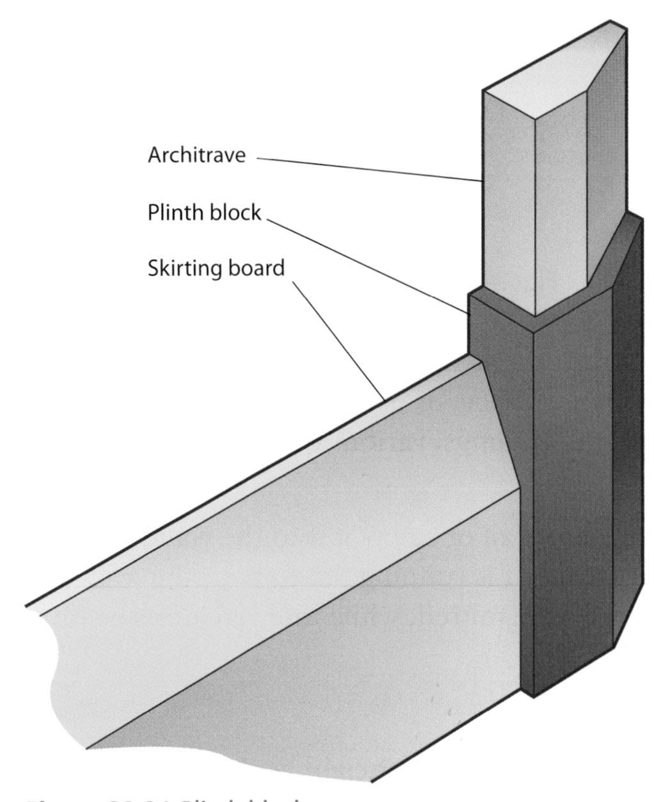

Figure 29.21 Plinth block

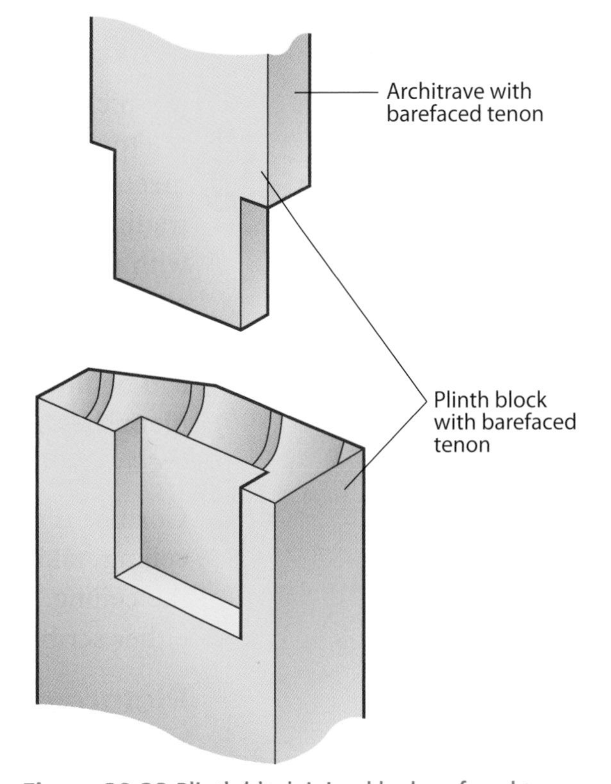

Figure 29.22 Plinth block joined by barefaced tenon

Key term

Butt joint – the simplest joint between two pieces of wood, with the end grain of one meeting the long grain of the other and glued, screwed or nailed together.

Plinth blocks should be approximately 15 mm taller than the skirting board and follow the basic profile of the architrave with at least a 6 mm excess. They should be **butt jointed** to the architrave or preferably fixed using a barefaced tenon (see Figure 29.22 on the previous page). The skirting board can also be butt jointed to the plinth block, although a housing joint would be a better option, as shown in Figure 29.23. Mortise and tenon joints are covered in more detail in Unit 3031, pages 299–301.

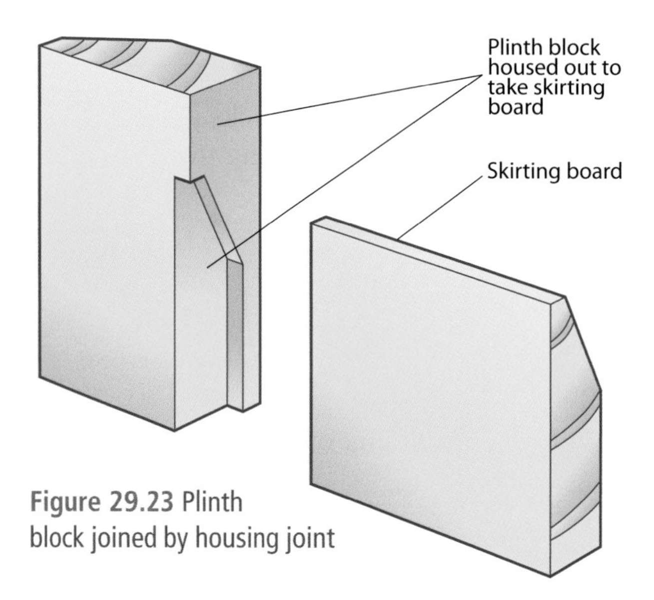

Figure 29.23 Plinth block joined by housing joint

Other types of moulding

Cornice

Cornice is fitted where the top of the wall meets the ceiling. It is traditionally associated with plasterers, although some cornice today is made from timber and is fitted by the carpenter. Cornice can be fitted as a decoration piece, or used to hide gaps or blemishes. As with most mouldings, various different designs are available.

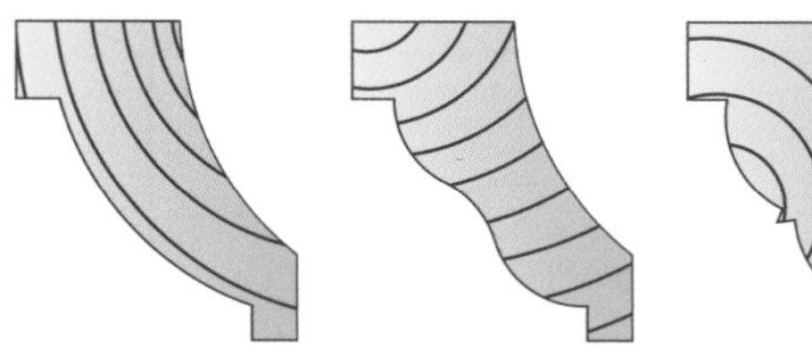

Figure 29.24 Cornice profiles

Cornice is simple to install. Nail or screw it into the wall along the ceiling, taking care to ensure it is running flat to both the wall and the ceiling. External joints are mitred, while internal joints can be either scribed or mitred.

Picture rail

Picture rail is usually fitted at the same height as the top architrave on a door, or just above or below this. Picture rail was used to hang paintings from, so the height used to be determined by the size of the pictures being hung. Today, picture rail is largely used for decoration. Picture rails also come in a variety of profiles.

Picture rail is slightly more difficult to fix than skirting or cornice, as the rail must be fixed level. First, mark a level line around the room to act as a guide for fitting the rail, then proceed as with any other moulding, with the joints scribed or mitred.

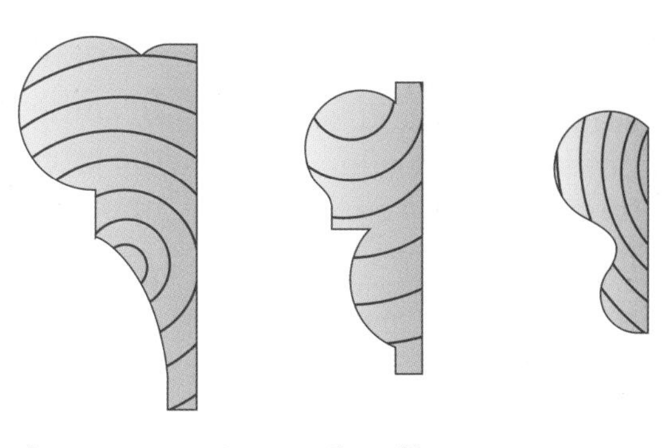

Figure 29.25 Picture rail profiles

Dado rail

Dado rail is fixed to the wall between the picture rail and the skirting, the exact height depending on how it is being used. Dado rails were generally used to guard the walls from damage from deck chairs, so would be set at the height of a chair. Today, dado rails in domestic dwellings are mainly for decorative purposes, so the height is up to the owner's preference. Dado rails also come in a variety of profiles.

Fix a dado rail in the same way as a picture rail, taking care to ensure that the rail is level and the joints are scribed and mitred.

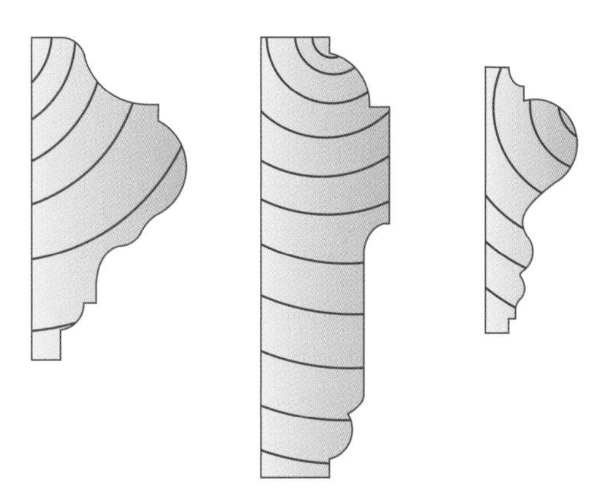

Figure 29.26 Dado rail profiles

Did you know?

A form of dado rail is still used today in places such as hospitals, to protect the walls from damage.

Find out

Sketch three different mouldings and then use the Internet and other resources to confirm where these might be used in a building and why.

K3. Know how to install service encasements and cladding

'Encasing services' usually refers to the carcass, framework and trim that cover BSBs, service pipes, cables, steel and concrete columns or, in some instances, unsightly spaces such as where a bath is situated or where the bulkhead for a staircase cuts into a room.

Whatever you are cladding, there is a variety of material to choose from:

- plasterboard
- man-made timber boards such as plywood or MDF
- tongued and grooved boarding
- melamine-faced boards.

The choice of materials depends on the finish wanted by the client and requested in the specification.

Did you know?

BSB stands for British Standard Beam, often used to provide load-bearing supports in a building because the shape is very resistant to bending.

Key terms

Nogging – a short length of timber, most often found fixed in a timber frame as a brace.

Halving joint – the same amount is removed from each piece of timber so that when fixed together the joint is the same thickness as the uncut timber.

Beams and columns

In some buildings BSBs have to be cladded to protect them from the effects of fire. The amount of protection required will depend upon the function and location of the beam.

How to clad a beam

If a beam to be clad is made of timber, or even concrete, it can have a frame fixed to it and then facing material put on to the frame, or the facing material can be fixed directly to it. The bigger problem is to clad a steel beam, often a BSB used as a load-bearing support.

Step-by-step: Cladding a beam

1

Fix noggings

Step 1 Fix **noggings** between the rolled edges of the beam. This can be done by accurately cutting lengths of 50 mm × 50 mm timber to match the vertical gap in the side of the BSB. Then drive them vertically into the gap on both sides of the beam so they wedge in place. They will provide attachment points for a cradle so should be no more than 600 mm apart.

Alternatively, timber supports can be fitted in the gap along the whole length of the beam and then bolted into place through holes in the beam (some come with them or they can be drilled). These similarly provide attachment points for a cradle.

2

Create a cradle

Step 2 Create a cradle using 50 mm × 25 mm treated softwood and use a simple **halving joint** (or housing or lap joint) to fix the corners together. Screw the cradle to the bearers. (If the noggings or timber supports come flush with the sides of the metal beam, it is possible to dispense with a cradle by fixing additional timber supports along the beam near the top and bottom as a support for facing material.)

Fitting soffits and facing material

Step 3 Run soffits along the length of the joist, fastening them into the cradle. The facing material can now be fixed to the soffits to conceal the sides of the beam.

How to clad a column

Concrete or steel columns are cladded by constructing a set of framed grounds (either two or four) slightly larger than the column itself. These are generally shaped like a ladder and made from rough sawn timber. They are assembled around the column and adjusted until plumb. Any slackness is taken up using wedges or packing pieces, with a screw or nail driven into them to stop them slipping. Facing material is then fixed to the framed grounds.

Service pipes and cables

Service pipes and cables are hidden from view whenever possible. However, there are occasions when it is not possible to do so; when the circuit of pipe work has to go from one storey to another, for instance. When this occurs they should be encased behind a timber stud frame. The method is very similar to cladding a column.

Did you know?

Plumb means vertical; hence, a plumb bob is used to check something is vertical.

Safety tip

One of the main dangers associated with drilling into walls or floors is the possibility of cutting hidden cables and pipes for services. Always check for these services before beginning work. If in any doubt, turn off the services at the mains.

Step-by-step: Encasing pipes and wires

1

Step 1 Measure the protrusion. The framework should not stand excessively far from the object to be encased.

2

Step 2 Construct two sets of framed grounds by using halving joints or butt nailing.

3

Step 3 Fix battens to the wall with plugs and screws or masonry nails.

4

Step 4 Fix the framed grounds to the battens.

5

Step 5 Clad with 6 mm ply, scribing the wall as necessary.

Remember

You will use variations on this technique for branch pipes, face panels, horizontal pipes and skirting board.

Baths

Baths are often panelled, but care should be taken to ensure that suitable materials are used that can cope with the heat and moisture associated with bathrooms. Also, the plumbing should be easily accessible.

Panelling a bath is similar to encasing pipe work, only the framed grounds will probably be wider. Figures 29.27–29.29 illustrate typical framework and panel fixing.

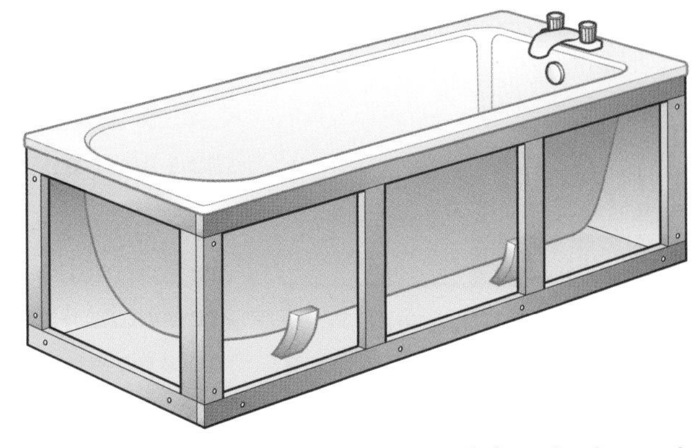

Figure 29.27 Framework for a bath panel

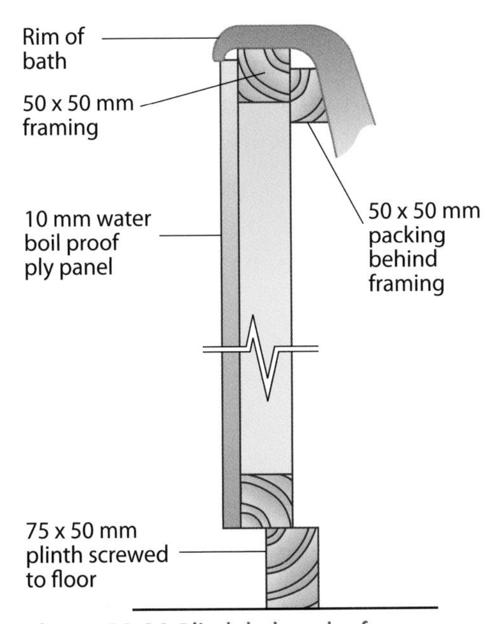

Figure 29.28 Plinth below the frame as a toe space

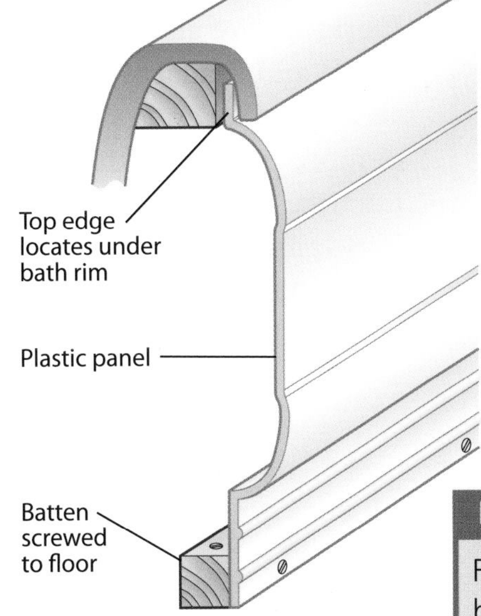

Figure 29.29 Detail of a standard panel

Did you know?

Fitting loft insulation behind bath panels helps keep the bath water warmer for longer and also provides some sound insulation.

K4. Know how to install wall and floor units and fitments

Normally you will be fitting units in kitchens. If you learn to follow these instructions, they will enable you to fit wall and floor units in a bedroom or anywhere else.

Kitchen units

The majority of kitchen units available today are constructed from melamine-faced particle board, and supplied as either base units (floor standing) or wall units.

Units are supplied ready built or flat-pack. When supplied as flat-packs, the assembly instructions must be followed with great care. Check whether shelves have to be fitted during assembly, rather than after installation, particularly in corner units.

Remember

It may be worth fixing wall units before base units and worktops go in.

Base units are available in widths of 300 mm to 1200 mm, and are normally 600 mm deep and 900 mm high. Corner units are available to match up with them. Most manufacturers incorporate adjustable height legs in their designs to allow for uneven floors.

Wall units are available in widths of 300 mm to 1000 mm; typically 600 mm, 720 mm or 900 mm high and 300 mm deep.

Remember

When using a spirit level to mark a datum line, always alternate the level along its length; this counteracts inaccuracies and prevents cumulative error.

Worktops are also mainly made from plastic-laminated particle board and available in widths of 600 mm to 900 mm, but granite, stainless steel and proprietary, hardwearing plastics are used in more expensive kitchens. The width fitted is dependent on the base unit used. Thickness is usually 40 mm, but can vary. Plastic-laminated particle boards have a square- or post-formed front edge.

Step-by-step: Fitting base units

1

Mark a horizontal datum line

Step 1 Carefully mark a horizontal datum line, approximately 1 m from the floor, on the walls that are to have base or wall units placed against them. You now have a point to work up or down from.

2

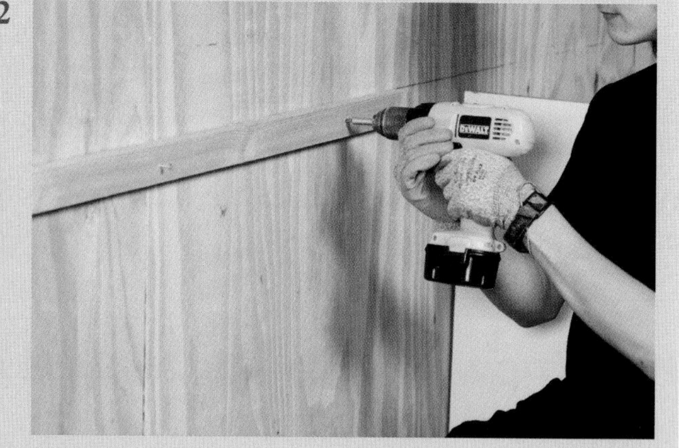

Fix a back rail

Step 2 Determine a user-friendly height for the units minus the worktop. Then, working down from the 1 m datum, fix a back rail to the wall. You now have a datum that allows you to level across and along the units.

3

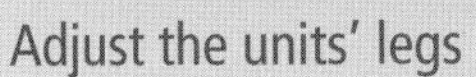
Adjust the units' legs

Step 3 Preferably working from a corner, place units where they should go, considering service pipes and cables. Adjust the legs so that units are level and in line with the back rail.

4

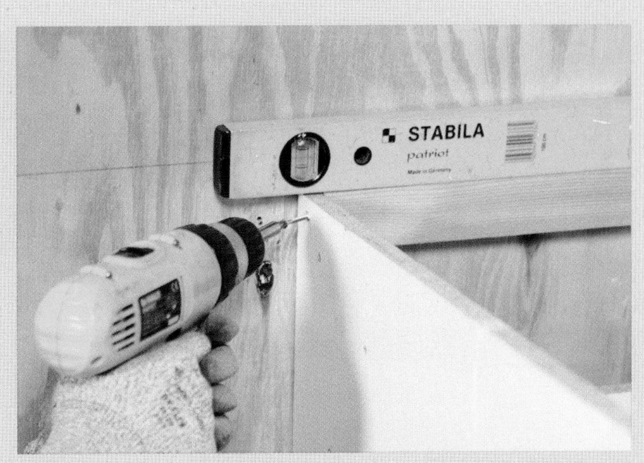

Fix base units into position

Step 4 Once all units are in place, they can be fixed together using connecting bolts, and fixed to the wall or floor using screws.

Step-by-step: Fitting worktops

1

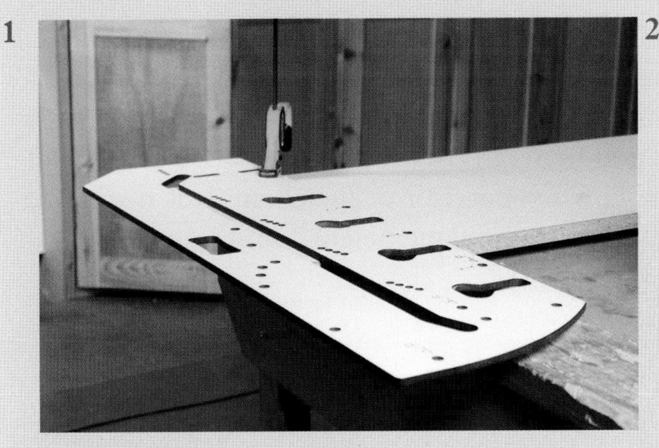

Cut lengths and internal corners

Step 1 Cut to length and form an internal corner using a purpose-made jig. These are produced by a variety of manufacturers and, when used in conjunction with a powerful router, a clean accurate cut is achievable.

2

Strengthen joints with biscuit jointer

Step 2 Strengthen joints using a biscuit jointer.

3

Cut worktops to house sink units

Step 3 Cut worktops to house sink units or appliances using a jigsaw with a downward cutting blade. This prevents the plastic-laminated surface from being chipped. The chipboard that is exposed by the cut should be coated in varnish to prevent any moisture penetration.

4

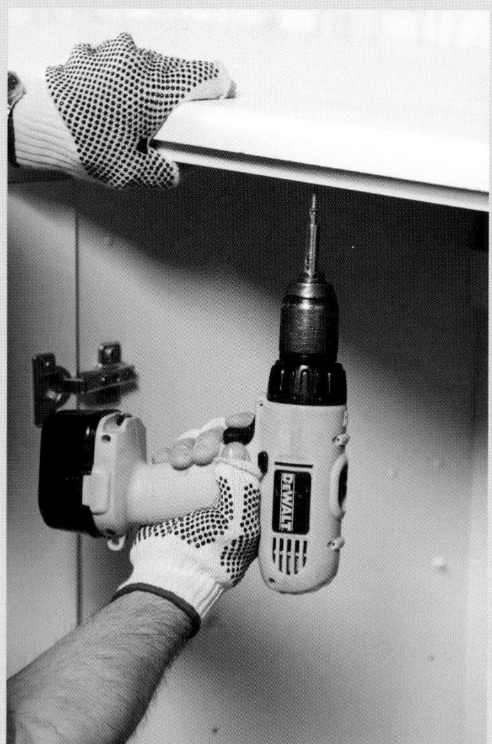

Fix worktops to base units

Step 4 Fix worktops to the base units with screws and connectors.

Remember

Carefully read all instructions supplied with jigs and routers, as these will give guidance on how to produce the cuts you want.

Fitting wall units

Most wall units are fastened to the wall by an adjustable bracket that hooks itself onto a steel hanger plate. The plans should show the clearance required between the worktop and base of the wall units. If not, check with a supervisor or the client. It is normal to have the tops of all units at the same height.

Step-by-step: Fitting wall units

1

Mark the tops of wall units

Step 1 Measure the height of units and add this to the clearance from the work surface, which gives the height where the tops of the wall units should go. Measure and mark these, working upwards from your datum line.

2

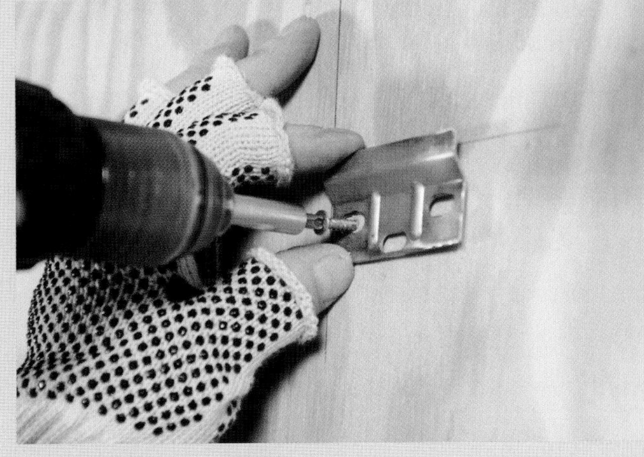

Mark and fix hanger plates

Step 2 Mark the locations for the hanger plates and fix in place. Often the manufacturer of the wall units provides a template to help position them.

3

Hang and adjust wall units

Step 3 Hang the wall units on the hanger plates and adjust for height by turning a screw housed within the adjustable bracket. Once done, another screw enables the bracket to be tightened onto the steel plate.

As with base units, wall units can be connected together by using connecting bolts.

Working life

Shauna has been asked to fit 300 mm × 30 mm skirting boards in a room. The architrave is 75 mm × 19 mm.

What might be a problem at the junction where the architrave and the skirting board meet? How could Shauna get around this? The room is also to have a dado rail fitted. The rail is 5 mm thicker than the architrave. How could Shauna resolve this problem?

Functional skills

When dealing with installation problems on site, you will be practising **FM 2.3.1** Judge whether findings answer the original problem and **FM 2.3.2** Communicate solutions to answer practical problems.

FAQ

What is a scribe?

A scribe is a copy of the surface it fits over.

What does bisecting an angle mean?

Cutting an angle equally in two, as when cutting a mitre to create an internal angle.

Do I scribe both ends of a length of skirting?

No. It looks neater if one end butts to the wall and the other end is scribed to the butted end of the next skirting.

Can I have a scribe and a mitre?

No. It would be very difficult to get the scribed edges to meet correctly.

Check it out

1. Name three examples of mouldings.
2. Explain how an architrave can be fitted to a surface that is not smooth.
3. Describe the different ways in which skirting can be fitted.
4. Briefly describe the two different methods of shaping skirting where it meets in a corner.
5. Explain why plinth blocks are used.
6. Describe in your own words the following types of door:
   - **a.** framed door
   - **b.** flush door
   - **c.** fire-resisting door.
7. Why should a door be stored flat for a few days prior to fitting?
8. If a door is too tall for an opening, where should you cut excess wood from?
9. State the size, type and number of hinges usually fitted to an external door.
10. Explain the purpose of a spy hole.
11. Explain what push plates are used for.
12. Explain why a door closer should be fitted to a door.
13. Explain the special feature of a helical or double-action hinge.
14. Why might a rolled steel joist (RSJ) need to be cladded?
15. What should you always consider when encasing service pipes or cables?
16. When fitting kitchen wall units, how do you know how much clearance is required between the worktop and the base of the unit?
17. Why are the doors and shelves of kitchen units fitted last?

Getting ready for assessment

The information contained in this unit, as well as continued practical assignments that you will carry out in your college or training centre, will help you with preparing for both your end of unit test and the diploma multiple-choice test. It will also aid you in preparing for the work that is required for the synoptic practical assignments.

The information contained within this unit will aid you in learning how to identify and calculate the materials and equipment required for second fixing.

You will need to be familiar with:

- preparing materials for a range of second fixing tasks
- installing side hung doors and ironmongery
- installing mouldings
- installing service encasements and cladding
- installing wall and floor units and fitments.

This unit will have made you familiar with the common second fixing tasks. For example, for learning outcome four you have seen the methods used for fixing wall and floor units, whilst protecting electric, gas and water pipes. You have also seen the importance of using drawings and specifications to complete work accurately. You will need to follow this information in order to successfully assemble and install wall and floor units. You will also need to select the correct tools to complete the work to specification.

The knowledge from this unit will also need to be used to install worktops, form openings for hobs and sinks and returned post-formed worktops using worktop jigs. This work will always need to be of a high standard, as it is client-facing and will leave the client with an overall impression of the quality of your work.

Before you start work on the synoptic practical test, it is important that you have had sufficient practice and that you feel you are capable of passing. It is best to have a plan of action and a work method that will help you. You will also need a copy of the required standards, any associated drawings and sufficient tools and materials. It is also wise to check your work at regular intervals. This will help you to be sure that you are working correctly and help you to avoid problems developing as you work.

Your speed at carrying out these tasks will also help you to prepare for the time limit that the synoptic practical task has. But remember, don't try to rush the job as speed will come with practice and it is important that you get the quality of workmanship right.

Make sure you are working to all the safety requirements given throughout the test and wear all appropriate personal protective equipment (PPE). When using tools, make sure you are using them correctly and safely.

Good luck!

CHECK YOUR KNOWLEDGE

1. Architraves should be kept back from the front edge of a frame by:

a 4–6 mm
b 6–10 mm
c 10–12 mm
d 12–14 mm

2. The moulding that is fixed at the bottom of an architrave where it meets the skirting is called a:

a plinth block
b pelmet block
c cornice block
d dado block

3. What is a door stile?

a The longest vertical timber in the door frame
b The longest horizontal timber in the door frame
c The shortest vertical timber in the door frame
d The shortest horizontal timber in the door frame

4. Which of these is not a standard door height?

a 1.97 m
b 2 m
c 2.03 m
d 2.17 m

5. What could cause a door to spring open when it is closed (binding)?

a The stops are not fitted correctly
b The door and lining are fitted out of level
c The screws in the hinges are not in straight
d All of the above

6. A lock block is contained in a:

a flush door
b fire door
c panel door
d framed door

7. Which hinge would you use if a door is to be fitted in an area with a floor that slopes upwards?

a Tee hinge
b Rising butt hinge
c Stormproof hinge
d Butt hinge

8. What ironmongery should be fitted to a fire escape door?

a Spy hole
b Letter plate
c Panic bar
d Dead lock

9. Ferrous nails should not be used outside because:

a they will rust
b they are weaker
c they are not long enough
d they are too long

10. Which type of screw head would be used for fixing metal components such as door handles?

a Mirror screw
b Flange head
c Countersunk
d Raised head

11. When it is not possible to hide service pipes and cables from view (for example, when the circuit of pipe work has to go from one storey to another), what should you encase the services behind?

a A tiled screen
b A timber stud frame
c An easily shattered covering
d Nothing – they must stay visible

12. The best way to cut a worktop joint is to use a:

a jigsaw
b router and template
c hand saw
d circular saw

Unit 3008

Know how to erect complex structural carcassing components

Structural carcassing covers all carpentry work associated with the structural elements of a building, such as floors and roofs. This unit is designed to help you identify the main activities associated with structural carcassing and to provide you with the knowledge and understanding required to carry them out.

You will need to be sure that you are working in accordance with the current health and safety and Building Regulations when working on new buildings or refurbishing old ones.

This unit contains material that supports the following NVQ units:

- VR 24 Erect structural carcassing
- VR 209 Confirm work activities and resources for work

This unit will cover the following learning outcomes:

- Know how to erect trussed rafter roofs
- Know how to construct verge and eaves finishes
- Know how to form dormer windows
- Know how to construct a traditional cut roof with hips and valleys

K1. Know how to erect trussed rafter roofs

Functional skills

While reading and understanding the text in this unit, you will be practising **FE 2.2.1** Select and use different types of texts to obtain relevant information, **FE 2.2.2** Read and summarise succinctly information/ideas from different sources and **FE 2.2.3** Identify the purposes of texts and comment on how effectively meaning is conveyed.

Most roofing on domestic dwellings consists of trusses, factory made from stress-graded PAR timber to a wide variety of designs, depending on requirements. Using trussed roofing saves on timber and makes the process of roofing easier and quicker than with a traditional pitched roof. Trussed roofs can also span greater distances without the need for support from intermediate walls.

In a trussed roof, each truss is composed of triangles, because of the structural stability of that shape. All joints are butt-jointed and held together with fixing plates, face-fixed on either side. These plates are usually made of galvanised steel, either nailed or factory pressed; they can also be gang-nailed gusset plates made of 12 mm resin-bonded plywood.

Handling and storing truss rafters

Truss rafters should be stored on raised bearers to avoid contact with the ground, and should be supported to prevent any distortion. Trusses should ideally be stored vertically on bearers located at the points of support, with suitable props to maintain them vertically. Alternatively, trusses can be stored horizontally, again on bearers and supported at each joint.

Remember

Standard trusses are strong enough to resist the eventual load of roofing materials, but they are not able to withstand pressure applied by lateral loading. This means that damage is most likely to occur during delivery, transport, site storage or lifting into position.

Manhandling trusses should be kept to a minimum due to the risk of manual handling injuries. When they have to be handled – for example, when they are transported – they must be handled with great care and supported correctly along any stress points.

Mechanical handling of trusses is usually done with a crane, when they should be lifted in banded sets and lowered onto suitable bearing supports. Suggested lifting points are at rafter or ceiling node points. A suitable spreader bar should be used to withstand any sling forces, and a guide rope should be attached to allow for precise movement.

The components of a roof were covered in Unit 3003, pages 92–99. The main components you will deal with when constructing trussed rafter roofs are hips, valleys, gable ladders, soffits, verges, wall plates, straps, bracing and truss clips.

Truss types

There are a variety of truss styles, the most common of which are shown in Figure 8.01.

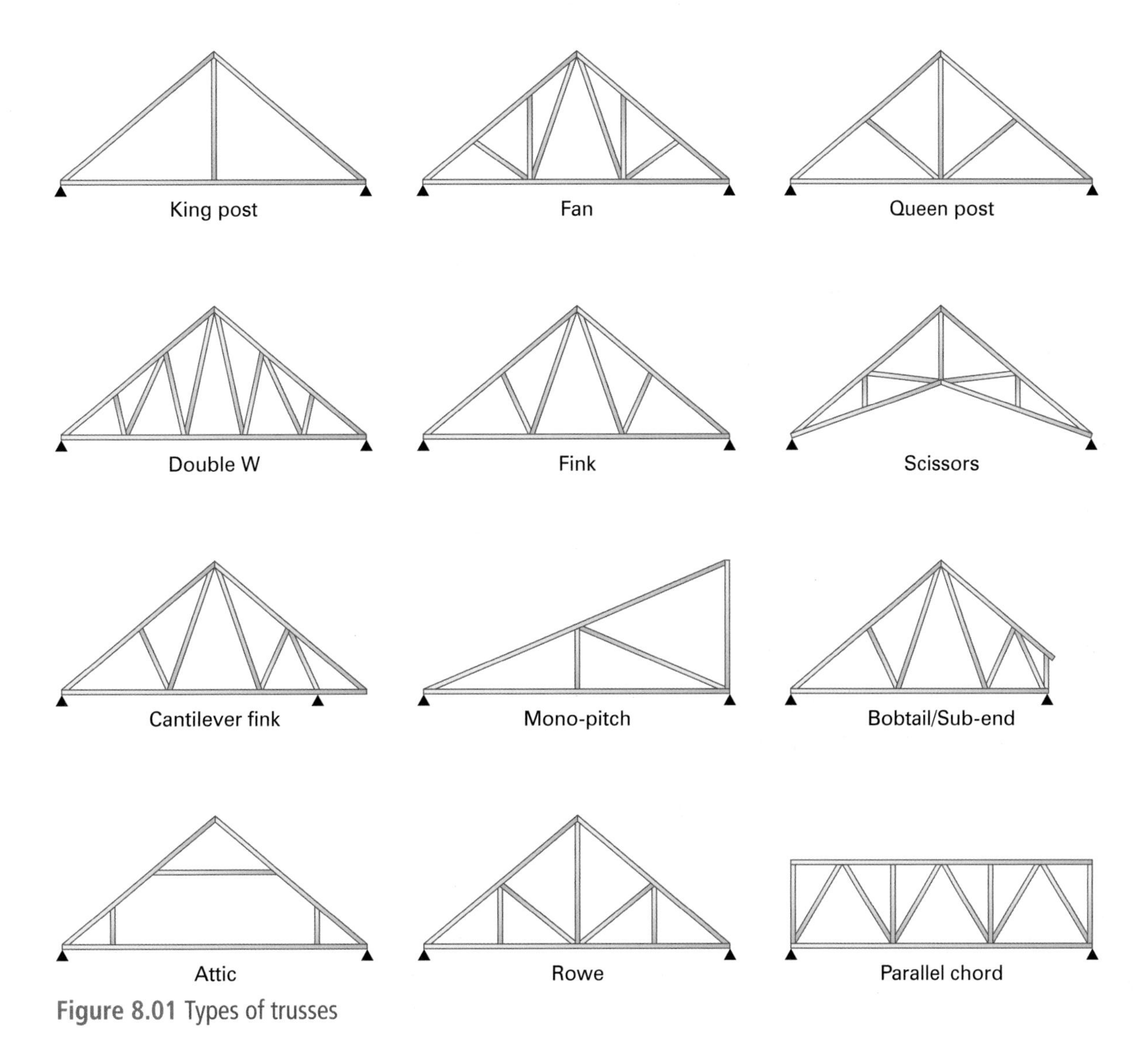

Figure 8.01 Types of trusses

Erecting trussed rafter roofs

Wall plates are bedded onto the wall and then onto the wall plate. The position of the trusses can be marked out at a maximum of 600 mm centres; then truss clips can be used or the trusses can be cheek-nailed (skew-nailed) into the wall plate.

Refer to Figure 8.02 on the next page when you are reading these instructions.

Step 1 Fix first truss using framing anchors (1).

Step 2 Stabilise and plumb first truss with temporary braces (D).

Step 3 Fix temporary battens on each side of ridge (A).

Step 4 Position next truss (2).

Step 5 Fix wall plate and temporary battens (A). Continue until last truss is positioned.

Remember

When marking out and fixing a trussed rafter roof, you will need to clearly set out and prepare an area to work in.

Remember

Never alter a trussed rafter without the structural designer's approval.

Step 6 Fix braces (B).

Step 7 Fix braces (C).

Step 8 Fix horizontal restraint straps at max 2.0 m centres across trusses onto the inner leaf of the gable walls.

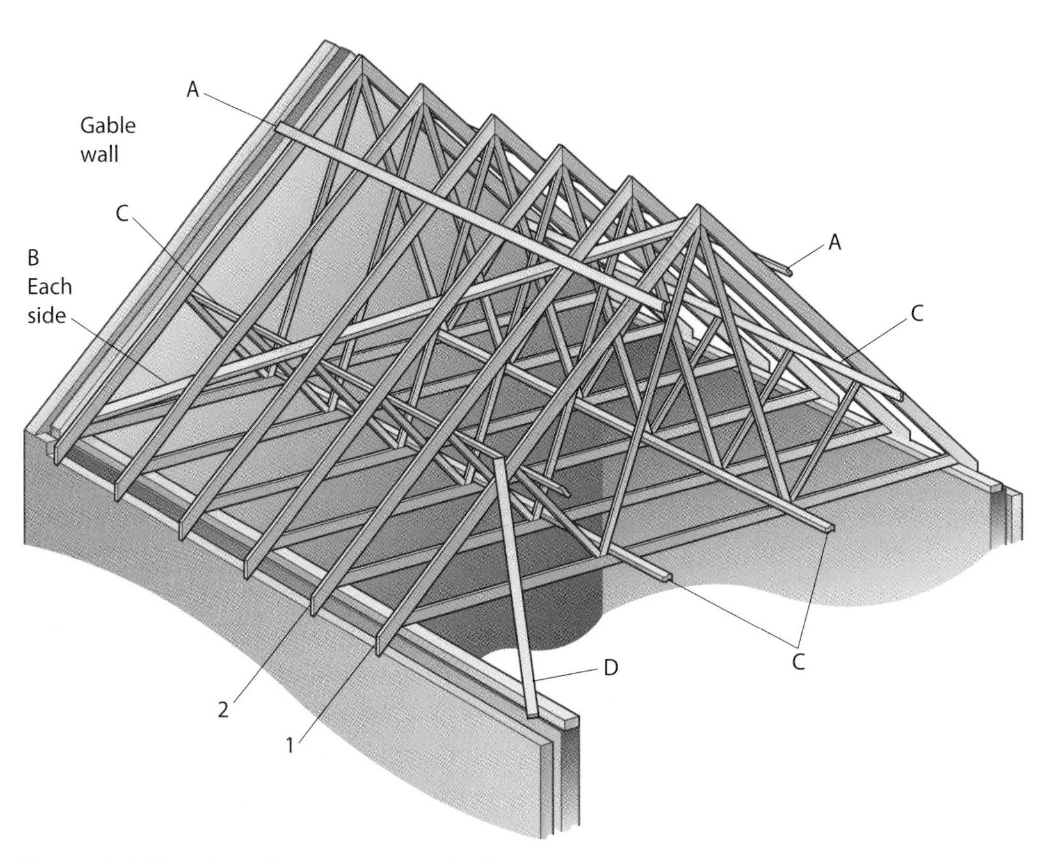

Figure 8.02 Erection of common trussed rafters

Fixing bracing

Fixing bracing is important for the safety and stability of any trussed roof. There are two main types of bracing:

- **Stability bracing** – this holds the trusses firmly in place and keeps them straight providing lateral stability, so that they can withstand all the loads on the roof, with the exception of wind loads.
- **Wind bracing** – this is additional bracing that is used to withstand the wind loads.

Gable ends

Setting out for a gable end

First, set out and fix the wall plate. The wall plate is set on the brick or blockwork and either bedded in by the bricklayer or temporarily fixed by nailing through the joints. Once secured it is held in place with restraint straps (see Figure 8.03). If the wall plate is to be joined in length, a halving joint is used (see Figure 8.04). It is vital that the wall plate is fixed level to avoid serious problems later.

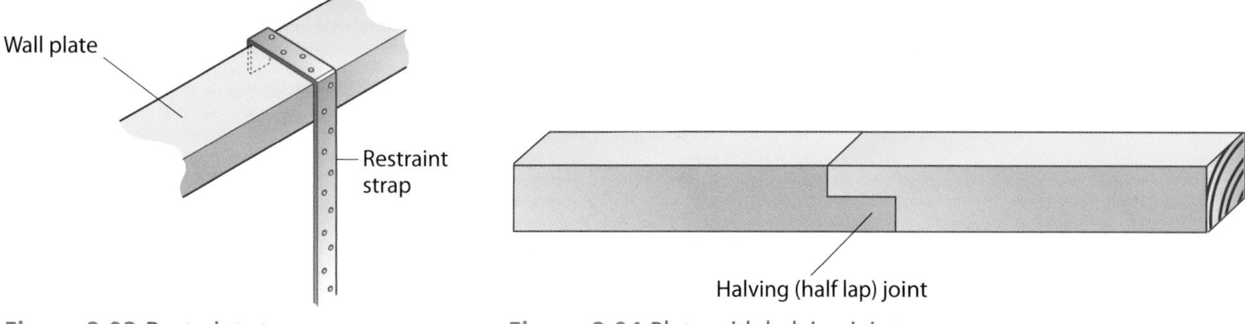

Figure 8.03 Restraint strap

Figure 8.04 Plate with halving joint

Once the wall plate is in place, you need to measure the span and the rise. You can use these measurements to work out the rafter length in different ways, using a roofing ready reckoner, **geometry** or **scale drawings**. Ready reckoners and geometry are covered later in this unit, so we will start with scale drawings.

For this example we will use a span of 5 m and a rise of 2.3 m.

Using a scrap piece of plywood or hardboard, we first draw the roof to a scale that will fit the scrap piece of plywood/hardboard (usually a scale of 1:20).

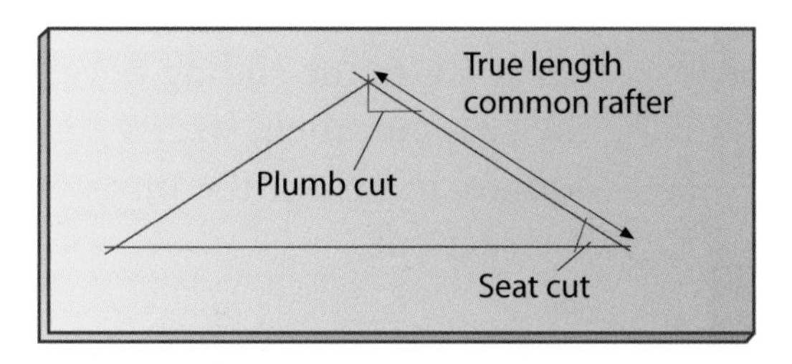

Figure 8.05 Sketch on a piece of scrap ply showing the true length, plumb and seat cut for a common rafter

From this drawing we can measure at scale and find the true length of the rafter. Then by using a **sliding bevel** we can work out the plumb cut and seat cut.

Making and using a pattern rafter

From our scale drawing we can mark out one rafter, which we will then use as a **pattern rafter**.

Key terms

Geometry – a form of mathematics using formulae, arithmetic and angles.

Scale drawing – a drawing of a building or component in the right proportions, but scaled down to fit on a piece of paper. On a drawing at a scale of 1:50, a line 10 mm long would represent 500 mm on the actual object.

Sliding bevel – a tool that can be set so that the user can mark out any angle.

Pattern rafter – a precisely cut rafter used as a template to mark out other rafters.

There are five easy steps to follow when marking out a pattern rafter.

Step-by-step: Marking out a pattern rafter

1

Step 1 Mark the pitch line one-third of the way up the width of the rafter.

2

Step 2 Set the sliding bevel to the plumb cut and mark the angle onto the top of the rafter.

3

Step 3 Mark the true length on the rafter, measuring along the line.

4

Step 4 Use the sliding bevel to mark out the seat cut; then, with a combination square, mark out the birdsmouth at 90° to the seat cut.

5

Step 5 Re-mark the plumb cut to allow for half the thickness of the ridge.

Once it has all been marked out, this can be cut and used as a pattern rafter.

The pattern rafter can be used to mark out all the remaining common rafters, although it is advisable to mark out and cut only four, then place two at each end of the roof to check whether the roof is going to be level.

Once all the rafters are cut, mark out the wall plate and fix the rafters. Rafters are normally placed at 400 mm centres, with the first and last rafter 50 mm away from the gable wall. The rafters are usually fixed at the foot by skew-nailing into the wall plate and at the head by nailing through the ridge board.

Did you know?

The first and last rafters are placed 50 mm away from the wall to prevent moisture that penetrates the outside wall coming into contact with the rafters, thus preventing rot.

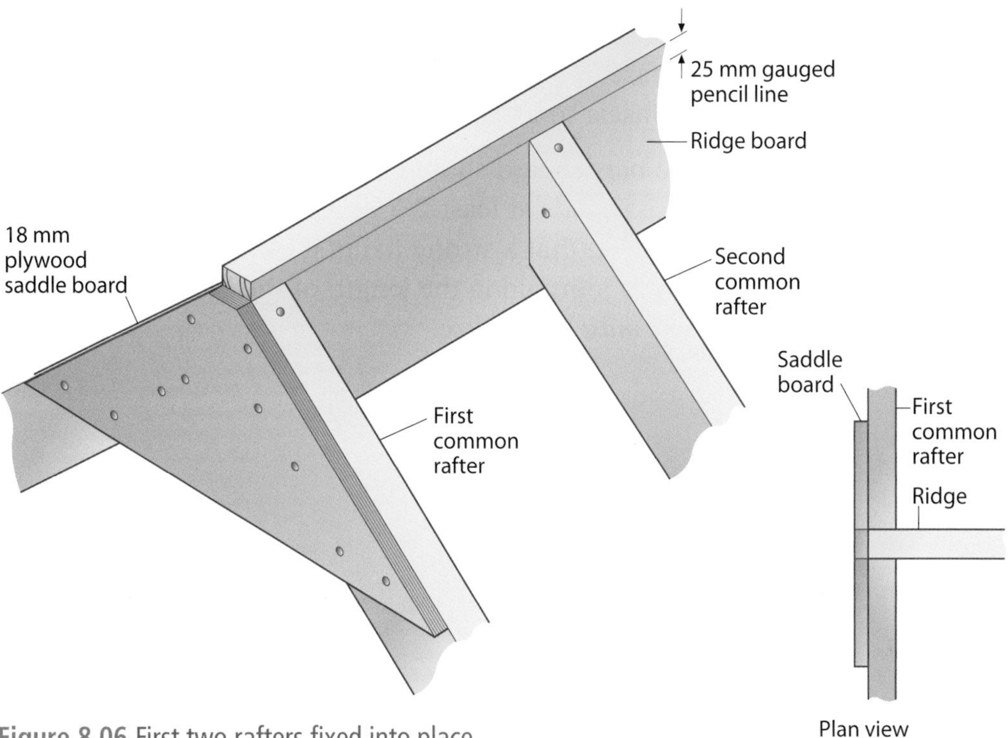

Figure 8.06 First two rafters fixed into place

If the roof requires a loft space, joists can be put in place and bolted to the rafters; if additional support is required, struts can be used.

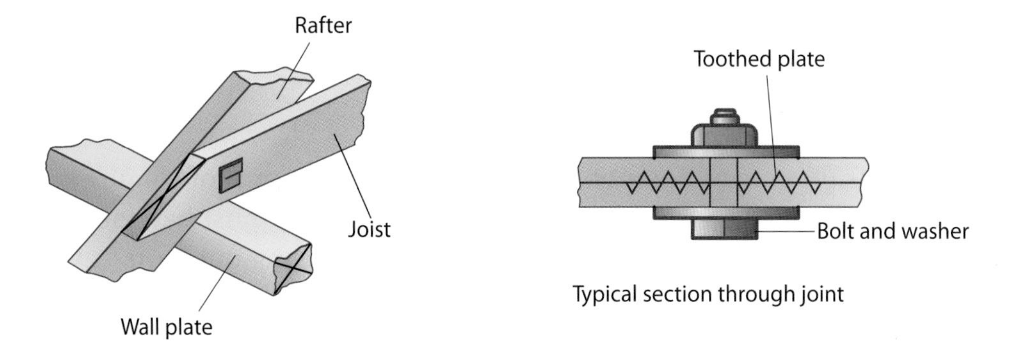

Figure 8.07 Rafter legs bolted to joists

Key term

Sand course – where the bricklayer beds in certain bricks with sand instead of cement so that they can be easily removed to accommodate things like purlins.

For roofs with a large span, purlins provide adequate support. Purlins are usually built into the brickwork: either the gable wall will not have finished being built yet or the bricklayer will have left a **sand course**. The same is true of roof ladders (see below).

Finishing a gable end (verge)

In Unit 3003, you saw how to use a roof ladder (page 101). Here we look at marking out and fixing a bargeboard.

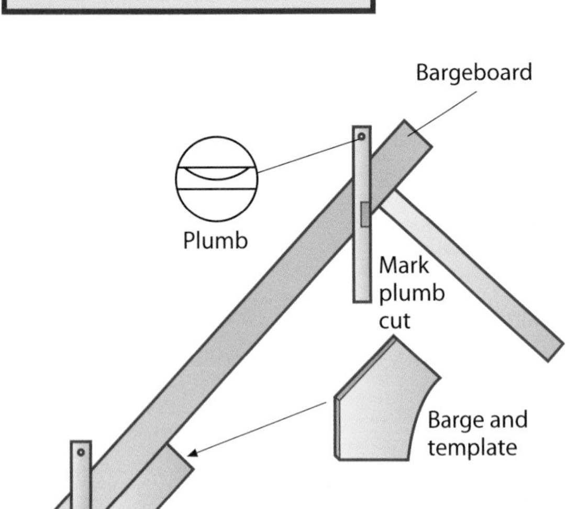

Figure 8.08 Marking out a bargeboard using a level

The simplest way of marking out the bargeboard is to temporarily fix it in place and use a level to mark the plumb and seat cut.

When fixing a bargeboard, the foot of the board may be mitred to the fascia, butted and finished flush with the fascia, or butted and extended slightly in front of the fascia to break the joint (see Figure 8.09).

The bargeboard should be fixed using oval nails or lost heads at least 2.5 times the thickness of the board, so that a strong fixing is obtained. If there is to be a joint along the length of the bargeboard, the joint *must* be a mitre.

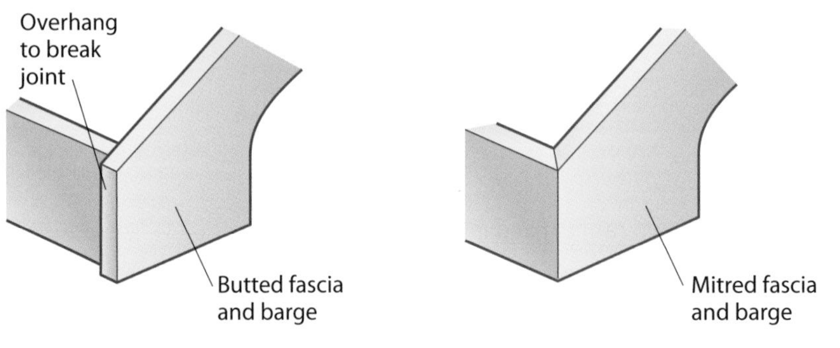

Figure 8.09 Fascia joined to bargeboard

Key term

Trimming an opening – removing structural timbers to allow a component such as a chimney or staircase to be fitted, and adding extra load-bearing timbers to spread the additional load.

Forming openings

Roofs often have components such as chimneys or roof windows. These components create extra work, as the roof must have an opening for them to be fitted. This involves cutting out parts of the rafter and putting in extra support to carry the weight of the roof over the missing rafters. This is called **trimming an opening**.

Chimneys

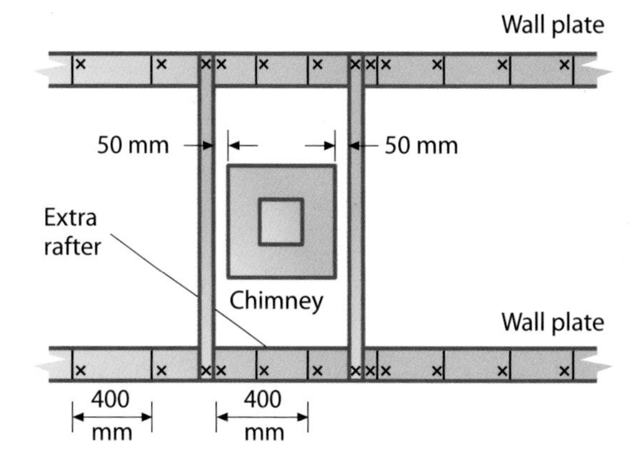

Figure 8.10 Wall plate marked out to allow for chimney

Chimneys are rarely used in new house construction as there are more efficient and environmental ways of heating, but many older houses will have chimneys and these roofs must be altered to suit.

When constructing such a roof, the chimney should already be in place, so you should cut and fit the rest of the roof, leaving out the rafters where the chimney is. When you mark out the wall plate, make sure that the rafters are positioned with a 50 mm gap between the chimney and the rafter. You may also need to put in extra rafters.

Next, fit the trimmer pieces between the rafters to bridge the gap, then fix the trimmed rafters – rafters running from wall plate to trimmer and from trimmer to ridge. The trimmed rafters are birdsmouthed at the bottom to sit over the wall plate, and the plumb and seat cut is the same as for common rafters.

If the chimney is at **mid-pitch** rather than at the ridge, you will need to fit a chimney/back gutter: this ensures the roof remains watertight by preventing the water gathering at one point. The chimney gutter should be fixed at the back of the stack (see Figure 8.12 on the next page).

Key term

Mid-pitch – in the middle of the pitch rather than at the apex or eaves.

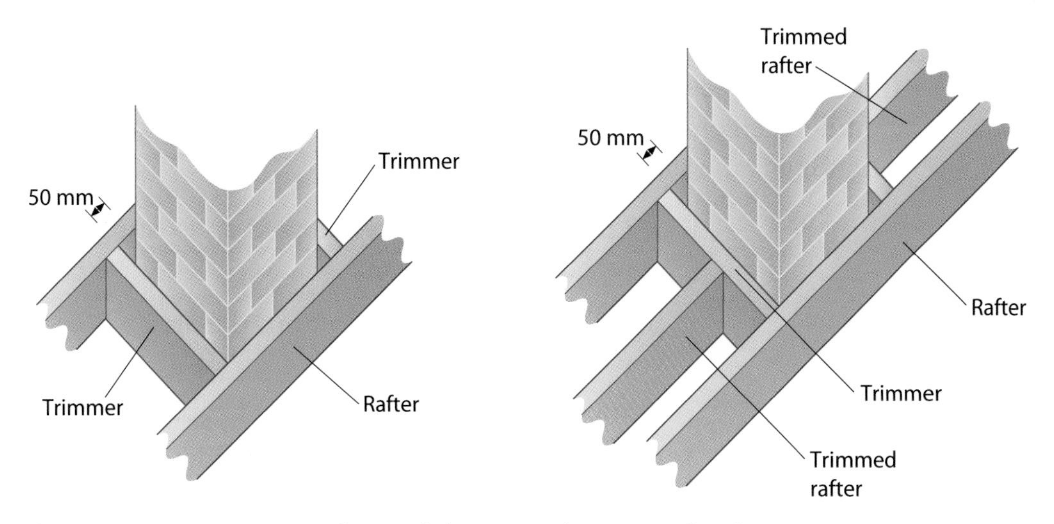

Figure 8.11 Opening trimmed around chimney with trimmers fitted

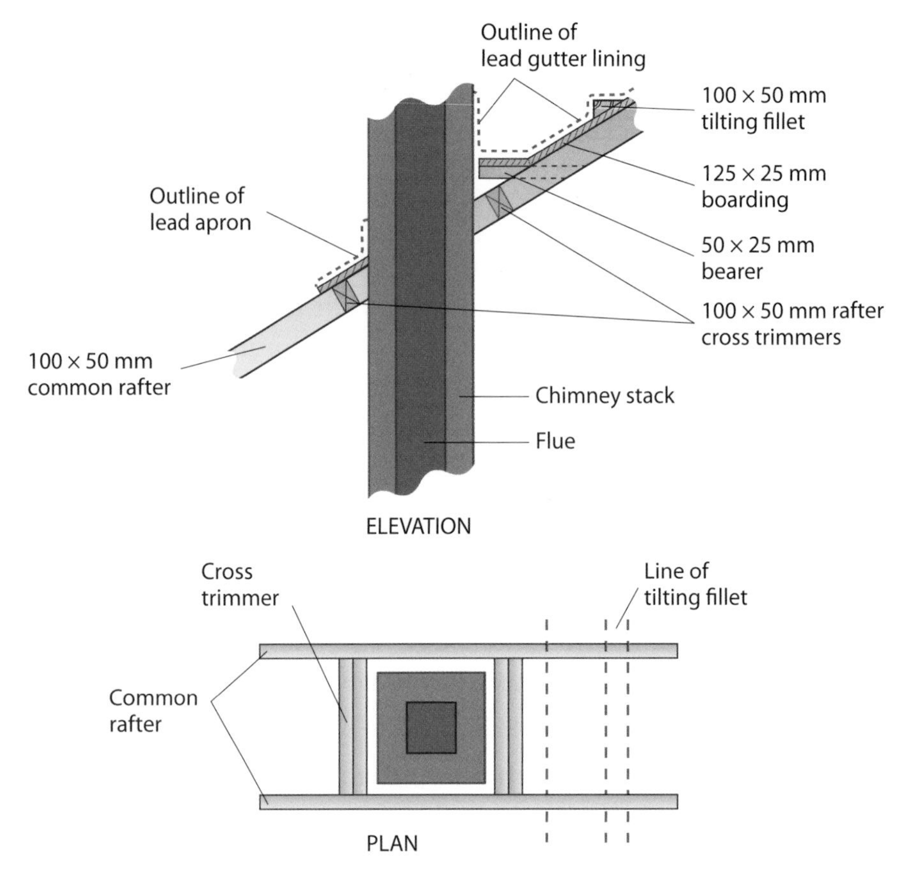

Figure 8.12 Gutter detail around chimney

Roof windows and skylights

You need to trim openings for roof windows and skylights too.

If the roof is new, you can plan it in the same way as for a chimney, remembering to make sure the area you leave for trimming is the same size as the window.

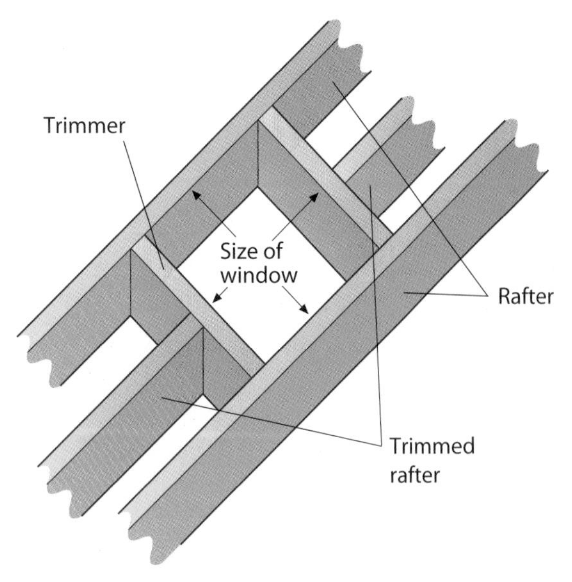

Figure 8.13 Roof trimmed for roof light

If you need to fit a roof window in an existing roof, the procedure is different:

1. Strip all the tiles or slates from the area where you want to fit the window.
2. Remove the tile battens, felt and roof cladding.
3. Mark out the position of the window and cut the rafters to suit.
4. Now fit the trimmer and trimmed rafters – you may need to double up the rafters for additional strength – and fit the window.
5. Re-fit the cladding and felt, and fix any flashings.
6. Cut and fit new tile battens, and finally re-tile or re-slate the roof.

Implications of constructing trussed rafter roofs at ground level

As well as the traditional method of constructing trussed roofs, more companies are looking at the method of constructing the roof at ground level. Constructing the roof at ground level can allow the roof to be constructed more safely and quickly. However, there are drawbacks to following this method.

In order to build the roof at ground level, there needs to be a large area on site for the construction to take place. Lifting the roof into place after completion is a large and complex operation that can create problems with safety. There can also be accuracy problems with constructing the roof, where slight errors in measurement prevent the roof from fitting accurately once in place.

The problems with size can be overcome by constructing the roof in a warehouse which has been specially designed for this use. Alternatively, a company can be hired to construct the roof and then deliver it on the back of a lorry.

Remember

The implications of constructing trussed rafter roofs at ground level are the same for all types of truss rafter, including fan, fink, king post and attic.

K2. Know how to construct verge and eaves finishes

As you will know from Unit 3003, the verge is the point in the roof where it overhangs at the gable. The construction of verges was covered in the section looking at finishing gable ends on page 101. You will have seen in this section how bargeboards, roof ladders and soffit boards are used in the construction.

Methods for finishing eaves

There are various ways of finishing a roof at the eaves; the four most common were covered in more detail on pages 102–103.

Ventilation

The soffit provides ventilation to the roof. Ventilation allows air to flow through the roof space. This helps to prevent the growth of damp and other problems in the roof. Without good ventilation, a roof could become infected with dry rot or similar mould and damp-based wood problems. This would lead to the roof becoming rotten, unstable and eventually unsafe.

Remember

Eaves are the lowest part of the roof surface. They are the point where the roof surface meets with the outside walls.

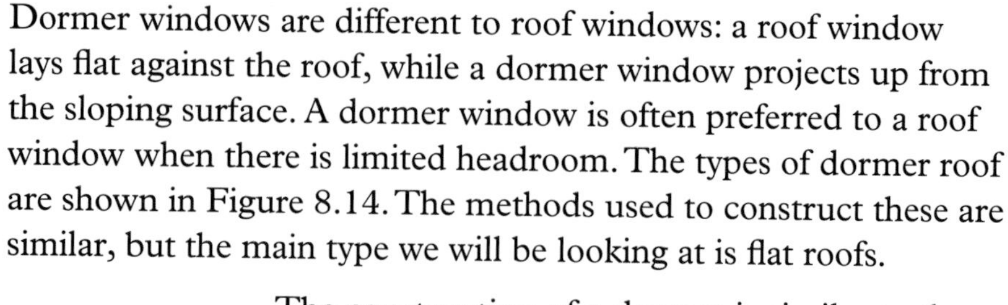

K3. Know how to form dormer windows

> **Remember**
>
> A segmental arch is also known as an elliptical arch.

Dormer windows are different to roof windows: a roof window lays flat against the roof, while a dormer window projects up from the sloping surface. A dormer window is often preferred to a roof window when there is limited headroom. The types of dormer roof are shown in Figure 8.14. The methods used to construct these are similar, but the main type we will be looking at is flat roofs.

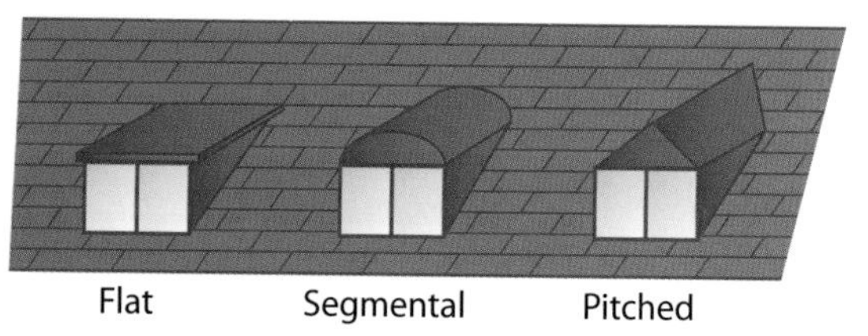

Figure 8.14 Dormer types

The construction of a dormer is similar to that of a roof window except that, once the opening is trimmed, you need to add a framework to give the dormer shape. For a dormer you usually double up the rafters on either side to give support for the extra load the dormer puts on the roof (see Figure 8.15).

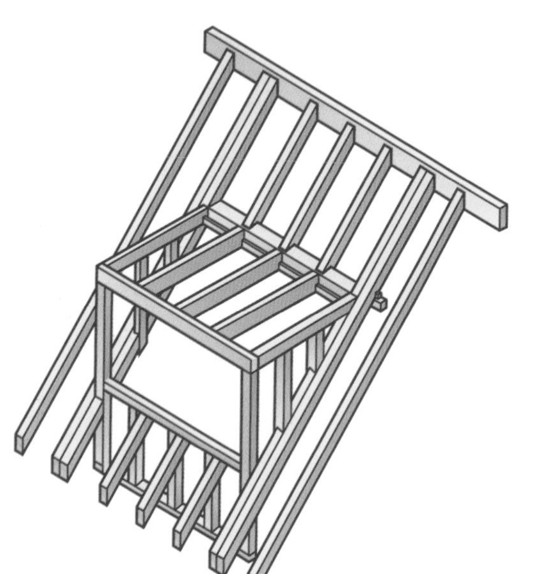

Figure 8.15 Framing for dormer

Forming dormer windows

Dormer windows are usually included within the original building but, with more people looking to convert their loft space, retrofitting dormers is becoming more common.

When forming dormer windows, the first job you need to do is to remove all the tiles, felt and batten from the existing roof in the area into which the new dormer will be constructed. This stripping back of the roof will have to be done from eaves to ridge to allow extra rafters to be fitted (see Figure 8.16).

The position of the dormer is then marked out on the wall plate and ridge board. Extra rafters may need to be fitted in place, or existing rafters may need to be doubled up, to give support for the extra load the dormer will place on the roof.

The position of the trimmer rafters can then be marked out and the trimmed rafters cut to allow the trimmers to be fitted.

With the roof opening for the rafter trimmed, the framing for the dormer can be constructed (see Figure 8.18).

The roof can then be re-felted and battened with the relevant flashings and tile finishes fitted to both the existing roof and new dormer roof. The vertical faces of the dormer can be finished with tiles, slates or cladding (see Figure 8.19).

Figure 8.16 Roof stripped to accommodate dormer

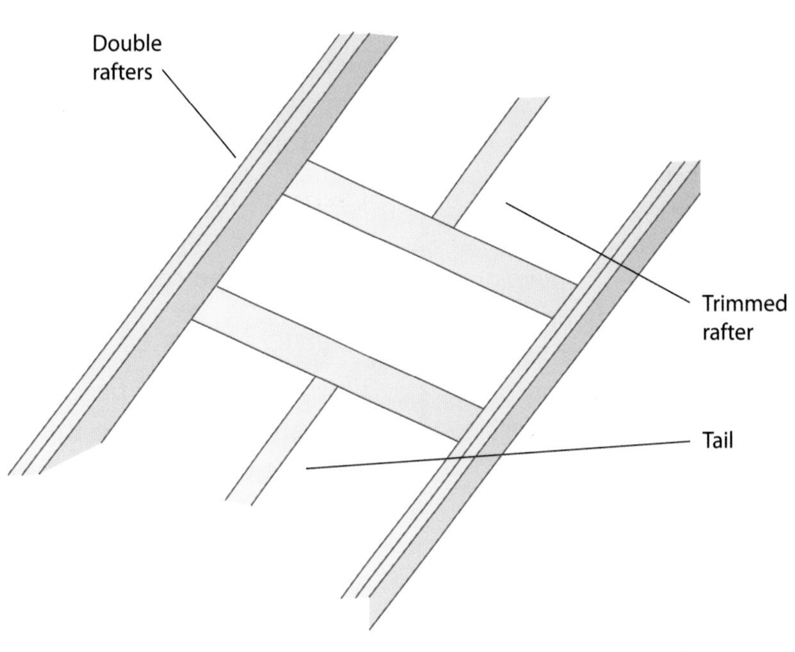

Figure 8.17 Extra rafters and trimmers being fitted

Figure 8.18 Constructed dormer

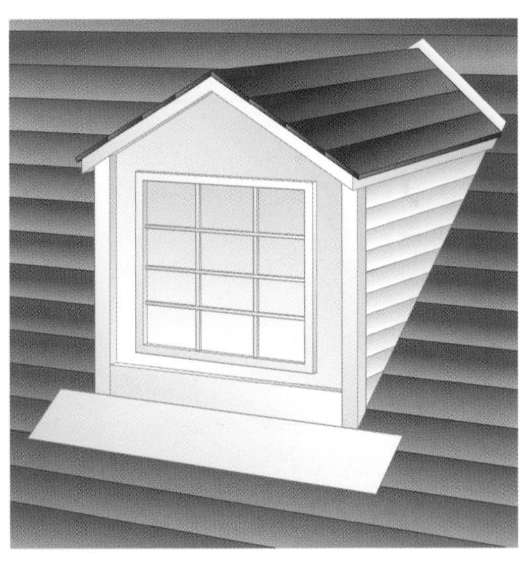

Figure 8.19 Completed dormer

Did you know?

If a flat roof had a guarantee for 50 years, the builder would be responsible for any maintenance work on the roof free of charge for 50 years. Since the average lifespan of a flat roof is between 12 and 15 years, the builder will only offer a guarantee for 10 years.

Constructing flat roofs

Flat roof joists are similar in construction to floor joists, but unless they are to be accessible they are not so heavily loaded. Joists are, therefore, of a smaller dimension than those used in flooring.

There are many ways to provide a fall on a flat roof. The method you choose depends on what the direction of fall is and where on a building the roof is situated.

Laying joists to a fall

This method is by far the easiest: all you have to do is ensure that the wall plate fixed to the wall is higher than the wall plate on the opposite wall, or vice versa. The problem is that this method will also give the interior of the roof a sloping ceiling. This may be fine for a room such as a garage, but for a room such as a kitchen extension the client might not want the ceiling to be sloped and another method would have to be used.

Key term

Firring piece – long wedge tapered at one end and fixed on top of a joist to create the fall on a flat roof.

Joists with firring pieces

Firring pieces provide a fall without disrupting the interior of the room, but involve more work. Firrings can be laid in two different ways, depending on the layout of the joists and the fall (see Figures 8.20 and 8.21).

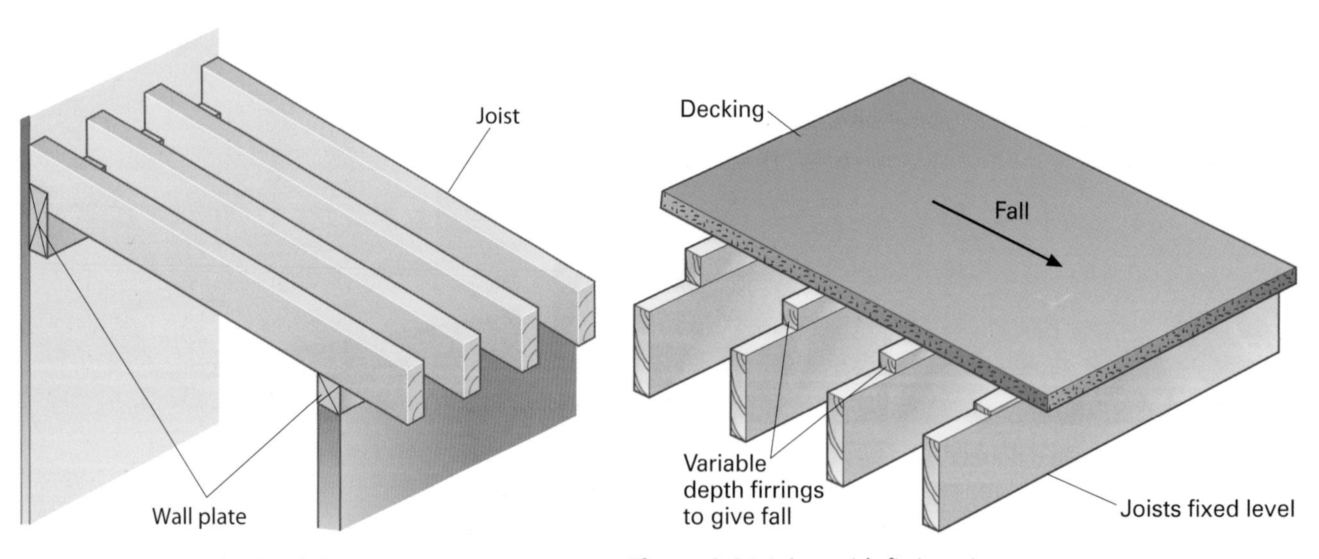

Figure 8.20 Joists laid to fall

Figure 8.21 Joists with firring pieces

Using firring pieces

The basic construction of a flat roof with firrings begins with the building of the exterior walls. Once the walls are in place at the correct height and level, the carpenter fits the wall plate on the eaves wall (there is no need for the wall plate to be fitted to the verge walls). This can be bedded down with cement, or nailed through the joints in the brick or blockwork with restraining straps fitted for extra strength. The carpenter then fixes the **header** to the existing wall.

The header can either be the same depth as the joists or have a smaller depth. If it is the same depth, the whole of the joist butts up to the header and the joists are fixed using joist hangers; if the header has a smaller depth, the joists can be notched to sit on top of the header, as well as using framing anchors.

Once the wall plate and header are fixed, they are marked out for the joists at the specified centres (300 mm, 400 mm or 600 mm). The joists are cut to length, checked for **camber (crown)** and fixed in place using joist hangers. Strutting or noggings are then fitted to help strengthen the joists.

Once the joists are fixed in place they must have restraining straps fitted. A strap must be fitted to a minimum of one joist per 2 m run, then firmly anchored to the wall to prevent movement in the roof under pressure from high winds.

The next step is to fix the firring pieces, which are either nailed or screwed down onto the top of the joists. Insulation is fitted between the joists, along with a vapour barrier to prevent the movement of moisture caused by condensation.

Key terms

Header – a piece of timber bolted to the wall to carry the weight of the joists.

Camber (crown) – the bow or round that occurs along the depth of some timbers.

Decking

Once the insulation and vapour barrier are fitted, it is time to fit the decking. As you may remember from Level 2, decking a flat roof can be done with a range of materials:

- **Tongued and grooved board** – these boards are usually made from pine and are not very moisture resistant, even when treated, so they are rarely used for decking these days. If used, the boards should be laid with, or diagonal to, the fall of the roof. Cupping of the boards laid across the fall could cause the roof covering to form hollows in which puddles could form.
- **Plywood** – only roofing boards stamped with WBP (weather/water boil proof) should be used. Boards must be supported and securely fixed on all edges in case there is any distortion, which could rip or tear the felt covering. A 1 mm gap must

Did you know?

There is no specific best or worst decking material, but some materials are better than others. All the materials stated serve a purpose, but only if they are finished correctly, e.g. a chipboard-covered roof that is poorly felted will leak, as will a metal roof if the screw or bolt holes are not sealed correctly.

be left between each board along all edges in case there is any movement caused by moisture, which again could cause damage to the felt.

- **Chipboard** – only the types with the required water resistance classified for this purpose must be used. Boards are available that have a covering of bituminous felt bonded to one surface, giving temporary protection against wet weather. Once laid, the edges and joints can be sealed. Edge support, laying and fixing are similar to floors. Moisture movement will be greater than with plywood as chipboard is more **porous**, so a 2 mm gap should be allowed along all joints, with at least a 10 mm gap around the roof edges. Tongued and grooved chipboard sheets should be fitted as per the manufacturer's instructions.
- **Oriented strand board (OSB)** – generally more stable than chipboard, but again only roofing grades must be used. Provision for moisture movement should be made, as with chipboard.
- **Cement-bonded chipboard** – strong and durable with high density (much heavier and greater moisture resistance than standard chipboard). Provisions for moisture movement should be the same as chipboard.
- **Metal decking** – profiled sheets of aluminium or galvanised steel, with a variety of factory-applied colour coatings and co-ordinated fixings available. Metal decking is more usually associated with large steel sub-structures and fixed by specialist installers, but it can be used on small roof spans to some effect. Sheets can be rolled to different profiles and cut to any reasonable length to suit individual requirements.
- **Translucent sheeting** – this might be corrugated or flat (e.g. polycarbonate twin wall), and must be installed as per the manufacturer's instructions.

Metal decking and translucent sheeting are supplied as finished products, but timber-based decking needs additional work to make it watertight, as explained in the next section.

Key term

Porous – full of pores or holes: if a board is porous, it may soak up moisture or water and swell.

Safety tip

Once the roof has been decked and is safe to walk on, the area around the roof (the verges and eaves) must be cordoned off with a suitable edge protection containing handrails and toeboards.

Key terms

Translucent – allowing some light through, but not clear like glass.

Bitumen – also known as pitch or tar, bitumen is a black sticky substance that turns into a liquid when heated, and is used on flat roofs to provide a waterproof seal.

Weatherproofing a flat roof

Once a flat roof has been constructed and decked, the next step is to make it watertight. The roof decking material can be covered using different methods and with different materials. One basic way of covering the decking is as follows.

The roof decking is covered in a layer of hot **bitumen** to seal any gaps in the joints. Then another layer of bitumen is poured over the top and felt is rolled onto it, sticking fast when the bitumen sets. A second roll of felt is stuck down with bitumen, but is laid at 90° to the first. Some people add more felt at this stage – sometimes up to five more layers.

The final step can also be done in a number of ways. A cold system is now available, which replaces the need to use hot melt pots and blowtorches. Some people put stones or chippings down on top of the final layer; some use felt that has stones or chippings embedded and a layer of dried bitumen on the back. The felt with stones embedded into it is laid by rolling the felt out and using a gas blowtorch to heat the back, which softens the bitumen, allowing it to stick.

Finishing a flat roof at the abutment and verge

Abutment

The abutment finish needs to take into consideration the existing wall as well as the flat roof. The abutment is finished by cutting a slot into the brick or blockwork and fixing lead to give a waterproof seal, which prevents water running down the face of the wall and into the room. **Tilt (angle) fillets** are used to help with the run of the water and to give a less severe angle for the lead to be dressed to.

Verge

Since a flat roof has such a shallow pitch, the verge needs some form of upstand to stop the water flowing over the sides at the verge instead of into the guttering at the eaves. This is done using tilt or angle fillets nailed to the decking down the full length of the verge, prior to the roof being felted or finished.

Eaves details

The eaves details can be finished in the same way as the eaves on a pitched roof, with the soffit fitted to the underside of the joists and the fascia fitted to the ends.

Working life

You have been tasked with fitting a flat roof. Your client has asked you for advice on building this, and the best way to carry out the work.

What method would you use? What type of decking would you use? What reasons do you have for your choices?

Safety tip

Bitumen in liquid form is very hot, and if it comes into contact with the skin, it will stick and cause severe burns. When working with bitumen you must wear gloves and goggles, and ensure that arms and legs are fully covered.

Did you know?

Bitumen is bought as a solid material. It is then broken up, placed in a cauldron and heated until it becomes a liquid, which can be spread over the flat roof. The bitumen will only stay in liquid form for a few minutes before it starts to set. Once set, it forms a waterproof barrier.

Key term

Tilt (angle) fillet – a triangular-shaped piece of wood used in flat roof construction.

Did you know?

Flat roofs have stones or chippings placed on top to reflect the heat from the sun, which can melt the bitumen, causing problems with leaks.

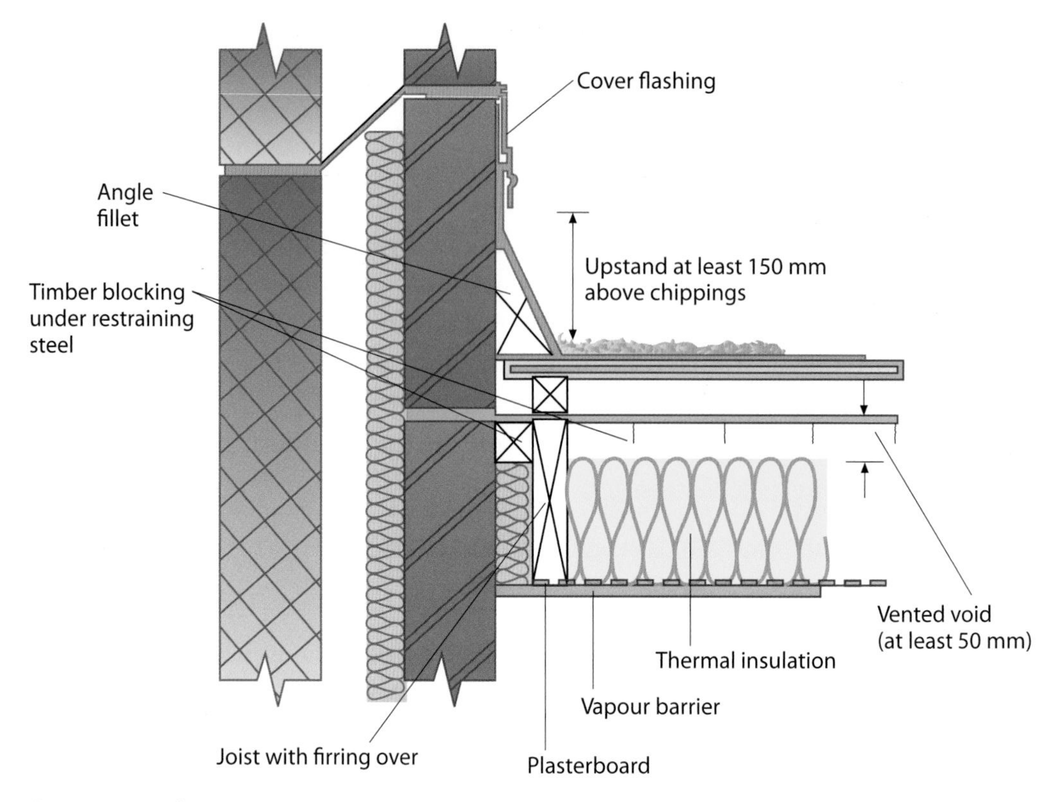

Figure 8.22 Abutment joint

Key term

Span – the distance measured in the direction of ceiling joists, from the outside of one wall plate to another; known as the overall (O/A) span.

Functional skills

While learning how to construct different types of roof, you will have the opportunity to practise **FM 2.3.1** Interpret and communicate solutions to multistage practical problems and **FM 2.3.2** Draw conclusions and provide mathematical justifications.

K4. Know how to construct a traditional cut roof with hips and valleys

Types of roof construction

A pitched roof can be constructed either as a single roof, where the rafters do not require any intermediate support, or a double roof, where the rafters are supported. Single roofs are used over a short **span** such as a garage; double roofs are used to span a longer distance such as a house or factory.

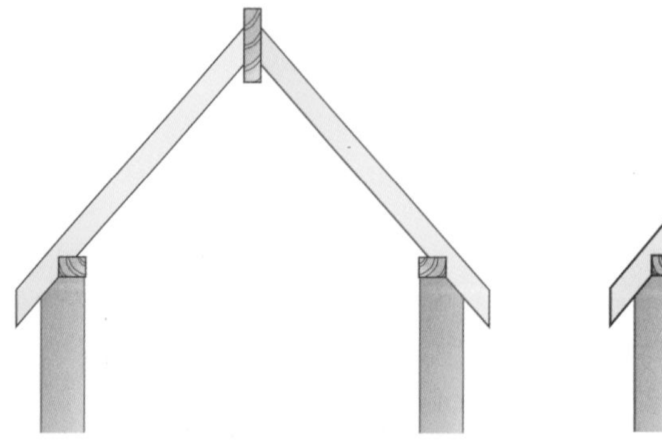

Figure 8.23 Single roof

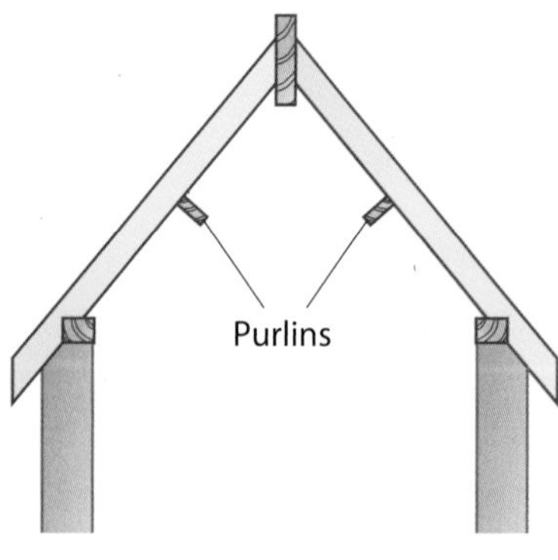

Figure 8.24 Double roof

There are many different types of pitched roof, including:

- **a hipped roof** – with hip ends incorporating crown, hip and jack rafters
- **a lean-to roof** – with a single pitch, which butts up to an existing building
- **a couple close roof** – with no ties and only suitable for short spans
- **a collar roof** – where the ceiling height is above the eaves and there are no ceiling joists.

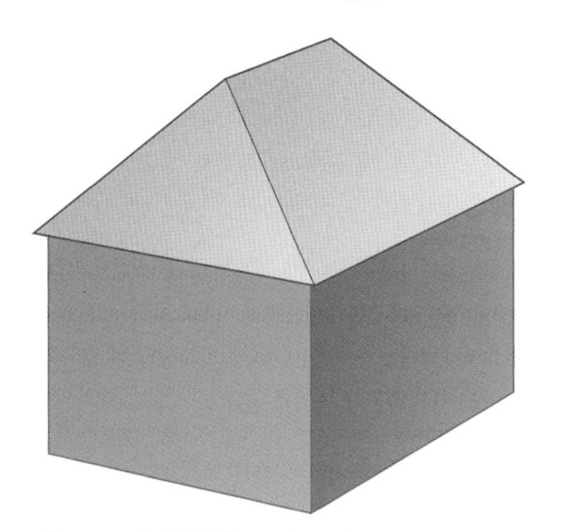

Figure 8.25 Hipped roof

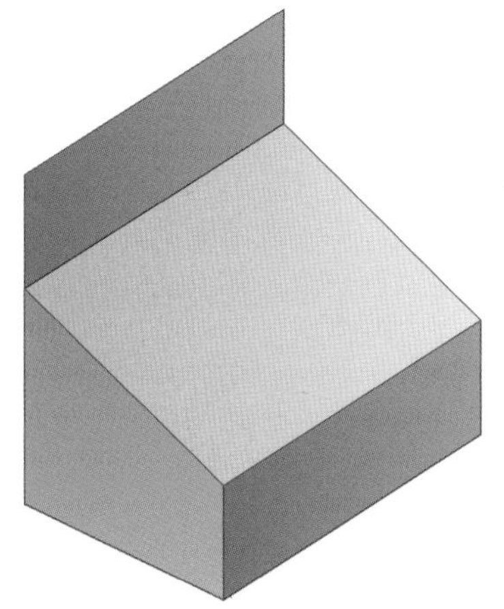

Figure 8.26 Lean-to roof

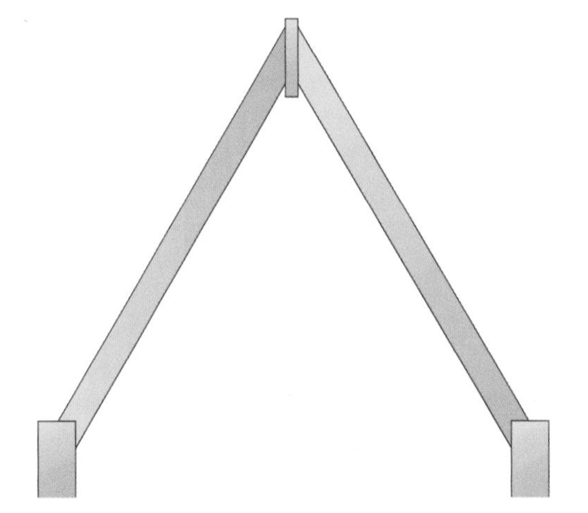

Figure 8.27 Couple close roof

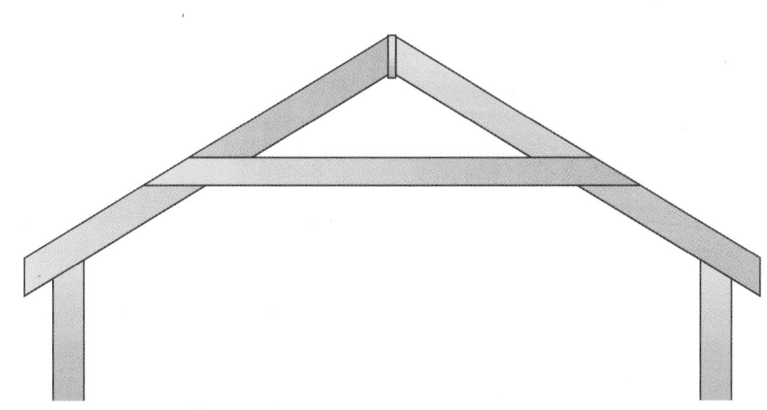

Figure 8.28 Collar roof

Constructing hipped roofs

In Unit 3003, you learned about marking out for a hipped roof, and how to find true lengths, using Pythagoras' theorem. Here we look at the rest of the marking and setting out process. This process will be used on all the types of roof we looked at earlier.

From this point, to make the geometrical drawings as clear as possible, abbreviated labels will be used (see Table 8.01 on page 224).

Safety tip

Never erect a roof on your own. It is a job for at least two people.

Table 8.01 Abbreviations

Abbreviation	Definition	Abbreviation	Definition
EC	Edge cut	SC	Seat cut
ECCrR	Edge cut cripple rafter	SCCR	Seat cut common rafter
ECHR	Edge cut hip rafter	SCHR	Seat cut hip rafter
ECJR	Edge cut jack rafter	SCVR	Seat cut valley rafter
ECVR	Edge cut valley rafter	TL	True length
PC	Plumb cut	TLCR	True length common rafter
PCCR	Plumb cut common rafter	TLCrR	True length cripple rafter
PCHR	Plumb cut hip rafter	TLHR	True length hip rafter
PCVR	Plumb cut valley rafter	TLJR	True length jack rafter

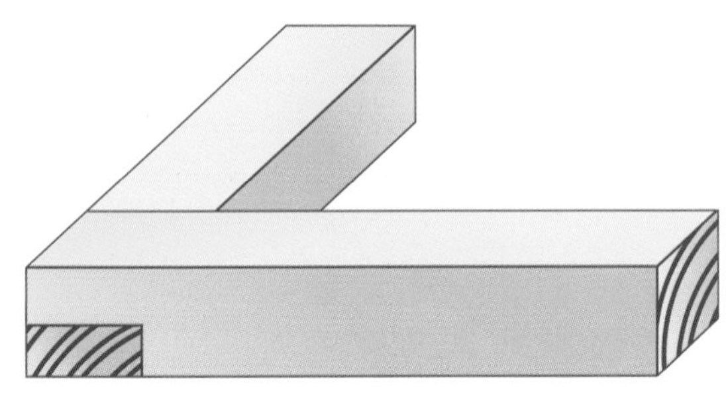

Figure 8.29 Corner halving

Setting out a hipped end

First, you need to fit the wall plate, then the ridge and common rafters. To know where to place the common rafters, you need to work out the true length of the rafter, then begin to mark out the wall plate.

The wall plate is joined at the corners and marked out, as shown in Figures 8.29 and 8.30.

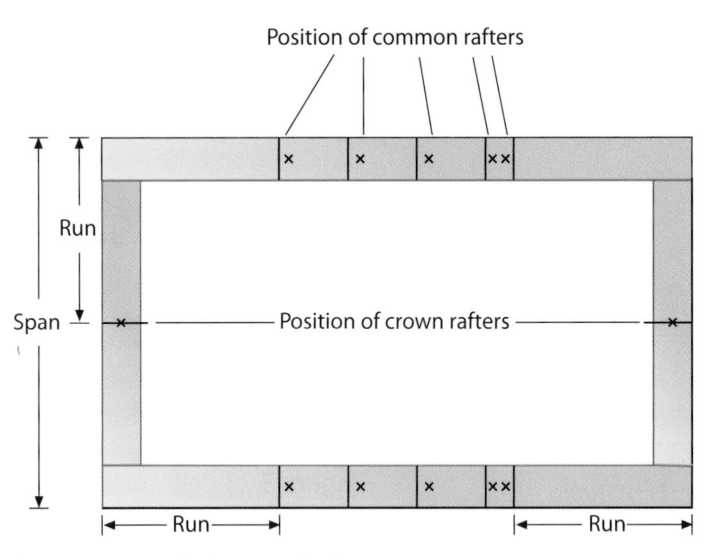

Figure 8.30 Wall plate marked out

Marking out for a hipped roof

To mark out for a hipped roof, follow these steps:

1. Measure the span and divide it by two to get the run, then mark this on the hipped ends – the centre of your crown rafter will line up with this mark.

2. Mark the run along the two longer wall plates – these marks will give you the position of your first and last common rafter.
3. The common rafter will sit to the side of this line, so a cross or other mark should be made to let you know on which side of the line the rafter will sit.
4. Mark positions for the rest of the rafters on the wall plate at the required centres, again using a cross or other mark to show you on which side of the line the rafter will sit. Note: the last two rafters may be closer together than the required centres, but must not be wider apart than the required centres.
5. Cut and fit the common rafters using the same method as used for a gable roof (see Figure 8.31).
6. Fit the **crown rafter**, which has the same plumb and seat cuts as the common rafter and is almost the same length – but here you should *not* remove half the ridge thickness (see Figure 8.31).

Key term

Crown rafter – the rafter that sits in the middle of a hip end, between the two hips.

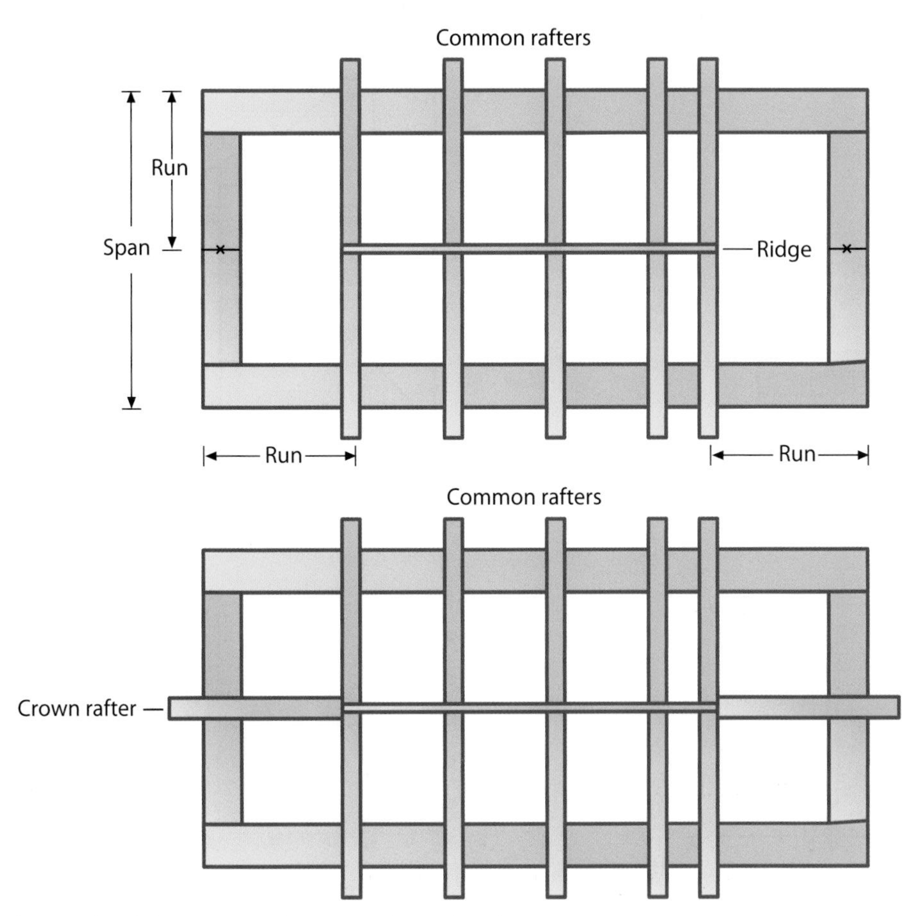

Figure 8.31 Common and crown rafters fitted

Remember

With a hip rafter it is important to remember that the pitch line is marked out differently. It is marked from the top of the rafter and is set at two-thirds of the depth of the *common* rafter. The best way to check your rafter before cutting is to measure from the point of the ridge down to the corner of the wall plate. This distance should be the same as marked on your rafter.

7. For the hip rafters, work out the true length, all the angles and bevels and mark out one hip as shown in Figure 8.32; then cut the hip and try it in the four corners. If the hip fits in all four corners, you can use it as a template to mark out the rest of the hips; if not, the roof is out of square or level, but you can still use this hip to help mark out the remaining three corners.
8. Cut in the jack rafters. Find out the true length and edge cut of all the jack rafters, then mark them out and cut them. As the jack rafters are of different sizes it is better to cut them individually to fit. They can still be used as template rafters on the opposite side of the hip (see Figure 8.35).

Valleys

Valleys are formed when two sloping parts of a pitched roof meet at an internal corner.

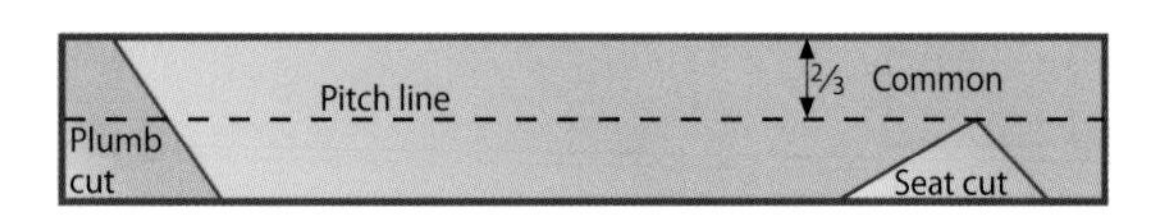

Figure 8.32 Pitch line marked on hip

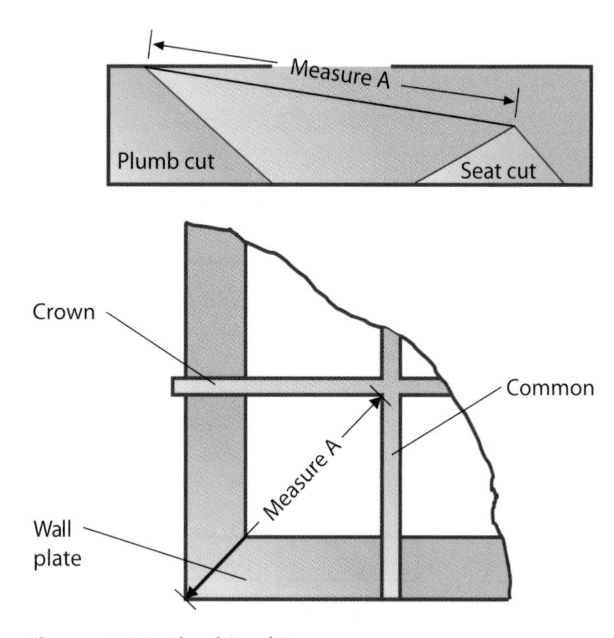

Figure 8.33 Checking hip measurements

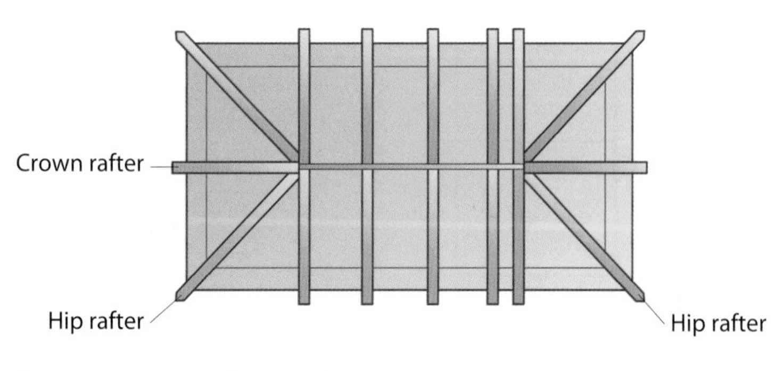

Figure 8.34 Hip rafters in place

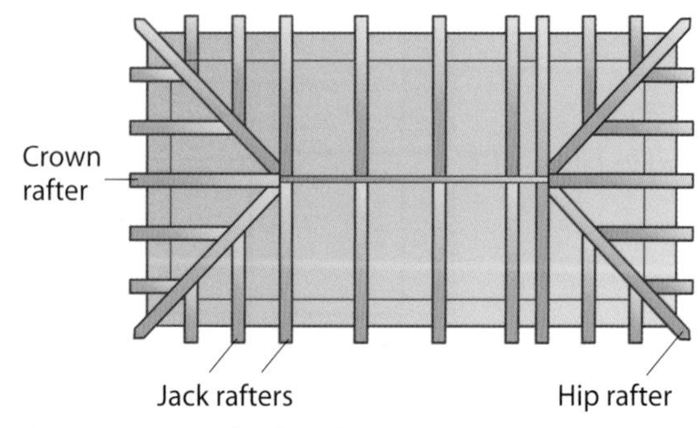

Figure 8.35 Jacks fitted

Marking out for valleys

Valleys can be worked out in the same way as hips, using either a ready reckoner or geometry. Here we will look at geometry.

Working out the angles for valleys is similar to doing so for hips, except that the key drawing is not a triangle but a plan drawing of the roof.

First, you need to find out the valley rafter true length, plumb and seat cut. Start by finding the rise of the roof and drawing a line this length at a right angle to the valley where it meets the ridge. Join this line to the point where the valley meets the wall plate. This will give you the true length of the valley rafter, as well as the plumb and seat cuts.

As with the hip rafter, there are two other angles to find for a valley rafter: the dihedral angle and the edge cut.

The dihedral angle for the valley is used in the same way as the hip dihedral and rarely in roofing today. Figure 8.38 on page 228 shows you how to work out the dihedral angle.

The final angle to find is the edge cut for the valley rafter, as follows:

1. Mark on the rise and true length of the valley rafter.
2. Draw a line at right angles to the valley where it meets the wall plate, and extend this line to touch the ridge at A.
3. Set your compass to the true length of the valley and **swing an arc** towards the ridge at B.

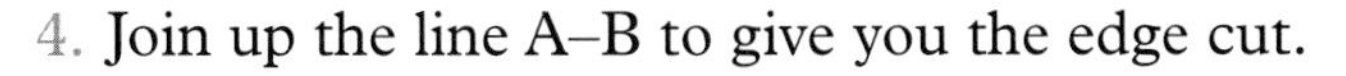

4. Join up the line A–B to give you the edge cut.

The final part of valley geometry is to find the true length and edge cut for the cripple rafters, as follows:

1. Draw out the roof plan as usual, then to the side of your plan draw out a section of the roof (see Figure 8.40).
2. Set your compass to the rafter length and swing an arc downwards.

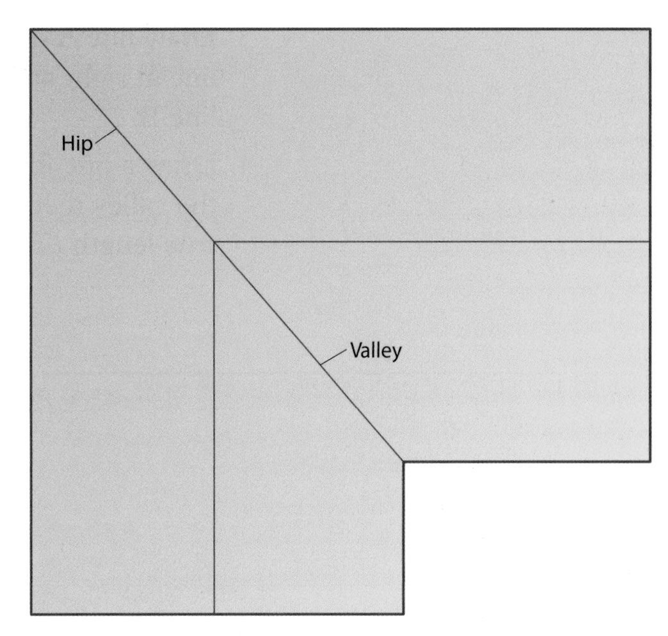

Figure 8.36 Plan of roof

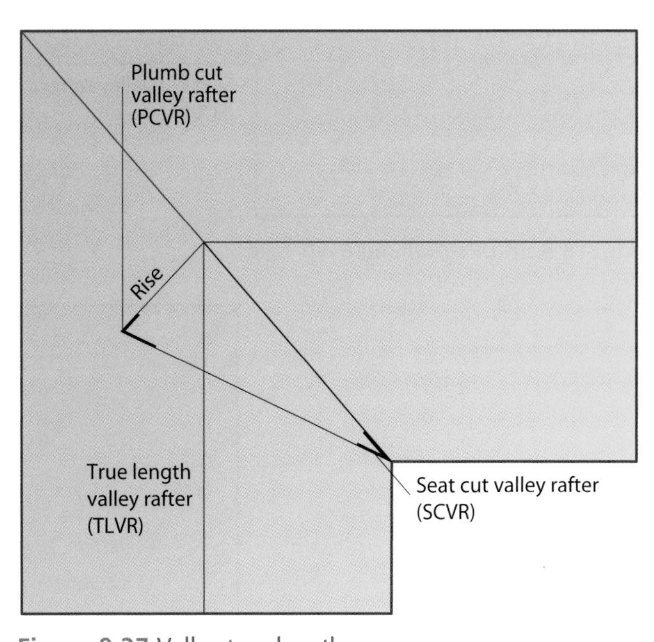

Figure 8.37 Valley true length

Key term

Swing an arc – set a compass to the required radius and lightly draw a circle or arc.

3. Draw line A downwards until it meets the arc; then draw a line at right angles to line A until it hits the wall plate, creating line B.
4. Draw a line from where line B hits the wall plate up to where the valley meets the ridge. This will give you the appropriate true length (TLCrR) and edge cut (ECCrR).

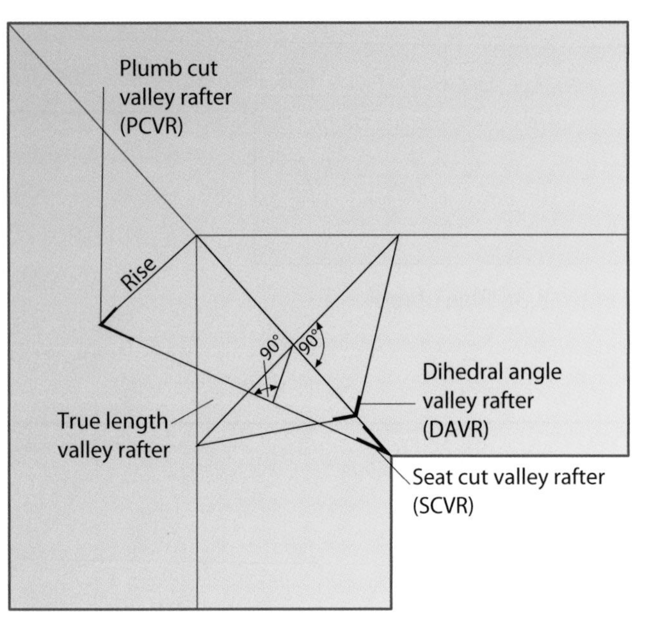

Figure 8.38 Dihedral angle hip

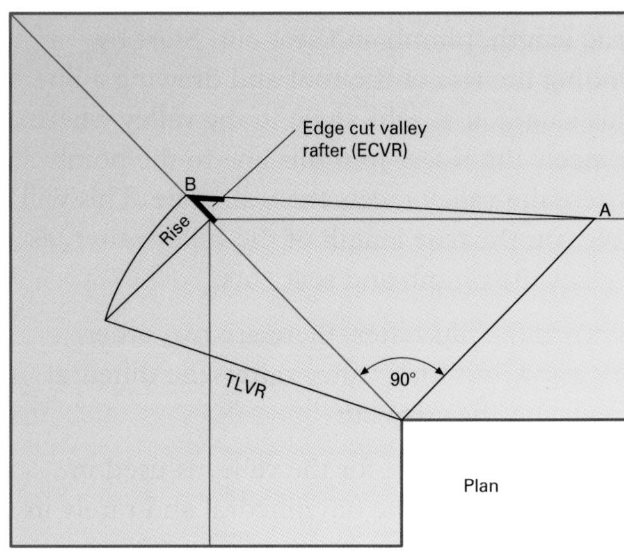

Figure 8.39 Edge cut

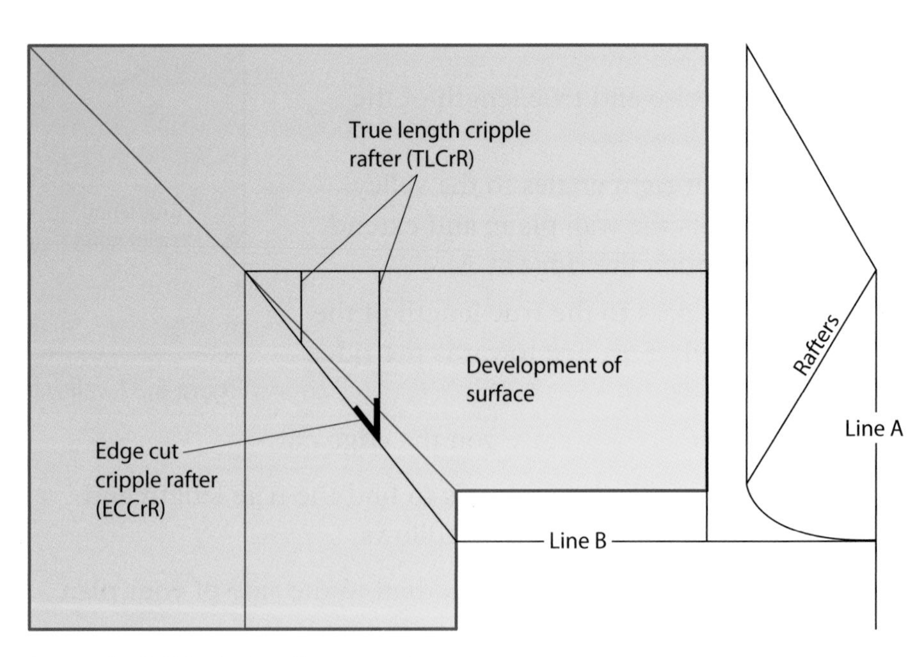

Figure 8.40 Cripple length and angle

Setting out a valley

There are four steps to follow when setting out a valley:

1. Fit the wall plate and mark it out with the position of the common rafters.
2. Fit the common rafters and ridge.
3. Fit the hip and jack rafters.
4. Fit the valley and cripple rafters, taking the true lengths and bevels from the drawings.

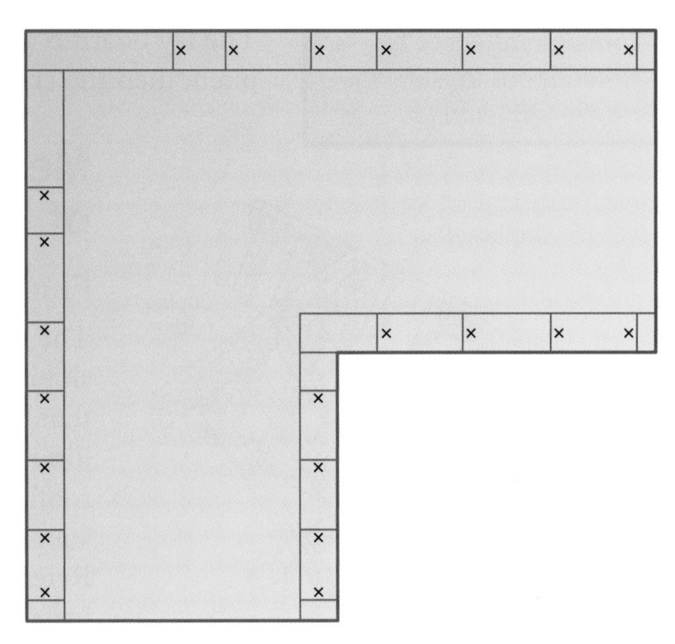

Figure 8.41 Common rafters fitted

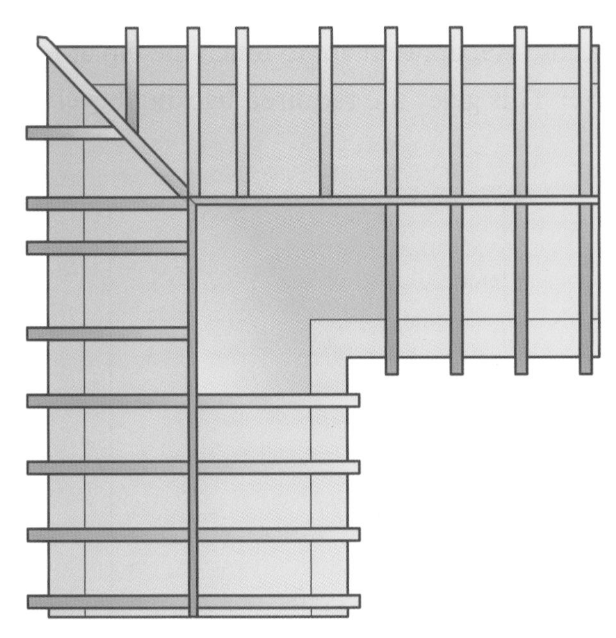

Figure 8.42 Hip and jack rafters fitted

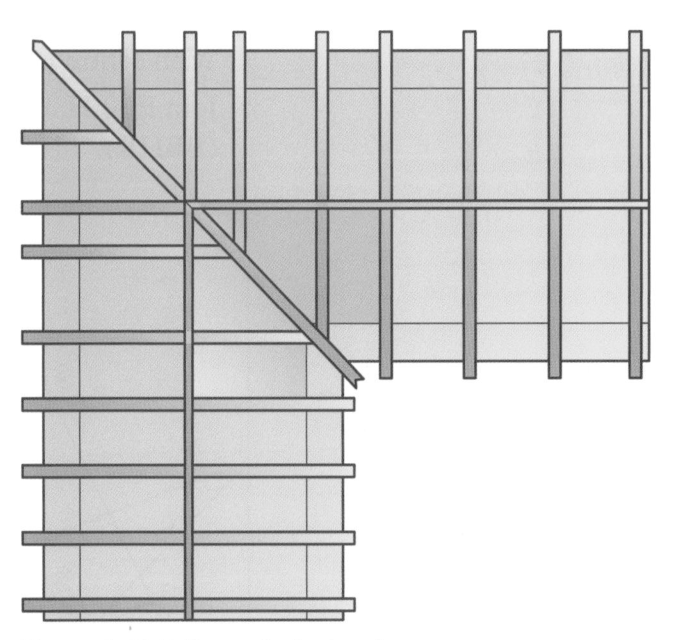

Figure 8.43 Valley and cripple rafters

Key term

Lay board – a piece of timber fitted to the common rafters of an existing roof to allow the cripple rafters to be fixed.

An alternative to using valley rafters is to use a **lay board**. Lay boards are most commonly used with extensions to existing roofs, or where there are dormer windows.

The lay board is fitted onto the existing rafters at the correct pitch; then the cripple rafters are cut and fixed to it.

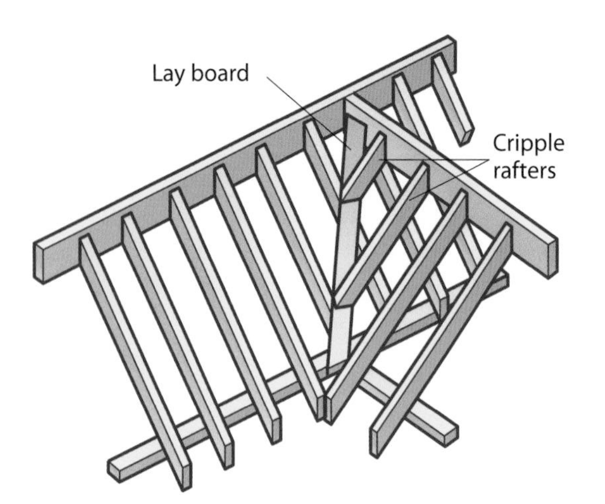

Figure 8.44 Valley lay board set-up

Methods of determining lengths of cuts

Finding the dihedral or backing bevel angle

The backing bevel angle is the angle between the two sloping roof surfaces. It provides a level surface so that the tile battens or roof boards can lie flat over the hip rafters. The backing bevel angle is rarely used in roofing today as the edge of the hip is usually worked square, but you should still know how to work it out. Refer to Figure 8.45 when you are reading these instructions.

1. Draw a plan of the roof and mark on the TLHR as before.
2. Draw a line at right angles to the hip on the plan at D, to touch the wall plates at E and F.
3. Draw a line at right angles to the TLHR at G, to touch point D.
4. With centre D and radius DG, draw an arc to touch the hip at H.
5. Join E to H and H to F. This gives the required backing bevel (BBHR).

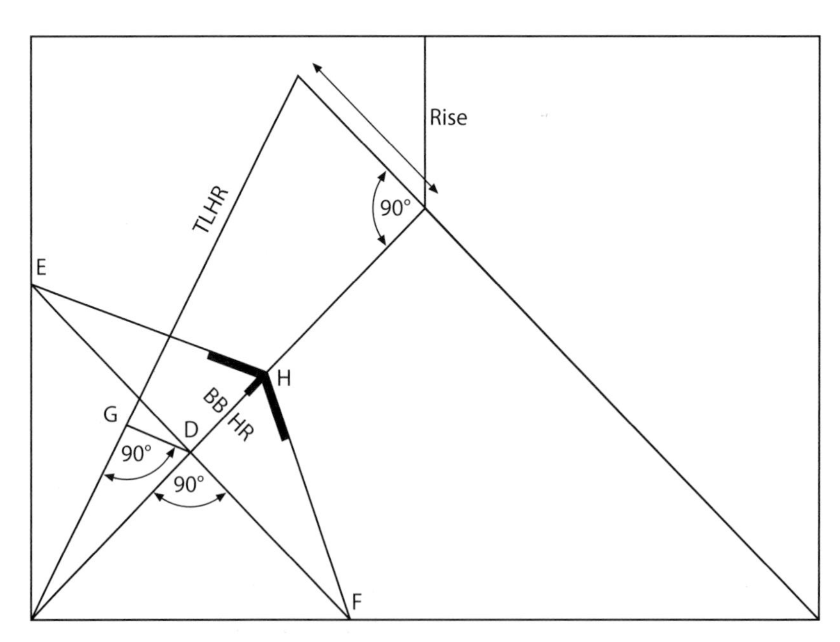

Figure 8.45 Dihedral angle for hip

Finding the edge cut

The edge cut is applied to both sides of the hip rafter at the plumb cut. It enables the hip to fit up to the ridge board between the crown and the common rafters.

1. Draw a plan of the roof and mark on the TLHR as before.
2. With centre I and radius IB, swing the TLHR down to J, making IJ, TLHR.
3. Draw lines at right angles from the ends of the hips and extend the ridge line. All three lines will intersect at K.
4. Join K to J. Angle IJK is the required edge cut (ECHR).

The jack rafter's plumb and seat cuts are the same as those used in the common rafters, so all you need to work out is the true length and edge cut (see Figure 8.48):

- Draw the plan and section of the roof. Mark on the plan the jack rafters. Develop roof surfaces by swinging TLCR down to L and projecting down to M¹.
- With centre N and radius NM¹, draw arc M10. Join points M¹ and 0 to ends of hips as shown.
- Continue jack rafters onto development.
- Mark the true length of jack rafter (TLJR) and edge cut for jack rafter (ECJR).

Ready reckoner

A ready reckoner is a book used as an alternative to the geometry method and is often the simplest way of working out lengths and bevels. The book consists of a series of tables that are easy to follow once you understand the basics.

To use the ready reckoner you must know the span and the pitch of the roof.

Example

Take a hipped roof with a 36° pitch and a span of 8.46 m. First you halve the span, getting a run of 4.23 m. Referring to the tables in the ready reckoner, you can work out the lengths of the common rafter as shown on the next page.

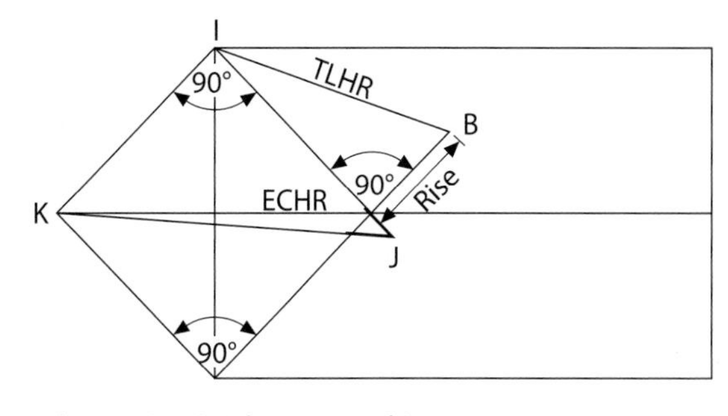

Figure 8.46 Edge cut on hip

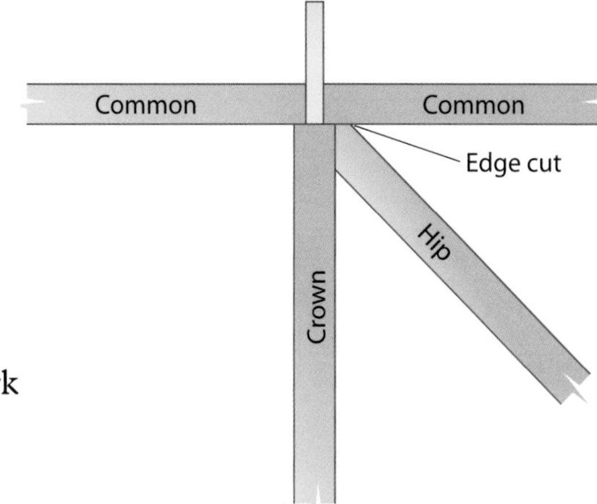

Figure 8.47 Edge cut on hip joined to ridge

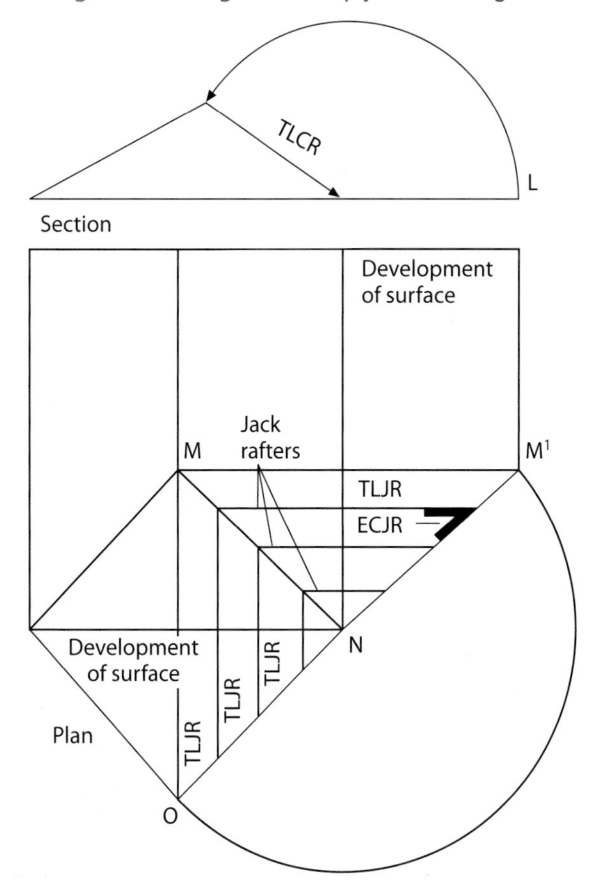

Figure 8.48 Jack rafter length and cuts

RISE OF COMMON RAFTER 0.727 m
PER METRE RUN PITCH 36°.

BEVELS	COMMON RAFTER	SEAT 36
	COMMON RAFTER	PLUMB 54
	HIP OR VALLEY	SEAT 27
	HIP OR VALLEY	RIDGE 63
	JACK RAFTER	EDGE 39

JACK RAFTERS	333 mm CENTRES DECREASE	412 mm
	400 mm CENTRES DECREASE	494 mm
	500 mm CENTRES DECREASE	618 mm
	600 mm CENTRES DECREASE	742 mm

Run of rafter	0.1	0.2	0.3	0.4	0.5	0.6	0.7	0.8	0.9	1.0
Length of rafter	0.124	0.247	0.371	0.494	0.618	0.742	0.865	0.989	1.112	1.236
Length of hip	0.159	0.318	0.477	0.636	0.795	0.954	1.113	1.272	1.431	1.590

You can see that the length of a rafter for a run of 0.1 m is 0.124 m, therefore for a run of 1 m, the rafter length will be 1.24 m. But you need to find the length of a rafter for a run of 4.23 m. This is how:

1.00 m = 1.24 m

So: 4.00 m = 4.944 m
0.20 m = 0.247 m
0.03 m = 0.037 m
total = 5.228 m

So the length of the common rafter is 5.228 m. However, there are a few adjustments you must make before finding the finished size.

You need to allow for an overhang and for the ridge, which can both easily be measured. For the purposes of this example, we will use an overhang of 556 mm and a ridge of 50 mm. The final calculation is:

Basic rafter 5.228 m
+ overhang 0.556 m
– half ridge 0.025 m
Total **5.759 m**

So the rafter length is 5.759 m.

Now you need to refer back to the table, which tells you that, for the common rafter, the seat cut is 36° and the plumb cut is 54°.

You can now mark out and cut your pattern rafter as before, remembering to mark on the pitch line and plumb cut. Measure the original size (5.228 m in our example) along the plumb cut, mark out the seat cut and finally cut.

The hip and valley rafters are worked out in the same way, but jack and cripple rafters are different.

Jack and cripple rafters are included in the table and are easy to work out. Continuing with the example, where the rafter length is 5.228 m, you look at the table and, if you are working at spacing the rafters at 400 mm centres, you reduce the length of the common rafter by 494 mm. The first jack/cripple rafter will be 5.228 m – 0.494 m = 4.734 m; the next jack/cripple rafter will be 4.734 m – 0.494 m = 4.240 m, and so on. The angles for all the rest of the cuts are shown in the table.

Remember

The plumb and seat cuts for the jack rafter are the same as for the common rafter.

Working life

You have been tasked with cutting in a garage roof with gable ends. You work out the angles and lengths, and cut the pattern rafter, which you use to mark out the other rafters. You cut and fit the rafters and are almost finished when you decide to put a level on the ridge. You find that the roof is out of level.

What could have caused this? What could have been done to prevent this? What needs to be done to make it right? What are the consequences of leaving the roof as it is? Who is to blame?

Functional skills

This exercise will allow you to practise the interpreting elements of functional skills, such as **FM 2.3.1** Interpret and communicate solutions to multistage practical problems and **FM 2.3.2** Draw conclusions and provide mathematical justifications.

FAQ

What is the best method to use when working out roof bevels and lengths?

There is no single best way: it all depends on the individual and what they find easiest to use.

When it comes to covering the roof, which method should I use?

The type of roof covering to be used depends on what the client and architect want, and on what is needed to meet planning and Building Regulations.

Why does there have to be a 50 mm gap between the rafters and the chimney?

The Building Regulations state that there must be a 50 mm gap to prevent the heat from the chimney combusting the timber.

How can I get the lead on a flat roof to fit into the brick/blockwork?

The lead is fitted into a channel or groove that is cut into the brick/blockwork by a bricklayer. Once the groove is cut, the lead is fed into the groove, wedged and sealed with a suitable mastic or silicon.

I have laid chipboard flooring and the floor is squeaking. What causes this and how can I stop it?

The squeaking is caused by the floorboards rubbing against the nails – something that happens after a while as the nails eventually work themselves loose. The best way to prevent this is to put a few screws into the floor to prevent movement – but be cautious: there may be wires or pipes under the floor.

Check it out

1. Describe the process for storing and handling trussed rafters, writing a method statement to explain the processes that should be followed.
2. State the distance the first and last common rafters must be from the wall, and explain why this distance is used. Use a sketch to help complete your answer.
3. Explain the process used to set out a gable end, including sketches to show the processes you will need to follow.
4. Explain why extra rafters need to be added when trimming an opening.
5. Describe the differences between the different types of eaves and explain when they may be used, sketching examples of each in situ.
6. Write a method statement describing the key steps for forming dormer windows. Illustrate each step with a diagram.
7. State and describe four different materials used when decking a roof. Explain how to weatherproof the roof once it is decked.
8. Describe two different methods of forming a valley.
9. Explain what the dihedral angle (backing bevel) is and what purpose it serves.
10. Sketch several plans of roofs and mark on the different sketches the backing bevel angle and edge cut.

Getting ready for assessment

The information contained in this unit, as well as continued practical assignments that you will carry out in your college or training centre, will help you with preparing for both your end of unit test and the diploma multiple-choice test. It will also aid you in preparing for the work that is required for the synoptic practical assignments.

The information contained within this unit will aid you in learning how to identify and calculate the materials and equipment required for erecting structural carcassing.

You will need to be familiar with:

- erecting trussed rafter roofs
- constructing gables, verges and eaves
- forming dormer windows
- constructing a traditional cut roof with hips and valleys.

This unit will have made you familiar with the methods of erecting structural carcassing. For example, learning outcome two will require you to construct and fit verge finishing, including gable ladders, bargeboards and soffit boards to specification. You will also need to construct and fix eaves finishing, including open, closed and flush and sprocketed eaves to specification. When working on roofs you will need to be able to work from scaffolding and at height.

Safety is a major issue when working at height, and you will need to be sure that you are avoiding the common hazards associated with this type of work. You will also need to use your knowledge of materials to select the best suited material for a particular job. To do this you will need to study the specifications and take into account the particular location you will be building in.

Before you start work on the synoptic practical test, it is important that you have had sufficient practice and that you feel that you are capable of passing. It is best to have a plan of action and a work method that will help you. You will also need a copy of the required standards, any associated drawings and sufficient tools and materials. It is also wise to check your work at regular intervals. This will help you to be sure that you are working correctly and help you to avoid problems developing as you work.

Your speed at carrying out these tasks will also help you to prepare for the time limit imposed by the synoptic practical task. But remember that you should not try to rush the job as speed will come with practice and it is important that you get the quality of workmanship right.

Make sure you are working to all the safety requirements given throughout the test and wear all appropriate personal protective equipment (PPE). When using tools, make sure you are using them correctly and safely.

Good luck!

CHECK YOUR KNOWLEDGE

1. In a trussed roof, what is each shape composed of?
- **a** Squares
- **b** Rectangles
- **c** Triangles
- **d** Pentagons

2. What is the maximum that trusses are marked out to?
- **a** 300 mm centres
- **b** 400 mm centres
- **c** 450 mm centres
- **d** 600 mm centres

3. Which of the following bracing is used to secure the trusses upright and in place?
- **a** Stability bracing
- **b** Wind bracing
- **c** Strength bracing
- **d** Security bracing

4. What is a precise cut rafter used as a template known as?
- **a** Pattern rafter
- **b** Common rafter
- **c** Hip rafter
- **d** Valley rafter

5. What is the main advantage of constructing a roof at ground level?
- **a** It is safer
- **b** It is a slower process, so easier to plan
- **c** It is less expensive
- **d** All of the above

6. When working with roofing geometry drawings, which of the following abbreviations relates to a hip rafter?
- **a** TLCR
- **b** TLHR
- **c** PCCR
- **d** SCCR

7. What are the diminishing rafters in a valley called?
- **a** Common
- **b** Hip
- **c** Jack
- **d** Cripple

8. What cuts are used on a hip rafter?
- **a** Plumb
- **b** Seat
- **c** Edge
- **d** All of the above

9. The lowest part of a roof where it meets the walls is called the:
- **a** eaves
- **b** verge
- **c** purlin
- **d** ridge

10. Where is a bargeboard fitted?
- **a** Eaves
- **b** Verge
- **c** Purlin
- **d** Ridge

Functional skills

In answering the Check it out and Check your knowledge questions, you will be practising **FE 2.2.1** Select and use different types of texts to obtain relevant information, **FE 2.2.2** Read and summarise succinctly information/ideas from different sources and **FE 2.2.3** Identify the purposes of texts and comment on how effectively meaning is conveyed.

You will also cover **FM 2.3.1** Interpret and communicate solutions to multistage practical problems.

Unit 3009

Know how to maintain non-structural and structural components

In time all components within a building will deteriorate. Timber will rot, brickwork will crumble and even the ironmongery will fail through wear and tear. These components need to be maintained or repaired to prevent the building falling into a state of disrepair. General building maintenance is now a recognised qualification, but traditionally this work is done by carpenters, as carpenters have the best links with other traders and the best understanding of what needs to be done.

The maintenance of a building can range from periodically re-painting the woodwork to replacing broken windows.

This unit contains material that supports the following NVQ unit:

- VR 25 Maintain non-structural and structural components

This unit will cover the following learning outcomes:

- Know how to maintain joist coverings
- Know how to maintain doors, windows and ironmongery
- Know how to maintain structural timbers
- Know how to replace broken glass
- Know how to maintain surface finishes

K1. Know how to maintain joist coverings

There are a range of joist coverings that need to be repaired across a building, both in the floor and in the roof. We have looked at some of the principles behind these structures in Unit 3003, pages 89–91.

We also looked at the types of covering for flat roofs in Unit 3008, pages 218–19 and types of fixings in Unit 2008, pages 156–58.

Functional skills

While reading and understanding the text in this unit, you will be practising **FE 2.2.1** Select and use different types of texts to obtain relevant information, **FE 2.2.2** Read and summarise succinctly information/ideas from different sources and **FE 2.2.3** Identify the purposes of texts and comment on how effectively meaning is conveyed.

Identifying and removing damaged floor coverings

There are a wide range of floor coverings you may encounter while working as a carpenter and joiner. These include:

- tongued and grooved boarding
- tongued and grooved chipboard
- plywood sheets.

It can be difficult to identify the defects in these floor coverings as they are often hidden away, for example by a carpet or a rug, etc. However, as floor coverings are a vital part of the structure and support a great deal of weight, it is vital that any defects be identified and repaired as soon as possible.

Remember

Flooring is intended to last a long time and stand up to constant use.

The main defects that can occur which would lead to replacement are problems connected with rot and insect attack. Identifying a damaged floorboard can be done in the same way that fungal attacks are identified, by a musty damp smell or by flight holes. When a floorboard that has been damaged in this way is identified, the first thing to do is to check the base of the skirting boards. If these show signs of fungal attack, then there is a good chance that the floor below will also be infected.

Did you know?

Any flooring that has been in place for a very long time may also need replacing due to wear and tear in areas of heavy traffic, but this is very rare.

It is also a good idea to occasionally check areas of high water use, such as the kitchen or bathrooms, as these are the areas that could most easily be affected.

Removing damaged floor boarding

The removal of floor boarding is a difficult job. This is because most boards, whether they are made from solid timber or sheet material, tend to be tongued and grooved. This means that they interlock and so are very difficult to lift.

Safety tip

Before beginning work on floorboards, you should always check inspection records to see if there are any services that could be damaged during your work. If there are any, you must take the correct precautions to ensure your safety.

You will also need to remember that there are a number of services, such as water, electricity and possibly even gas, which have supply cables and pipes under the floorboards. Unless care is used when removing floorboards, it is very easy to damage these cables/pipes and cause a serious accident, which could be fatal.

In order to remove the boards, you will need to cut them out. This can be done by using a flooring saw. A flooring saw is specially designed to cut through floorboards before drilling through first, and it causes less damage to neighbouring boards. There are also teeth on an angled front edge, which allows cutting into skirting boards.

Figure 9.01 Damaged flooring and skirting board

Once the damaged boards have been removed and disposed of correctly, the cause of the damage must be identified. Simply replacing the boards without fixing the cause will lead to them being infected and needing replacement again. The causes can range from small leaks in pipe work to blocked ventilation, or even carelessness with surface water being not mopped up. The joists should also be checked at this stage to ensure that they are not damaged. It is important that any replacement flooring should be fit for purpose and match the existing floorboards.

Figure 9.02 Flooring saw

Identifying and replacing damaged roof coverings

The covering of a roof might comprise simple felt and tiles/slates, or the roofing could be clad in some form of synthetic board such as OSB (oriented strand board), which would be clad on the outside with felt and then tiled.

Damaged roof felt can cause major problems and can easily be identified in the roof space as damaged or torn. Even if the roof is clad with boarding, the damage can be seen from the inside because if the felt has been damaged then the boarding will also show signs of damage (see Figure 9.03 on the next page).

Other defects which can affect a roof include broken or missing tiles (see Figure 9.04). These can be a result of severe weather conditions.

Did you know?

You can also use a circular saw to remove flooring boards, with the blade set to exactly the same thickness as the boarding.

Figure 9.03 Damaged roof felt

Figure 9.04 Broken or missing tiles

Figure 9.05 Stripping tiles from damaged area of roof

Replacing roof coverings

Once the damaged area has been identified, the tiles will need to be stripped from the damaged area of the roof. You have to remove the tile battens and then the damaged felt and roof boarding.

Replacement boards or felt should then be applied. When replacing them, ensure that you start at the bottom. This is because the felt must be overlapped at the top to prevent water penetration.

New tile battens then need to be fixed. Finally, the tiles or slates should be replaced with new ones, making sure that the pattern matches that of the other original tiles/slates.

Figure 9.06 Applying new felt to roof

Figure 9.07 Replacing roof slates

K2. Know how to maintain doors, windows and ironmongery

You will need to be familiar with the different types of damage that can affect doors, windows and ironmongery in order to maintain them correctly.

Types of fungal attack

There are many forms of fungi which, under suitable conditions, will attack timber until it is destroyed. They mainly fall into two groups:

- dry rot
- wet rot.

This is known as fungal attack. Both dry and wet rot can be a serious hazard to constructional timbers.

Dry rot

This is the most common and most serious of wood-destroying fungi. The fungus does most damage to softwoods but also attacks hardwoods, particularly when they are close to softwoods already infected. If left undetected or untreated, it can destroy much of the timber in a building.

Dry rot is so-called because of the dry, crumbly appearance of the infected timber. It is, however, excessive moisture that is the main cause of the decay. If the wood is kept dry and well ventilated there should be little chance of dry rot occurring.

The main conditions for an attack of dry rot are:

- damp timber, with a moisture content above 20 per cent (which exceeds the **dry rot safety line**)
- poor, or no, ventilation (i.e. no circulation of fresh air).

The attack of dry rot takes place in three stages, which are shown in Figures 9.08–9.10:

1. The spores (seeds) of the fungus germinate and send out hyphae (roots) which bore into the timber.
2. The hyphae branch out and spread through the timber. A fruiting body now starts to grow.
3. The fruiting body, which resembles a large, fleshy pancake, starts to ripen. When fully ripened it discharges millions of red spores into the air. The spores attach to fresh timber and the cycle starts again.

Did you know?

Both hardwood and softwood timbers are equally affected by fungal attack, although some timbers such as western red cedar have naturally occurring chemicals that make them more resistant to attack.

Key term

Dry rot safety line – when the moisture content of damp timber reaches 20 per cent. Dry rot is likely if the moisture content exceeds this.

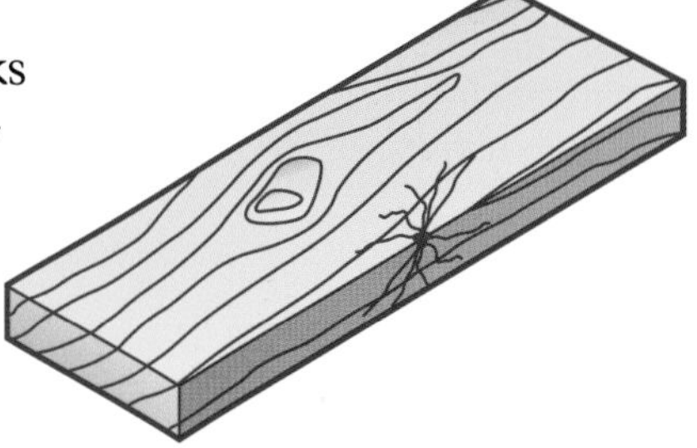

Figure 9.08 Dry rot stage one

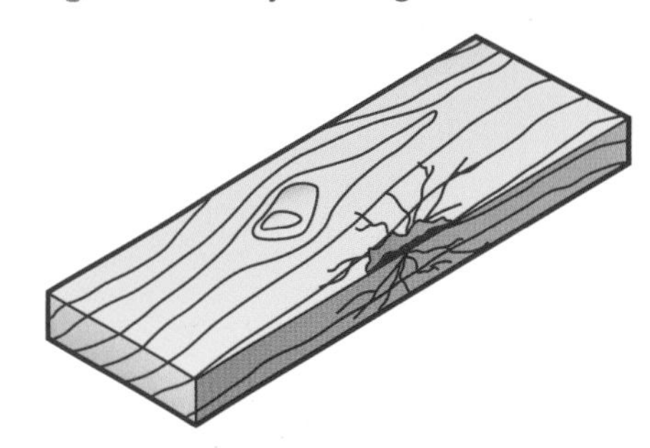

Figure 9.09 Dry rot stage two

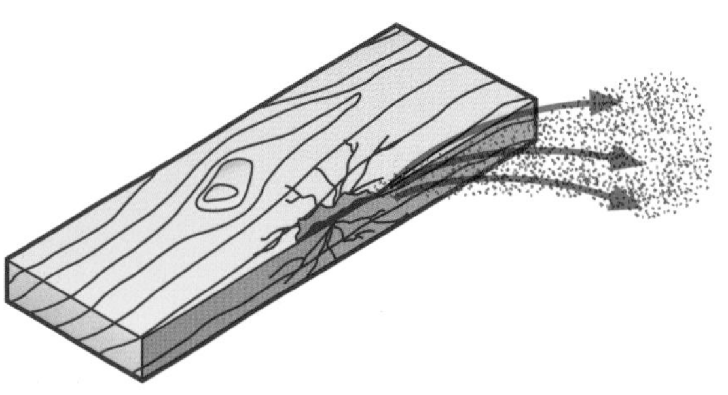

Figure 9.10 Dry rot stage three

Identification of dry rot

Dry rot can be identified by:

- an unpleasant, musty smell
- visible distortion of infected timber; warped, sunken and/or shrinkage cracks
- probing to test the timber for softening or crumbling
- the appearance of fruiting bodies
- the presence of fine, orange-red dust on the floorboards and other parts of the structure
- the presence of whitish-grey strands on the surface of the timber.

Did you know?

Fungal fruiting bodies are the sign of advanced decay in timber.

Eradication of dry rot

Dry rot is eradicated by carrying out the following actions:

1. Eliminate all possible sources of dampness, such as blocked airbricks, bridged damp proof course, leaking pipes, etc.
2. Determine the extent of the attack.
3. Remove all infected timber.
4. Clean and treat surrounding walls, floors, etc. with a suitable fungicide.
5. Treat any remaining timber with a preservative.
6. Replace rotted timber with new treated timber.
7. Monitor completed work for signs of further attack.

Wet rot

This is a general name given to another type of wood-destroying fungus. The conditions where wet rot is found are usually wet rather than damp. Although wet rot is capable of destroying timber it is not as serious a problem as dry rot and, if the source of wetness is found, the wet rot can be halted.

The most likely places to find wet rot are:

- badly maintained external joinery, where water has penetrated
- ends of rafters and floor joists
- fences and gate posts
- under leaking sinks or baths, etc.

Identification of wet rot

The signs for identifying wet rot include:

- timber becoming darker in colour, with cracks along the grain
- decay occuring internally, leaving a thin layer of relatively sound timber on the outside

- localised areas of decay close to wetness
- a musty, damp smell.

Eradication of wet rot

As wet rot is not as serious a problem as dry rot, less extreme measures are normally involved. It is usually sufficient to remove the rotted timber, treat the remaining timber with a fungicide and replace any rotted timber with treated timber. Lastly, if possible, the source of any wetness should be rectified.

Insect infestations

Many insects attack or eat wood, causing structural damage. There are four main types of insects found in the UK, all classed as beetles.

Remember

Timbers affected by dry rot, wet rot or insect attack should be cut back at least 600 mm.

Common furniture beetle

This wood-boring insect can damage both softwoods and hardwoods. The larvae of the beetle bore through the wood, digesting the cellulose. After about three years, the beetle forms a pupal chamber near the surface, where it changes into an adult beetle. In the summer, the beetle bites its way out to the surface, forming the characteristic round flight hole, measuring about 1.5 mm in diameter. After mating, the females lay their eggs (up to 80) in cracks, crevices or old flight holes. The eggs hatch and a new generation begins a fresh lifecycle.

Death watch beetle

This wood-boring insect is related to the common furniture beetle, but is much larger, with a flight hole of about 3 mm in diameter, usually found in decaying oak. The female lays up to 200 eggs. While generally attacking hardwoods only, this wood-boring insect has been known to feed on decaying softwood timbers.

Did you know?

The death watch beetle is well known for making a tapping sound, caused by the head of the male during the mating season.

Powder post beetle

This beetle gets its name from the way it can reduce timber to a fine powder. Powder post beetles generally attack timber with a high moisture content. As with all other beetles, the female lays eggs and the larvae do the damage.

House longhorn beetle

This wood-destroying insect attacks seasoned softwoods, laying its eggs in the cracks and crevices of wood. In Great Britain, this insect is found mainly in Surrey and Hampshire.

Find out

Using the Internet and other sources, find out about the lifecycle of one of the wood-boring insects in this section, and draw a sketch or diagram illustrating it.

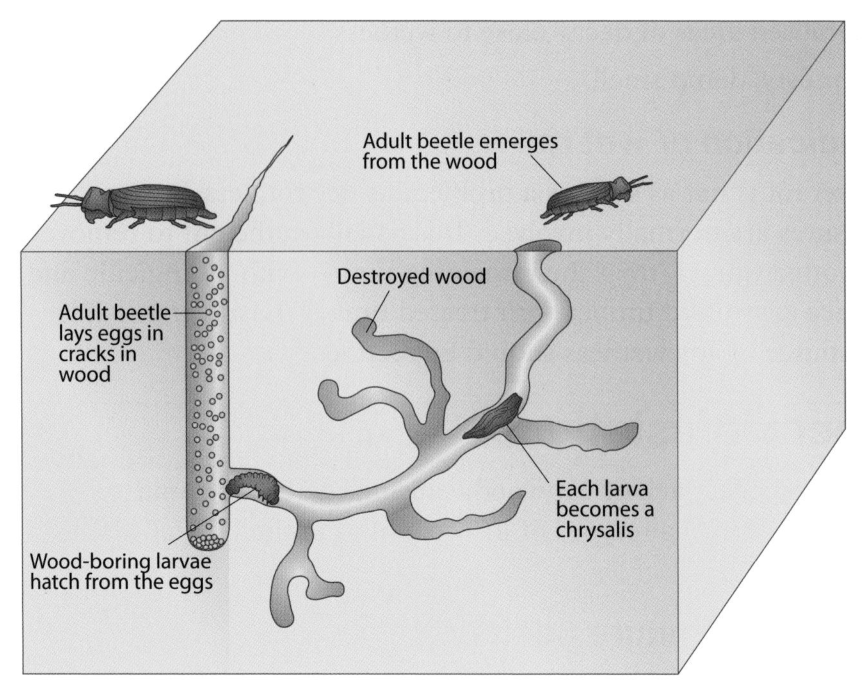

Figure 9.11 Insects can cause significant damage

As well as beetles, weevils and wood wasps can also cause a problem.

Wood-boring weevil

This is a wood-boring insect similar in appearance and size to the common furniture beetle. It differs in that it will only attack timber that is already decayed by wood-rotting fungi. There are over 50,000 species of weevil, all of which have long snouts. The wood-boring weevil is prolific, and is known to have up to two complete lifecycles in one year. Its presence may therefore be accompanied by serious structural collapse of timber.

Wood wasp

Remember

The good news is that wood wasps only lay their eggs in weakened forest trees, and will not re-infest buildings or household structures.

This insect is also known as a horntail. Wood wasps are wasp-like insects and may be seen or heard buzzing and causing cosmetic damage around the home. The body is 2–5 cm long, and the wasp is multi-coloured, blue or black, with wings of matching colour. Walls, wood floors, doors and other wooden surfaces may have holes that are approximately 3 mm in diameter. These holes may also appear on items covering wooden structures, such as carpeting, wallpaper and linoleum. Wood wasps lay their eggs in forest trees that are either damaged or weak. It can take up to five years for adult wood wasps to emerge, so they can be found in trees that have been felled for construction purposes – often even after the wood has been stored. However, most wood wasp holes will appear within the first two years after use of the cut wood.

Eradication of insects

If the insect attack is confined to a small area, the infected timbers can be cut out and replaced. If the attack is over a large area, it is better to pass the work to a specialist firm.

Working life

You have been tasked with re-insulating an attic space. When you remove the old insulation, you notice that the joists show signs of insect infestation. You inform the client, who thinks you are just trying to get extra work.

How can you prove to the client that the infestation is real? What needs to be done to correct the infestation? You will need to be able to identify exactly what type of insect has infested the wood to be certain you can correctly repair the damage. What could the consequences be if the client ignores your advice?

Remember

Contact a professional for advice where load-bearing timbers have been attacked, as there may be a need to prop or shore-up to prevent collapse of the structure.

Methods of applying preservatives to timber

Timber needs to be protected from fungal and insect attack and in addition to this, where timber is exposed, it needs protecting from the weather. Preservation extends the life of timber and greatly reduces the cost of maintenance.

Types of preservative

Timber preservatives are divided into three groups:

- tar oils
- water-borne preservatives
- organic solvents.

Tar oils

Tar oils are derived from coal tar, are very effective and relatively cheap. However, they give off a very strong odour, which may contaminate other materials.

Water-borne preservatives

Water-borne preservatives are mainly solutions of copper, zinc, mercury or chrome. Water is used to carry the chemical into the timber and then allowed to evaporate, leaving the chemical in the timber. They are very effective against fungi and insects, and are able to penetrate into the timber. They are also easily painted over and relatively inexpensive.

Organic solvents

Organic solvents are the most effective, but also the most costly of the preservatives. They have excellent penetrating qualities and dry out rapidly. Many of this type are proprietary brands such as those manufactured by Cuprinol™.

Safety tip

All preservatives are toxic and care should be taken at all times. Protective clothing should always be worn when using preservatives.

Methods of application

Preservatives can be applied in two ways:

- **non-pressure methods** – brushing, spraying, dipping or steeping
- **pressure methods** – empty cell, full cell or double vacuum.

Non-pressure methods

Although satisfactory results can be achieved, there are disadvantages with using non-pressure methods. The depth of penetration is uneven and, with certain timbers, impregnation is insufficient to prevent leaking out (leaching). Table 9.01 lists non-pressure methods, and how and where to employ them.

Table 9.01 Non-pressure methods

Method	How and where to employ it
Brushing	This is the most commonly used method of applying preservative; it is important to apply the preservative liberally and allow it to soak in.
Spraying	Usually used where brushing is difficult to carry out, in areas such as roof spaces.
Dipping	Timbers are submerged in a bath of preservative for up to 15 minutes.
Steeping	Similar to dipping but the timber is left submerged for up to two weeks.

Pressure methods

Pressure methods generally give better results, with deeper penetration and less leaching. Table 9.02 lists pressure methods, and how and where to employ them.

Table 9.02 Pressure methods

Method	How and where to employ it
Empty cell	Preservative is forced into the timber under pressure. When the pressure is released the air within the cells expands and blows out the surplus for re-use. This method is suitable for water-borne and organic solvent preservatives.
Full cell	Similar to empty cell, but prior to impregnation a vacuum is applied to the timber. The preservative is then introduced under pressure to fill the cells completely. Suitable for tar oils and water-borne preservatives.
Double vacuum	A vacuum is applied to remove air from the cells, the preservative is introduced, the vacuum released and pressure applied. The pressure is released and a second vacuum applied to recover surplus preservative. This method is used for organic solvent preservatives.

Disposing of affected timber

All affected timber must be disposed of to ensure that it does not affect other pieces of timber. The affected timber should be moved to a safe location and burnt as soon as it has been removed from the structure.

Maintaining doors and windows

Repairing a door that is binding

One of the most common problems that occur with doors is that they **bind**. A door can bind at several different points, as you can see in Figure 9.12.

> **Key term**
>
> **Bind** – a door 'binds' when it will not close properly and it springs open slightly when it is pushed closed.

These binding problems are simple to fix:

- If the door is binding at the hinges, it is usually caused by either a screw head sticking out too far or by a screw that has been put in squint. In these cases, screw the screw in fully or replace it straight. If the hinge is bent, fit a new one.
- If the door is binding on the hanging stile, it may not have been back bevelled when hung. In this case, take the door off, remove the hinges, plane the door with a back bevel and then re-hang it. Alternatively, it may be expansion or swelling due to changes in temperature that is causing the problem. In this case, take off the door, plane it and re-hang it.
- If the door is binding at the stop on the hanging side, the fitter may not have left a 1–2 mm gap between the stop and the door to allow for paint, in which case the stop will have to be moved. If there is a door frame rather than a door lining, there is no stop to remove: simply plane the rebate using a rebate plane.

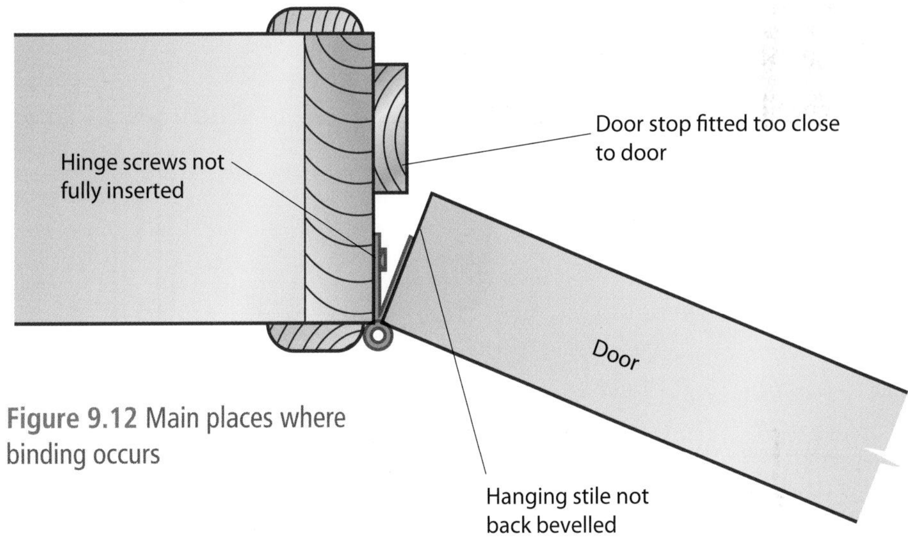

Figure 9.12 Main places where binding occurs

Repairing damaged windowsills and door frames

Provided they are painted or treated from time to time, exterior door frames and windows should last a long time, but certain areas of frames and windows are more susceptible to damage than others. The base of a door frame is where water can be absorbed into the frame; with windows, the sill is the most likely to suffer damage as water can sit on the sill and slowly penetrate it.

Remember

The choice usually comes down to a balance of cost and longevity: as the experienced tradesperson, you must help the client choose the best course of action.

With damaged areas like these, the first thing to consider is whether to repair the damage or replace the component. With door frames, replacement may be the best option, but you must take into account the extra work required to make good: repairing the frame may cost less and require less work. With a rotten windowsill, replacement involves taking out the entire window so that the new sill can be fitted, so repair may be the better option.

In this section we will look at repairing the base of a door frame and a windowsill.

Repairing a door frame

The base of a door frame is most susceptible to rot as the end grain of the timber acts like a sponge, drawing water up into the timber (this is why you should always treat cut ends before fixing).

First, check that the timber is rotten. Push a blunt instrument like a screwdriver into the timber: if it pushes into the timber, the rot is evident; if it does not, the frame is all right. Rot is also indicated by the paint or finish flaking off, or a musty smell.

If there is rot, repair it by using a splice, as follows.

Step-by-step: Repairing a door frame using a splice

1

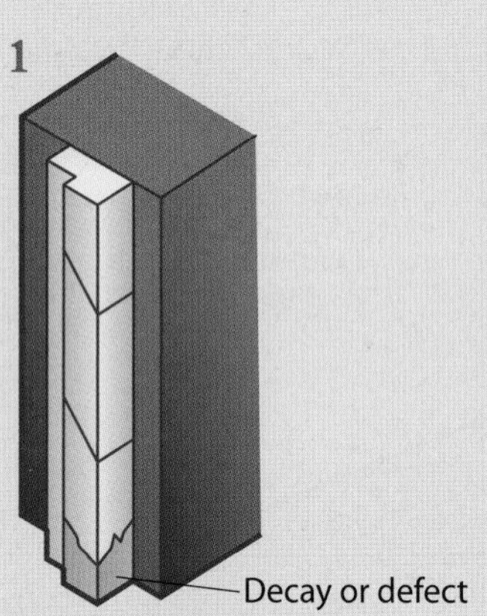

Make cuts in the frame

Step 1 Make 45° cuts in the frame, with the cuts clopping down outwards, to stop surface water running into the joints.

2

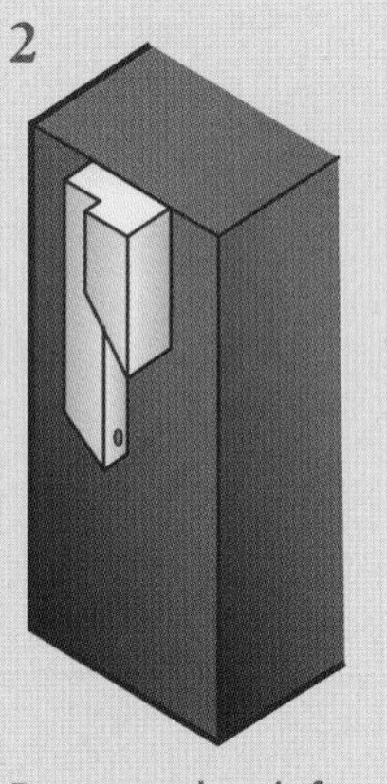

Remove the defect

Step 2 Remove the defective piece and use a sharp chisel to chop away the waste, forming the scarf.

3

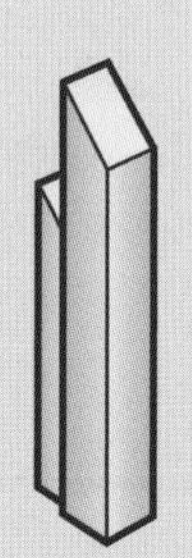

Scarfed splice

Step 3 Use either a piece of similar stock or, if none is available, a square piece planed to the same size and shape as the original, making sure the ends are thoroughly treated before fitting to prevent a re-occurrence.

4

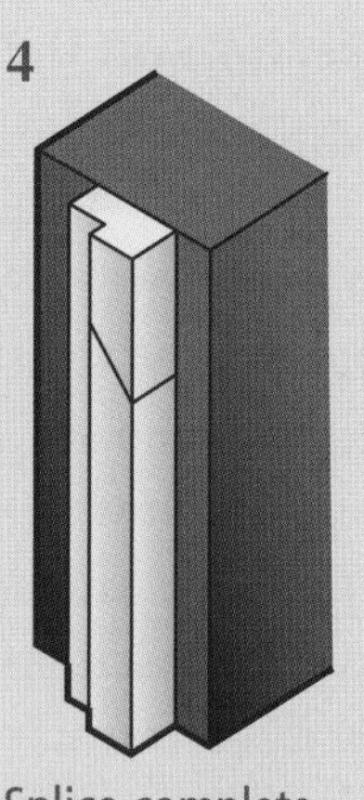

Splice complete

Step 4 Fix the splice in place, dress the joint using sharp planes, and make good any plaster or render.

This method can also be used to repair door frames damaged by having large furniture moved through them, but the higher up the damage on the frame, the more likely the frame will need to be replaced rather than repaired.

Repairing a windowsill

As with a door frame, this method uses a splice, as follows.

Step-by-step: Repairing a windowsill using a splice

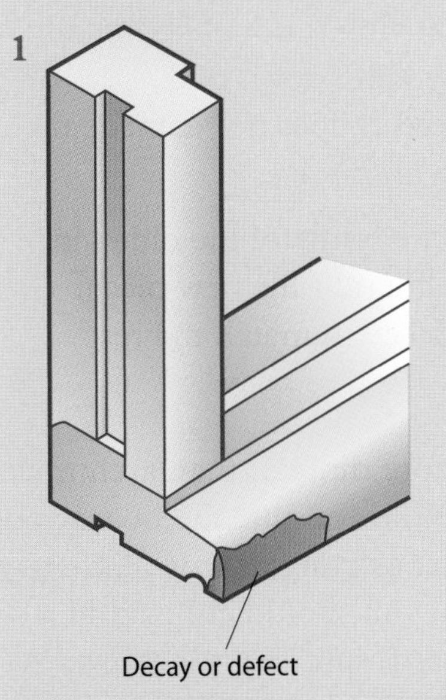

Remove the defect

Step 1 Make a cut at 45°, then carefully saw along the front of the windowsill to remove the rotten area.

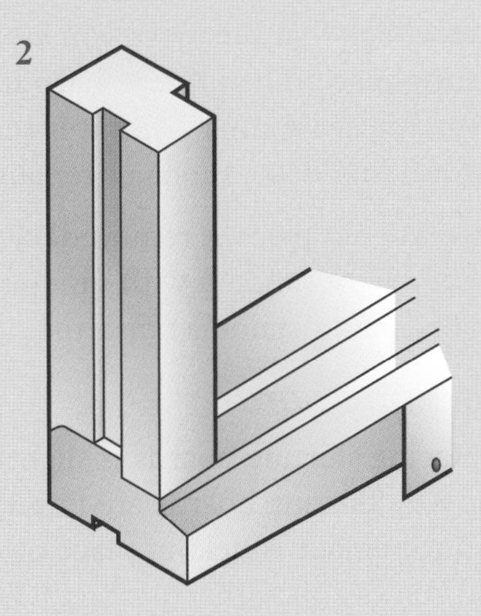

Clean the area

Step 2 Clean the removed area using a sharp chisel and then, as before, use either a piece of similar stock or, if none is available, a square piece planed to the same size and shape as the removed area.

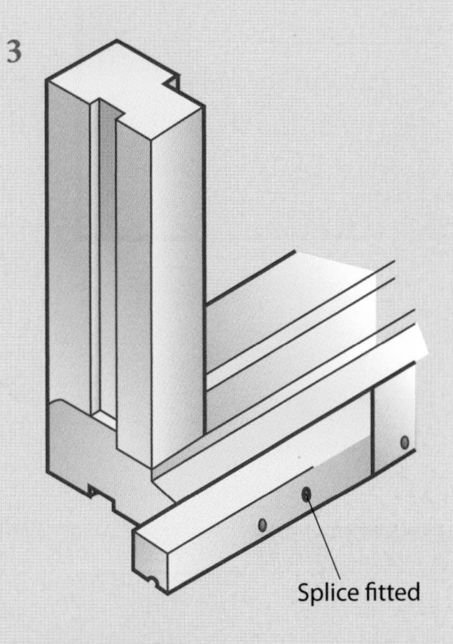

Attach the splice

Step 3 Give all cut areas a thorough treating with a preservative; then attach the splice to match the existing windowsill shape and paint or finish to match the rest.

Working life

You have been asked to give a quote for repairing the base of a door frame. The client asks you what is best to do: repair or replace it?

What advice would you give? What factors do you need to take into consideration? You will need to think about the work you will need to do and what the long-term effects would be for the client. What do you think the client will be most concerned about?

Functional skills

This exercise will allow you to practise the interpreting elements of functional skills, such as **FM 2.3.1** Interpret and communicate solutions to multistage practical problems.

Repairing and replacing mouldings

Mouldings such as skirting or architrave rarely need repairing – usually only because of damp or through damage when moving furniture – and in most cases it is easier to replace them rather than repair them.

Architrave

Remember

When replacing mouldings, you may need to rub down the existing moulding work and re-apply a finish to both the old and the new to get a good match.

To replace architrave, you simply remove the damaged piece and fit a new piece.

First, check that the piece you are removing is not nailed to an existing piece. Then run a sharp utility knife down both sides of the architrave so that when it is removed it does not damage the surrounding decorations or remaining architrave.

Once the old piece is removed, clean the frame of the old paint, give it a light sanding with sandpaper, then fit the new piece, Finally, paint, stain or finish the new piece to match the rest.

Skirting board

Safety tip

Take care if gripper rods are fitted in front of the skirting, as this will make replacing the skirting more difficult and more dangerous.

Replacing skirting boards is slightly more difficult than architrave. The way skirting boards are fitted could mean that the board you wish to replace is held in place by other skirting boards. Rather than remove the other skirting boards, you should cut or drill a series of holes in the middle of the board, splitting it in two so that you can remove it that way. Again, running a utility knife along the top of the skirting board will avoid unnecessary damage to surrounding decorations. Once the old skirting has been removed, the new piece can be fitted and finished to match the existing skirting.

You can replace other mouldings, such as picture and dado rails, in the same way as skirting boards, taking care not to damage the existing decorations.

Replacing window sash cords

As you saw in Unit 2008, box sash windows use weights attached to cords that run over a pulley system, to hold the sashes open and closed. The sash cord will eventually break through wear and tear, and will need to be replaced. This is not a large job so, if one cord needs attention, it is most cost-effective to replace them all at the same time.

There are various ways to replace sash cords. The method outlined on the next three pages is quick and easy, and is done from the inside.

Step-by-step: Replacing sash window cords

Step 1 Remove the staff bead, taking care not to damage the rest of the window. Cut the cords supporting the bottom sash, lowering the weight gently to avoid damaging the case. Take out the bottom sash, removing the old nails and bits of cord, and put to one side.

Step 2 Pull the top sash down and cut the cords carefully. Remove the parting bead and then the top sash, again removing the old nails and bits of cord, and put to one side.

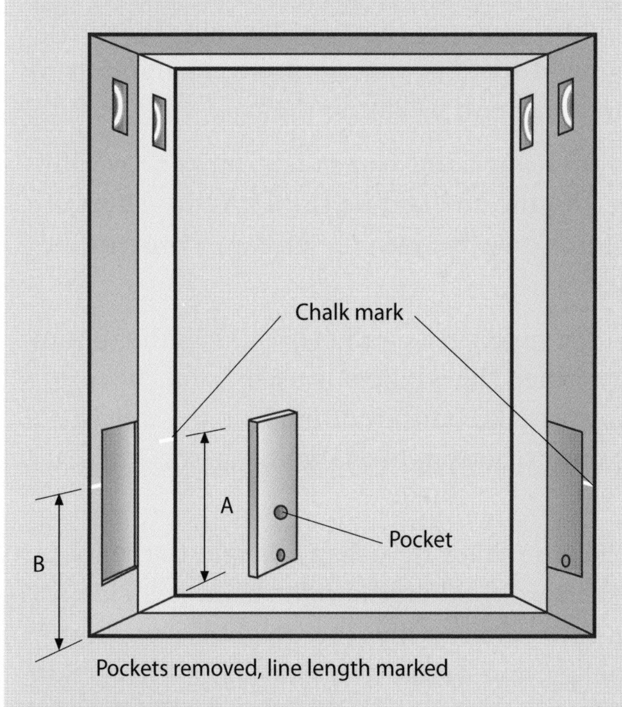

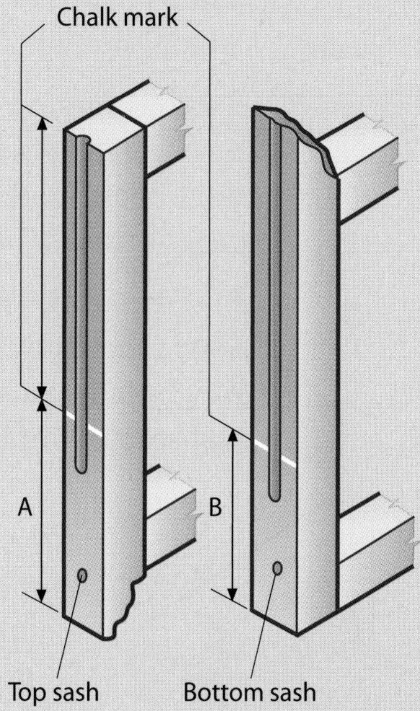

Figure 9.13 Marking sash lengths

Step 3 Remove the pockets and take out the weights, removing the cord from them and laying them at the appropriate side of the window.

Step 4 Slide the top sash into place at the bottom of the frame and, using chalk, mark the position of the sash cord onto the face of the pulley stiles (see distance A in Figure 9.13). Mark the bottom sash in the same way (distance B). Put the sashes to one side.

Step 5 Next, attach a **mouse** to the cord end, making sure it is long enough for the weighted end to reach the pocket before the sash cord reaches the pulley.

Step 6 Now cord the window. If only one cord is to be replaced, you can do this by feeding one pulley, then cutting

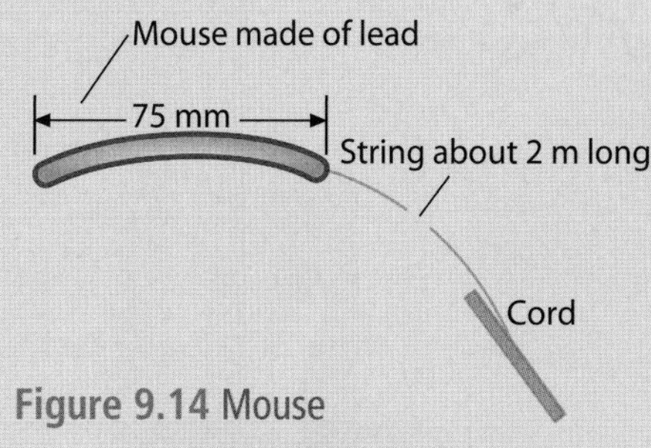

Figure 9.14 Mouse

Key term

Mouse – a piece of strong, thin wire or rope with a weight such as a piece of lead attached to one end; used to feed the sash cord over the pulleys in a box sash window.

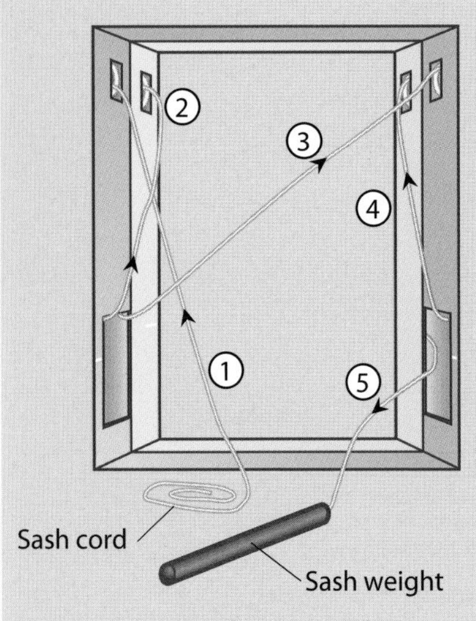

Figure 9.15 Cording a window

the sash to length, re-attaching the end of the mouse to the sash cord and then feeding the next pulley. If you are replacing more than one cord, it is more efficient to feed the pulleys in succession and cut the cords after. In Figure 9.15 you can see one method for cording a window.

Feed the cord through the top left nearside pulley (1), out through the left-hand pocket, in through the top left far side pulley (2), then back through the same pocket. Then feed it through the top right nearside pulley (3), out through the right-hand pocket, in through the top right far side pulley (4) and back out of the right-hand pocket. Remove the mouse and attach the right-hand rear sash weight to the cord end (5).

Step 7 Working on the right-hand rear pulley, pull one weight through the pocket into the box and up until it is just short of the pulley. Lightly force a wedge into the pulley to prevent the weight from falling.

Pull the cord down the pulley stile and cut it to length, 50 mm above the chalk line (when the window is closed and the weight falls back down to the bottom of the box, the weight will stop 50 mm from the bottom of the frame as shown in Figure 9.16).

Step 8 Tie the other right-hand weight to the loose end of the cord. Still on the right-hand nearside, pull the cord again so that the weight is just short of the pulley, and wedge it in place. Cut the cord to length, this time 30 mm longer than the chalk line (when the window is closed the bottom sash weight will be 30 mm away from the pulley, as shown in Figure 9.17).

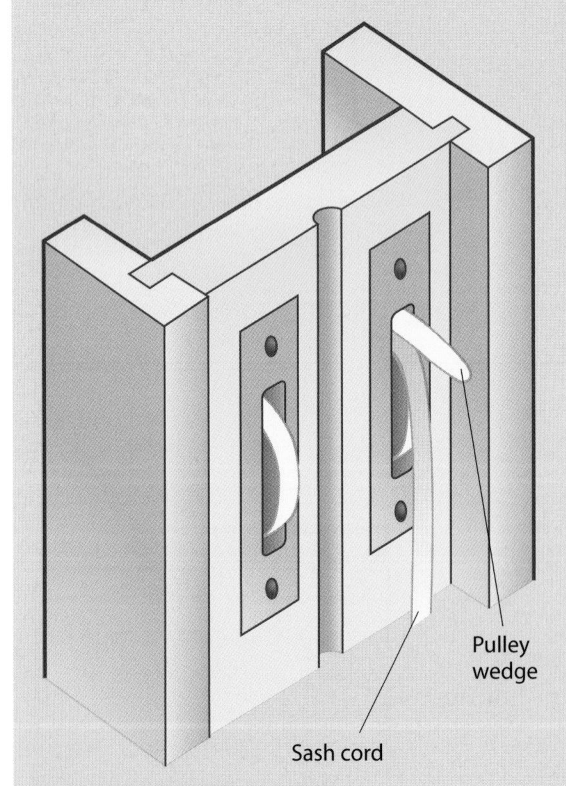

Figure 9.16 Wedged pulley

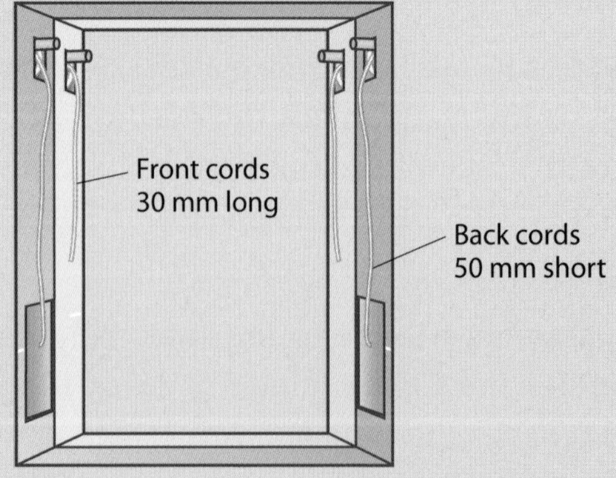

Figure 9.17 All pulleys wedged and cords cut to length

Step 9 Fix the right-hand pocket back in place, then repeat Steps 7 and 8 for the left-hand side. Now all the pulleys are wedged and all the cords cut to length.

Step 10 Next fit the sashes, starting with the top sash. Fix the cord to the sash by either using a knot with a tack driven through it, or a series of tacks driven through the cord. Take care not to hamper the opening of the window, and do not use long nails as they will drive through the sash stile and damage the glass.

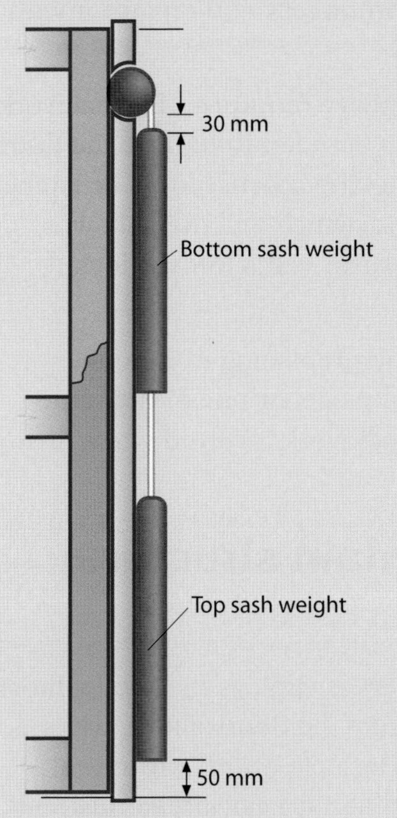

Figure 9.18 Sash weights at 50 mm and 30 mm

Step 11 Once the cord is fitted to both sides of the sash, slide the sash into place and carefully remove the wedges. Now test the top sash for movement and then re-fit the parting bead to keep the top sash in place. Now fit the bottom sash as in Step 10, test it and re-fit the staff beads, to secure the bottom sash in place.

Now test the whole window by sliding both sashes up and down. Finally, touch up any minor damage to the staff and parting beads.

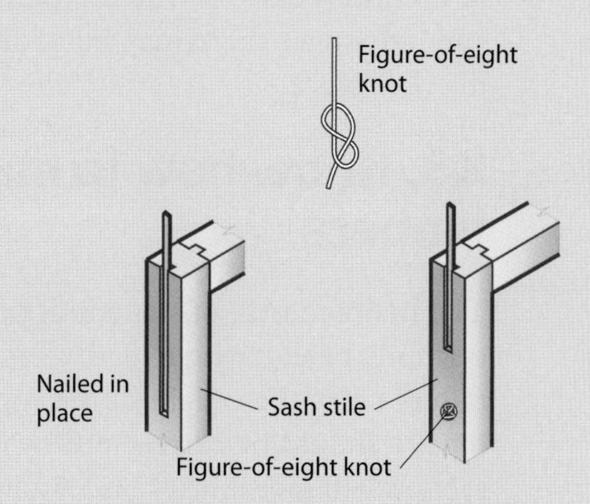

Figure 9.19 Two methods for fixing cord to sashes

Working life

Alice is working as trainee for a construction company and has been sent to carry out some maintenance work in a house recently constructed by the company. She has been asked to repair a box sash window that has not been opening properly.

What might the cause be? Alice will need to think about the likely parts of the window that could have been damaged and are preventing it from opening. What can be done to rectify it? If there are major problems, what is it best to do: repair or replace?

Methods of replacing ironmongery

As well as making repairs to components such as windows and doors, there will also come a time when the ironmongery such as handles, hinges, locks, latches, etc. may need to be replaced.

Normally when replacing these, the first job is to remove the screws that attach the ironmongery to the component. You will then need to remove the item of ironmongery and replace it with the same type.

However, some examples of ironmongery require additional work. For example, when replacing hinges you need to remove the door or window in the process. When removing mortise locks or latches a handle will need to be removed first, which will then allow you to remove the spindle which, in turn, will allow the lock to be removed.

An important thing to remember when replacing ironmongery is to try and replace it with an identical piece, or one of the same size; otherwise there may be additional work involved when fitting the new piece.

K3. Know how to maintain structural timbers

The maintenance of structural timbers is vital, as they will almost certainly be carrying a load: joists carry the floors above, while rafters carry the weight of the roof. Because of this, structural repairs should be carried out by qualified specialists, so this section will give a brief understanding of the work that is involved.

Joists

Joist ends are susceptible to rot: they are close to the exterior walls and can be affected in areas like bathrooms if the floor gets soaked and does not dry out. Joists can also be attacked by wood-boring insects. For both rot and insect damage, the repair method is the same.

Shore the area, with the weight spread over the props, then lift the floorboards to see what the problem is (for the purposes of this example, we will use dry rot) and throw the old floorboards away.

Before making any repairs, you need to find and fix the cause of the problem – otherwise the same problems will keep on arising. Rot at the ends of joists is usually caused by poor ventilation: the cavity or the airbricks may be blocked, for example.

Safety tip

When repairing or replacing joists, you must first shore up the ceiling to carry the weight while you work. Follow the instructions found in Unit 3003 and do not alter or remove the shoring until the job is complete.

Next, cut away the joists (see Figure 9.20), allowing 600 mm into sound timber, and treat the cut ends with a suitable preservative.

New pre-treated timbers should be laid and bolted onto the existing joists with at least 1 m overlap (see Figure 9.21). New timbers should ideally be placed either side of the existing joist, and in place of the removed timber.

Any new untreated timbers should be treated with a suitable preservative, before or after laying. Now slowly remove the props and make good as necessary.

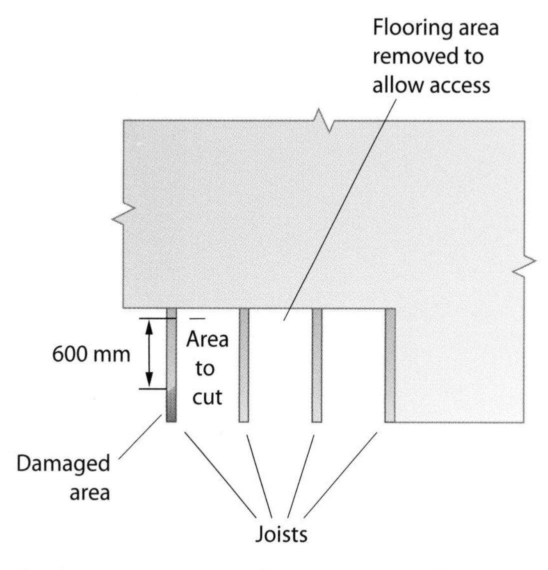

Figure 9.20 Cut away the rotten area of the joist

Rafters

Rafters are susceptible to the same problems as joists, and especially to rot caused by a lack of ventilation in the roof space. Again, the problem needs to be remedied before any repairs start.

Rafters are replaced or repaired in the same way as joists. They are easier to access, but shoring rafters can be difficult. In some cases it is best to strip the tiles and felt from the affected portion of the roof before starting.

For trussed roofs you should contact the manufacturer: trusses are stress graded to carry a certain weight, so attempting to modify them without expert advice could be disastrous.

Repairs to other roofing components, such as purlins and ridge boards, will follow a similar pattern, with the roof having to be supported for repairs to purlins and stripped for ridge repairs.

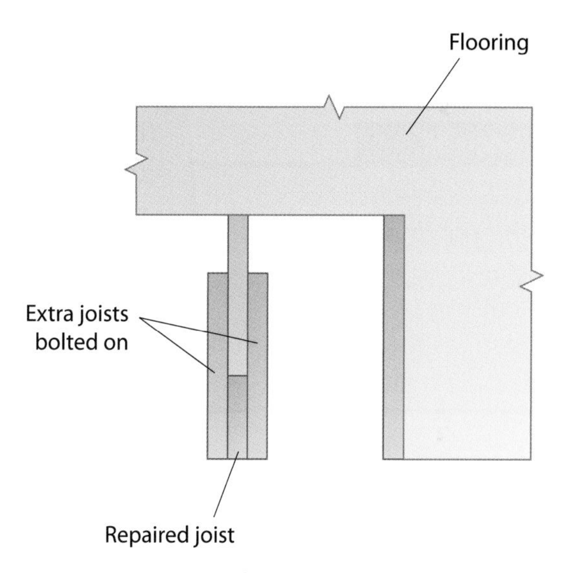

Figure 9.21 Repaired joist

Working life

You have been asked to look at joists that appear to be damaged by rot. The client wants to know what has caused it.

What do you think has caused it? How could you confirm what has caused the problem? What needs to be done to rectify it?

Remember

The same processes will be used for both hip and valley rafters. You will need to remember, however, that hip rafters are fitted on the corner of a hip roof (the external corner), while valley rafters are fitted on the corner of a valley (the external corner).

K4. Know how to replace broken glass

This is a job done largely by specialist glazing companies, but there is still a call for a carpenter to replace broken glazing.

This section will look at replacing both single and double-glazed windowpanes.

Removing and disposing of broken glass

The first thing to do when replacing a pane of glass is to remove the old pane. If the glass is smashed, you must remove the remaining pieces very carefully, wearing suitable gloves and other necessary personal protective equipment (PPE). If the glass is only cracked, it is safer to tape the glass to prevent it shattering, then tap it gently from the inside until it falls out.

Next, you need to clean out the rebate, removing all the old putty so that the new pane of glass can be fitted properly. Use a sharp chisel, taking care not to damage the rebate.

For double glazing, if the unit is just cracked, use tape to stop it shattering. Then remove the timber beads that keep the window in place, taking care not to damage them or the rubber if they are to be used again.

Safety tip

Glass can easily cut through most ordinary gloves, so it is vital that you wear the correct gloves when handling glass.

Now, wearing the correct protective clothing, carefully take out the glass. If the glass was installed with double-sided sticky rubber tape, run a sharp utility knife around the inside to free the glass.

Broken glass causes an obvious cutting hazard, so it is important that it is disposed of correctly. At the site all glass should be placed into a plastic or metal bin and covered until it can be disposed of properly. This can be done either by placing it into a recycling area or ensuring that it is disposed of safely in a skip, with no sharp edges protruding.

Types of glass

When replacing a broken piece of glass, there are several different types of glass available from which to select the replacement piece. You will need to know the proposed use of the glass and the likely conditions it will encounter when making your choice about the type of glass you use to replace the broken piece.

Single glazing

Single glazing is mainly found in older buildings such as factories or country houses. The glass is fixed using putty and held in place with either putty tooled to a smooth finish or timber beading.

Double glazing

Double glazing can be found in either timber or uPVC windows. With uPVC double glazing, there are many different ways to install the glass using different types of beads and rubber components, so it is best to refer to the manufacturer when changing a unit. Here we will concentrate on replacing double-glazed units on timber windows.

Laminated glass

Laminated glass is a type of safety glass that holds together even when it is broken. It is made from two or more pieces of glass sealed together with a thin chemical sheet between the pieces. This keeps the layers of glass bonded even when broken. Its high strength prevents the glass from breaking up into large sharp pieces. When the glass is hit with a force not strong enough to completely pierce the glass, it produces a characteristic 'spider web' cracking pattern.

Laminated glass is normally used when there is a possibility of human impact, or where the glass could fall if shattered. Skylight glazing, car windscreens and shop fronts are typical places where it is used.

Toughened glass

Toughened or tempered glass is glass that has been made with chemical treatments to increase its strength compared with normal glass. It will usually shatter into small fragments instead of sharp shards. This makes it less likely to cause severe injury or deep lacerations.

As a result of its safety and strength, tempered glass is used in a variety of demanding applications, including passenger vehicle windows, glass doors and tables, telephone box windows and various types of plates and cookware.

Methods of securing glass into frames

The next task you will need to carry out, once you have removed a piece of glass, is to fit the new pane of glass. There are several methods for doing this: use the method that matches the existing glazing in the building.

Method 1: Putty only

This is the traditional method, best done with a putty knife.

First, make the putty useable – when it is first taken out of the tub, putty is very sticky and oily, and must be kneaded like dough to make it workable. Then push the putty into the rebate, taking a small amount of putty in your hand and feeding it into the rebate using your forefinger and thumb.

Did you know?

Double glazing consists of two single panes of glass together within a frame, with the air sucked out of the gap between the two panes of glass. The vacuum created helps with both heat and sound insulation.

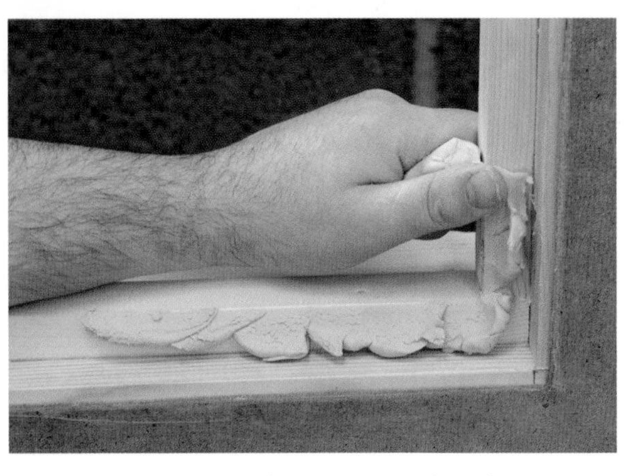

Figure 9.22 Pushing the putty into the rebate

Once the putty is all around the rebate, offer the glass into place starting at the bottom, then apply a little pressure to ensure that the glass is squashed up against the putty. Do not apply too much pressure as this will break the glass.

Now, insert glazing sprigs or panel pins to keep the glass in place until the putty sets. To avoid breaking the glass, place the edge of the hammer against the glass and slide it along the glass when driving in the pins or sprigs. If the hammer stays in contact with the glass, breakages are less likely. Drive the pins or sprigs in far enough that they are not in view, but not so far that they catch the edge of the glass and break it. Small panes need only one pin or sprig; larger panes will need more.

Did you know?

When using a putty knife, it can be useful to dip the knife into water to prevent the putty sticking to the blade and dragging the putty.

Take another ball of putty and feed it around the frame. Take the putty knife and, starting at the top, draw the knife down the window, squeezing the putty and shaping it at the same time. Repeat several times if necessary to get the desired finish. Remove any excess putty, including any on the inside that has been squeezed out from the rebate when the glass was pushed into place.

Figure 9.23 Sprigs being fitted

Figure 9.24 Putty being tooled

Method 2: Putty and timber beading

This is similar to Method 1, following the same steps until the glass has been pushed into place. At this point, instead of the putty being tooled along the outside, a timber bead is used.

First, place putty on the face of the bead that will be pushed against the glass, again using your forefinger and thumb.

Next, fit the beads, which are usually secured using panel pins or small finish nails such as brads or lost heads. The beads should be fitted in sequence: top bead first, then the bottom and finally the sides.

Press the beads into place with enough force to squeeze the putty out, leaving just a small amount, then drive the pins or nails in as in Method 1. It can be difficult to put pressure on the bead and nail it at the same time, so ideally you need two people. Alternatively, you can drive the nail into the bead a little way, but not far enough to penetrate the other side, then push the bead into place with one hand and drive the nail home with the other.

Figure 9.25 Nailing a bead into place

Once all the beads are fitted, remove the excess putty from both the inside and outside of the glass.

Method 3: Silicone and beads

This is identical to Method 2, but uses silicone instead of putty. It is often preferred as it is quicker and not as messy.

Other methods

When replacing double glazing, silicone and putty are rarely used, as a pre-made rubber bead, tape or strip is preferred. Various types of bead, tape and strip are available. Just a few examples are shown in Figure 9.26.

After removing the glass from double-glazed windows, clean the beads and the rebate and, if you are using double-sided sticky rubber tape, apply this to the rebate. Now the new double-glazed unit can be fitted and the beads pinned back into place.

Take care as some manufacturers have special insulated units that require the glass to go in a specific way. Such units carry a sticker that states something like 'this surface to face the outside'.

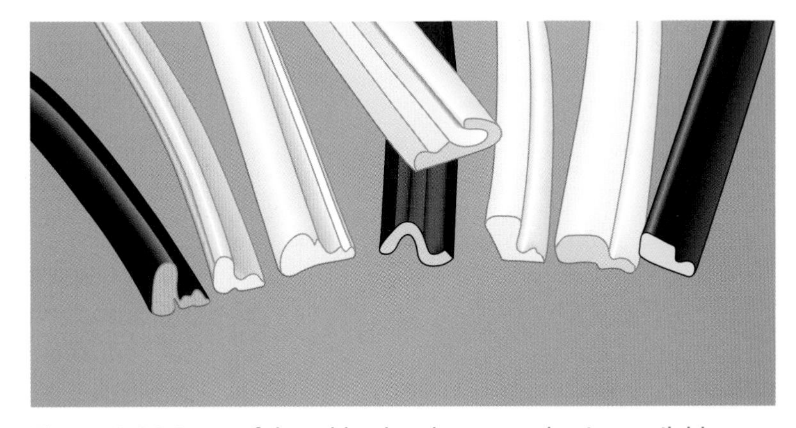

Figure 9.26 Some of the rubber beads, tape and strips available

Remember

With silicone, take care not to put too much on: once the glass or the timber bead is pushed into place, the silicone spreads more than putty but does not clean off glass as easily.

Did you know?

Double-sided sticky rubber tape is a security device to prevent burglars removing the beads and taking the glass out.

Working life

You have been tasked with replacing a double-glazed unit. The original unit has condensation between the two panes of glass. What could have caused this?

You remove the beading, but the unit still won't come out. What needs to be done? What should be done before fitting the new unit?

K5. Know how to maintain surface finishes

Remember

Attempt only minor repairs unless you are fully trained. Otherwise you could end up doing more harm than good.

During repair work there is always a chance that the interior plaster or brickwork may get damaged. Rather than call in a specialist, most tradespeople will repair the damage themselves.

Preparing old surfaces

Repairing plasterwork

Plaster damage most often occurs when windows or door frames are removed – the plaster cracks or comes loose – and is simple to fix with either ready-mixed plaster (ideal for such tasks) or traditional bagged plaster, which is cheaper.

Bagged plaster needs to be mixed with water and stirred until it is the right consistency. Some people prefer a thinner mix, as it can be worked for longer, while others prefer a consistency more like thick custard, as it can be easier to use.

Whichever type of plaster you use, the method of application is the same. The repair style will also not alter greatly for many of the types of surface onto which the plaster is applied. The technique will be similar for brick, block or stone. Lath and plaster may require building up in layers with a few coats of bonding plaster, and plasterboard may require a patched piece of plasterboard to be inserted first.

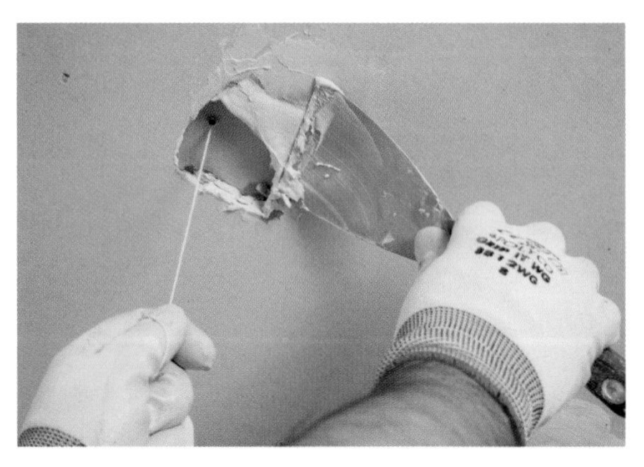

Figure 9.27 Plaster being applied and trowelled

First, brush the affected area with a PVA mix, watered down so it can be applied by brush. This acts as a bonding agent to help the plaster adhere to the wall.

Next, use a trowel or float to force the plaster into the damaged area. If it is quite deep you may have to part-fill the area and leave it to set, then put a finish skim over it. Filling a deep cavity in one go may result in the plaster running, leaving a bad finish.

When the plaster is almost dry, dampen the surface with water and use a wet trowel to skim over the plastered area, leaving a smooth finish. Once the plaster is dry, the area can be re-decorated.

Repairs to exterior render are done in the same way, using cement instead of plaster.

Repairing brickwork

With repairs to brickwork, one of the first problems is finding bricks that match the existing ones, especially in older buildings. After that, the method is as follows.

Step-by-step: Repairing brickwork

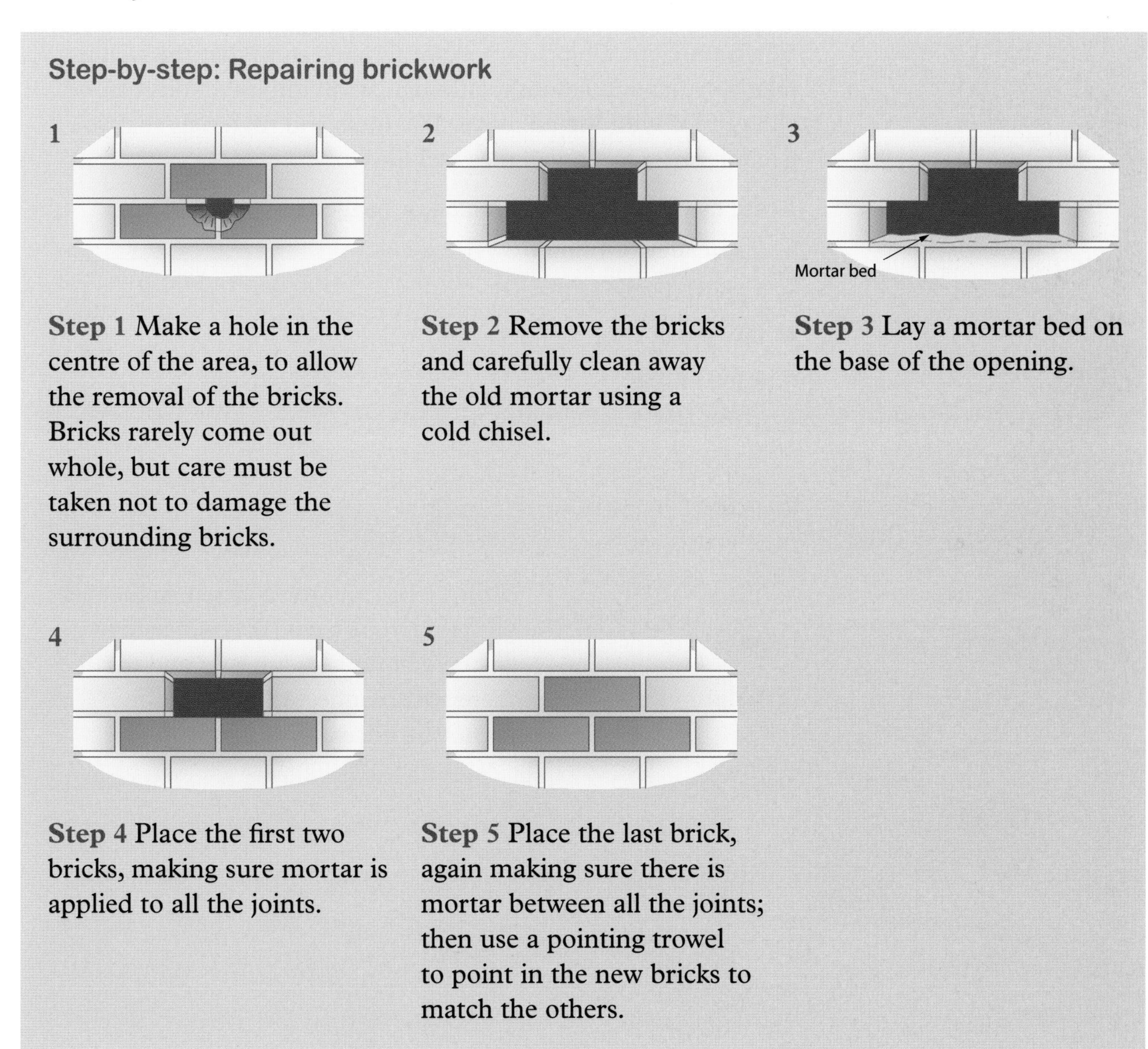

Step 1 Make a hole in the centre of the area, to allow the removal of the bricks. Bricks rarely come out whole, but care must be taken not to damage the surrounding bricks.

Step 2 Remove the bricks and carefully clean away the old mortar using a cold chisel.

Step 3 Lay a mortar bed on the base of the opening.

Step 4 Place the first two bricks, making sure mortar is applied to all the joints.

Step 5 Place the last brick, again making sure there is mortar between all the joints; then use a pointing trowel to point in the new bricks to match the others.

Working life

You have been asked to make plaster repairs around a new window that has been fitted. The plaster keeps slipping and you can't get an even finish.

What can be causing this? You will need to look at the process that might have been used to prepare and plaster, and the surface area. What can be done to rectify this?

Method for laying and replacing tiles

When tiling a wall it is important to make sure that the wall is relatively flat and smooth, although small imperfections can be allowed for with the adhesive. The tools and materials that you need in order to fit wall tiles include:

- adhesive (either ready mixed or bagged)
- a spreader or tiling trowel.
- a spirit level
- tile spacers
- a tile cutter
- grout and a grouting tool.

Step-by-step: Tiling a wall

1

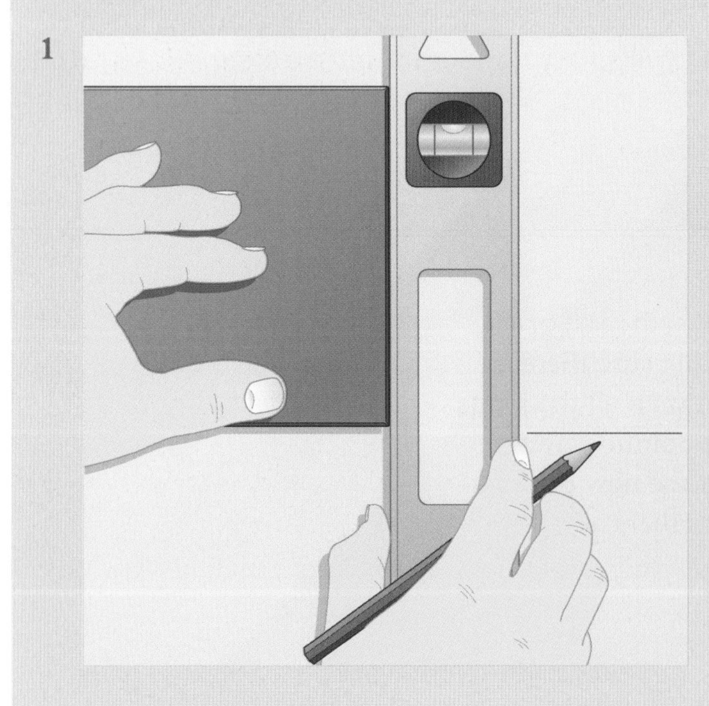

Step 1 Mark out the area where the tiles are going to be fixed. This will help to ensure that the tiles are level and that there are no unnecessary cuts in the corners or edges.

2

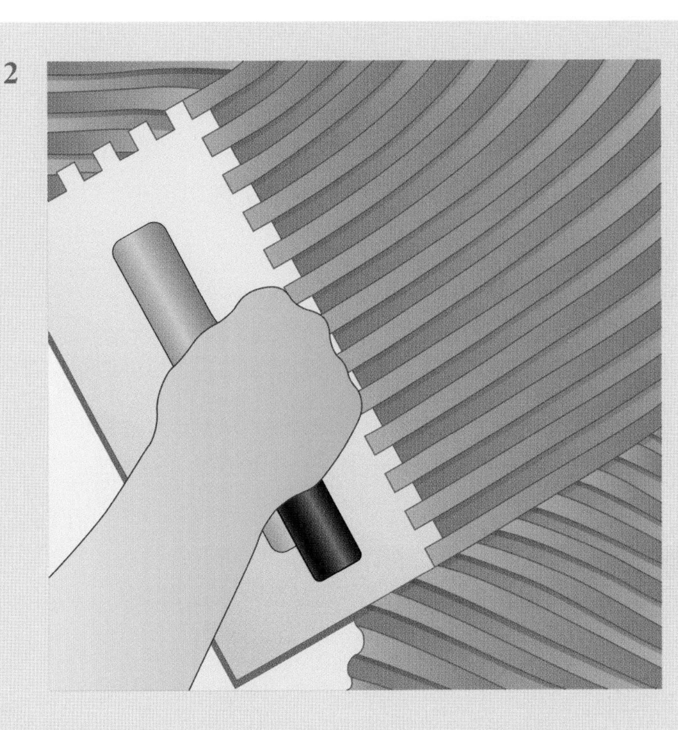

Step 2 Using a tile adhesive, spread an even amount of adhesive over the area to be tiled, ensuring that you don't cover over the lines you used to mark out for level.

3

Step 3 Starting at the bottom, apply the tiles. Remember to place the spacers between the tiles to ensure uniformity and to check that the tiles are running in line. Spacers can be removed once the adhesive is set. An alternative is to insert the spacers flat and leave them in place on the wall.

4

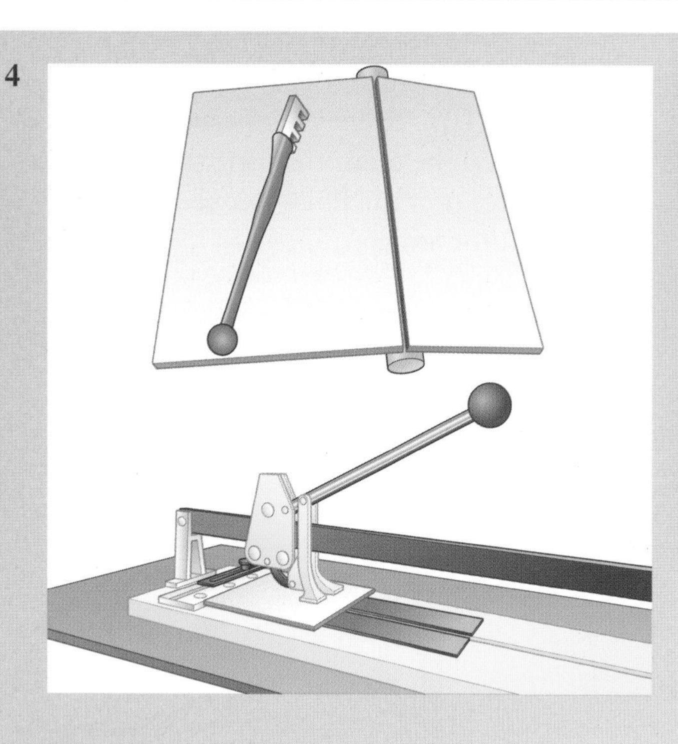

Step 4 You may need to cut down some tiles in order to fit in particular spaces. If cuts are required then mark out the tile and cut using an appropriate cutter.

5

Step 5 Once the whole area is tiled, it must be left until the adhesive has cured. The spacers can then be removed, if necessary, and the area grouted.

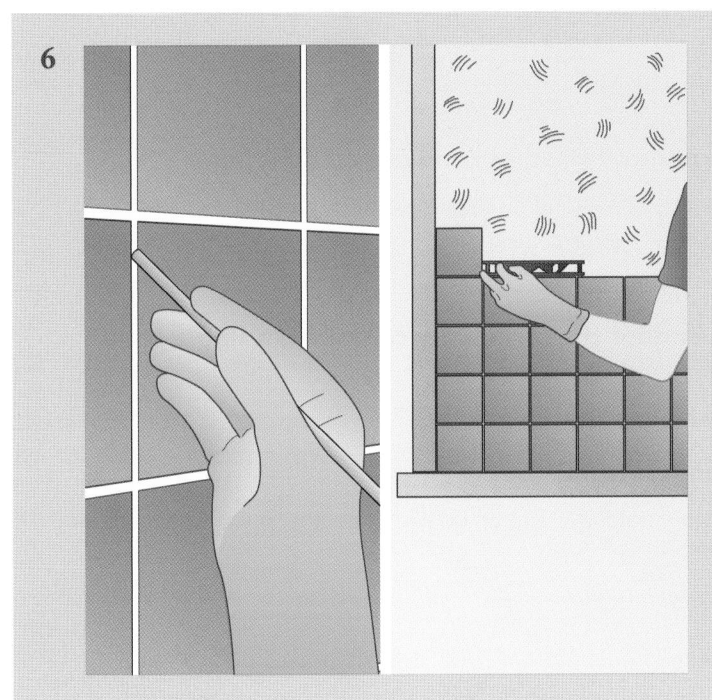

Step 6 The grout should be applied with a spreader and forced into the gaps between the tiles. Excess grout must be removed before it has set, otherwise it can be difficult to remove. The grout finish, similar to pointing in brickwork, can be achieved by using a grouting tool or even the end of a pencil. Areas where the tiles are not sat on the floor or a work surface will have to be supported with some form of temporary support.

Laying floor tiles

The method of tiling for floors is the same as tiling walls, except it is vital that the floor is flat and level. Problems with this can be overcome by applying a self-levelling compound before tile application. This will level the floor up. The other point to remember is that you should tile from one corner towards the door – otherwise you may end up backing yourself into a corner and have to stand on the tiles to get out. Doing this could cause the tiles to crack or, if the adhesive has not set, set them off level.

Removing old tiles

If a tile gets damaged then it will need to be replaced. This can be done by first removing all the grout that surrounds the damaged tile. Next, drill a series of holes within the tile using a masonry drill bit. Remove the damaged tile carefully, ensuring that protection is placed on the floor or work surface to prevent any damage that could be caused by the tiles falling.

Scrape away the old adhesive, apply new adhesive and insert a new tile, using spacers if necessary. Then, once the adhesive is dry, grout the area around the new tile to match the existing tiles on the wall.

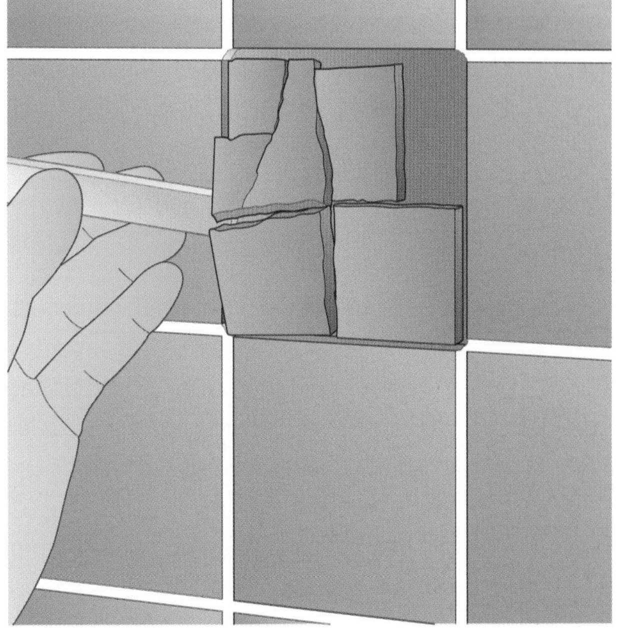

Figure 9.28 Removing a damaged tile

FAQ

Do I have to use putty or silicone to re-fix a pane of glass, or can I use an adhesive?

Like other substances, glass expands and contracts slightly depending on the weather. This means that silicone or putty must be used, as they remain flexible even when set. Adhesive sets rigid so, if any expansion or contraction occurs, the glass will break.

Can I put a double-glazed unit in a window or door that originally contained a single pane of glass?

Yes, as long as the rebate is wide enough to accept the unit and beading.

Should I go on a plastering course to help with repairs?

This is not essential as the repairs you would be doing are not large, but do go if you feel that it would benefit you.

When replacing tiles, are all tiles the same?

No. Like most components each manufacturer has their own designs and specific shading of colours, so it is important to ensure the replacement tiles match. Remember, however, that new tiles will never exactly match existing tiles, as these will have faded due to extremes of weather. If you need to search for a particular type of tile, this can often be done at reclamation yards.

When replacing a single wall tile, do you need to re-grout the entire area?

This depends on how the new grout will match any existing grout.

Check it out

1. Write a method statement describing the process used to identify and remove damaged floor coverings and roof coverings, showing clearly the different techniques that are needed for each of these processes.
2. Explain the methods used to eradicate wet and dry rot.
3. Describe the four types of insect that attack timber, and explain how to eradicate the infestations caused by these insects.
4. Name the four non-pressure methods of applying preservatives, and describe how a preservative is applied by double vacuum.
5. Define the term 'binding' in relation to doors and explain the process used to check if a door frame is rotten.
6. Complete method statements, with accompanying diagrams and sketches for each stage, to describe the two methods used to fix sash cords to sashes.
7. Explain why trussed rafters should only be repaired by a specialist, and describe the process they would use.
8. Describe the three methods which can be used for single glazing and explain what factors would lead you to choose a particular method.
9. What is the first step you need to follow when applying plaster? Explain why this is.
10. Explain the method used for laying tiles, preparing a method statement with accompanying diagrams to illustrate the process you should use.

Getting ready for assessment

The information contained in this unit, as well as continued practical assignments that you will carry out in your college or training centre, will help you with preparing for both your end of unit test and the diploma multiple-choice test. It will also aid you in preparing for the work that is required for the synoptic practical assignments.

The information contained within this unit will aid you in learning how to identify and calculate the materials and equipment required to maintain non-structural and structural components.

You will need to be familiar with:

- maintaining joist coverings
- maintaining doors, windows and ironmongery
- maintaining structural timbers
- replacing broken glass
- maintaining surface finishes.

This unit will have helped you to become familiar with the methods used to maintain different parts of a building. For example, for learning outcome three you have learned about the different methods used for maintaining structural timbers, the materials that can be used to repair these and how they can become damaged. You will need to be able to replace damaged and rotten structural timbers and splice new sections into structural timbers. You will also need to be able to treat new timber and protect the surrounding areas from damage. You will need to use this information on site to quickly see and identify problems with structural carcassing and decide the best method to use in replacing or repairing them. Again, you will need to work safely from height and follow the best working methods.

Before you start work on the synoptic practical test, it is important that you have had sufficient practice and that you feel that you are capable of passing. It is best to have a plan of action and a work method that will help you. You will also need a copy of the required standards, any associated drawings and sufficient tools and materials. It is also wise to check your work at regular intervals. This will help you to be sure that you are working correctly and help you to avoid problems developing as you work.

Your speed at carrying out these tasks will also help you to prepare for the time limit of the synoptic practical task. But remember that you should not try to rush the job as speed will come with practice and it is important that you get the quality of workmanship right.

Make sure you are working to all the safety requirements given throughout the test and wear all appropriate personal protective equipment (PPE). When using tools, make sure you are using them correctly and safely.

Good luck!

CHECK YOUR KNOWLEDGE

1. Why are floor coverings vital?
   - **a** They spread the load
   - **b** They hide the joists
   - **c** They allow carpets to be fitted
   - **d** All of the above
2. Why are floorboards difficult to lift?
   - **a** Long nails are used
   - **b** They are often interlocked
   - **c** It is difficult to get leverage
   - **d** The screws are difficult to get out
3. How should felt be laid on a roof?
   - **a** Vertically overlapped on the left
   - **b** Horizontally overlapped at the top
   - **c** Vertically overlapped on the right
   - **d** Horizontally overlapped on the bottom
4. Which timbers are affected more by rot?
   - **a** Softwood
   - **b** Hardwood
   - **c** Sheet materials
   - **d** They are all affected equally
5. What is the first step in eradicating rot?
   - **a** Remove the damaged area
   - **b** Treat the affected area with fungicide
   - **c** Treat the cause
   - **d** Fit new treated timber
6. How many steps are involved in splicing a door frame?
   - **a** 3
   - **b** 4
   - **c** 5
   - **d** 6
7. What do you need to be aware of when replacing ironmongery?
   - **a** All screws must be replaced
   - **b** The component should be checked for faults
   - **c** The component should be removed before ironmongery is replaced
   - **d** Replace it with an identical piece
8. What is glass called when it is made from two or more pieces stuck together with a thin chemical sheet?
   - **a** Laminated glass
   - **b** Toughened glass
   - **c** Double glazing
   - **d** Single glazing
9. When tiling, where is it important to start?
   - **a** At the bottom
   - **b** Near an opening
   - **c** Far away from an opening
   - **d** At the top
10. What is the first important job to carry out when wall tiling?
    - **a** Marking the tile positions
    - **b** Spreading the adhesive
    - **c** Cutting the tiles
    - **d** Applying grout

Functional skills

In answering the Check it out and Check your knowledge questions, you will be practising **FE 2.2.1** Select and use different types of texts to obtain relevant information and **FE 2.2.2** Read and summarise succinctly information/ideas from different sources.

You will also cover **FM 2.3.1** Interpret and communicate solutions to multistage practical problems.

Unit 3030

Know how to set up and use fixed and transportable machinery

Woodworking machines are vital in producing components accurately and safely. The machining of timber has become more prevalent in recent years. Technology has allowed the building of machines that can transform stock timber into any size or shape, making the woodworker's job far easier. Every carpenter or joiner will at some stage come across a woodworking machine, whether it is a combination planer in a workshop, or a table saw on site. It is important that you understand the principal uses, safety considerations and set-up of woodworking machines.

This unit contains material that supports the following NVQ unit:

- VR 26 Set up and use fixed or transportable machinery

This unit will cover the following learning outcomes:

- Know how to inspect and maintain fixed and transportable power machinery and equipment
- Know how to use machinery efficiently and safely

K1. Know how to inspect and maintain fixed and transportable power machinery and equipment

Functional skills

While reading and understanding the text in this unit, you will be practising **FE 2.2.1** Select and use different types of texts to obtain relevant information, **FE 2.2.2** Read and summarise succinctly information/ideas from different sources and **FM 2.3.1** Interpret and communicate solutions to multistage practical problems.

Did you know?

A common error is to think that the level of voltage is directly related to the level of injury or danger of death. A small shock from static electricity may contain thousands of volts, but has very little current behind it.

Remember

Electricity is a killer. A fatal electric shock can be received from working too close to live overhead cables, plastering a wall with electric sockets, carrying out maintenance on a floor and drilling into a wall.

Whenever you are working with machinery as a carpenter or joiner, you will need to be sure that the machines are working accurately and correctly. As such it is vital that you are familiar, and confident, with all methods used to inspect, fix and maintain machines, and that you follow best practice for keeping machines in their prime condition.

Analyse the manufacturer's literature for servicing and maintenance standards relating to machinery

When looking at woodworking machine maintenance, the most important thing is the manufacturer's literature. This will tell you all the maintenance requirements of the particular machine.

General safety rules

There are several key general safety rules that the manufacturer's instructions will provide you with when working with machinery. Never operate a tool until it has been fully installed and checked, according to the manufacturer's instructions. Make sure you are wearing full personal protective equipment (PPE), and keep hands and fingers well away from any cutters.

- Read the instruction manual before operating the tool and learn the tool's application and limitations, as well as any specific hazards peculiar to it.
- Keep guards in place and always make sure that they are in working order.
- Always wear eye protection.
- Remove adjusting keys and wrenches before use.
- Keep the work area clean.
- Don't use tools in dangerous environments. Never use power tools in damp or wet locations, or expose them to rain. Always keep the work area well lit.

The following points are also important to remember when using tools:

- Keep children and visitors away. Use padlocks, master switches and remove starter keys to make workshops childproof.

- Never force a tool. A tool will always do the job better and be safer if you use it at the rate for which it was designed.
- Never force a tool or attachment to do a job for which it was not designed.
- Wear the proper clothing. Never wear any loose clothing, gloves, neckties, rings, bracelets, or other jewellery which could get caught in moving parts. Non-slip footwear is recommended. Wear protective hair covering to contain long hair.
- Always use safety glasses. You should also use a face or dust mask if any cutting is likely to create dust.
- Use clamps or a vice to hold your work in place. This will also free up both your hands to operate the tool.
- Don't over-reach. Always make sure that you keep a proper footing and balance at all times while using machinery.
- Always maintain tools in a top condition. Keep tools sharp and clean to have the best and safest performance. Always make sure you follow instructions for lubricating and changing accessories.
- Always disconnect tools before servicing and when changing accessories such as blades, bits, cutters, etc.
- Always use recommended accessories. The use of non-recommended accessories or attachments may cause hazards or risk of injury.
- Always prepare the piece before work begins – never perform layout, assembly or set-up as you work – and allow adequate support to the rear and sides for long pieces.

If you are not entirely familiar with the operation of tools, obtain advice from your supervisor, instructor or another qualified person before beginning work. If the machine is damaged, fails, or in some way malfunctions, shut if off and remove the plug. Replace any parts and run a full check before continuing work.

Other information contained in the manufacturer's instructions

The other information you receive in the manufacturer's instructions will differ according to the type of machinery you are using. The information will include advice on:

- connecting the machine to its power supply
- assembling the machine and any components such as adjusting guards and fences
- changing any tooling such as blades or cutters

- maintenance, including changing any belts
- service intervals
- additional components that can be purchased and used with the machinery.

Failure to comply with the maintenance and service schedule as laid out by the manufacturer will invalidate any warranty on the machine, as well as possibly making the machine unsafe.

Regulations applicable to the use of machinery

The only regulations that deal with power tools or machinery are PUWER (see page 3 for more information on the duties and requirements of the regulations). These cover all new or existing work equipment – leased, hired or second-hand. They apply in most working environments where HASAWA applies, including all industrial, off-shore and service operations. PUWER covers the starting, stopping, regular use, transport, repair, modification, servicing and cleaning of equipment.

'Work equipment' includes all machinery, appliances, apparatus or tools, and any assembly of components that are used in non-domestic premises.

Good practice for working with machinery

Compared to other industries, woodworking accounts for a large proportion of accidents. Woodworking machines often have high-speed cutters, and many cannot be fully enclosed owing to the nature of the work they do.

Table 30.01 on the next page lists a few of the requirements relating to the safe use of all woodworking machines. Safety regulations relating to a particular machine are noted in dedicated sections in this unit.

Safety tip

Machines must always be switched off when not in use and never left unattended until the cutter has come to a complete standstill.

Working life

You are using a circular saw when a strange noise starts up. You switch the machine off and isolate it. Your boss tells you to fix it yourself.

What should you do? Are you qualified to carry out these repairs? What information would you need before carrying out any work? What could the consequences be? You will need to think about the impact on yourself, and anyone else who might use the equipment later.

Table 30.01 Requirements for safe use of woodworking machines

Safety issue	Requirements
Safety appliance	Push/sticks, jigs, etc. must be designed to keep operator's hands safe. Modern machines use power feed systems to keep the operator away from the cutting action, and these should be used if available.
Working area	An unobstructed area is vital. The positioning of the machine must be carefully thought out. When there are several machines in a work site, layout should be arranged so materials follow a logical path. Access routes must be kept clear, with suitable storage for materials.
Floors	Floors must be flat, in good condition and free from debris. Electricity supply, ductwork, etc. must be run overhead. Avoid polished services and clean up spills immediately.
Lighting	Adequate light is essential. All dials and gauges should be visible with a good view of the machine, without glare or light shining in the operator's eyes.
Heating	The temperature should not be too warm or too cold. 16°C is suitable.
Controls	All machines should have a means of isolation, separate from the on/off buttons. The isolator should be easily accessible. Ideally, a second cut-off switch accessible by others should be in place. The stop/start mechanism must be efficient and in easy reach.
Braking	All new machines must have an automatic braking system (ABS), which stops within ten seconds of the machine being switched off. A risk assessment must be carried out to see if an ABS needs to be installed. Contact the Health and Safety Executive (HSE) if unsure.
Dust collection	An efficient means of collecting dust and chippings must be installed.
Training	Never use a machine unless you have been suitably trained and deemed competent.
Maintenance	Follow the manufacturer's instructions and check them prior to use, with inspections and any findings recorded in a log.

K2. Know how to use machinery efficiently and safely

For all types of power tools you will need to know several key things:

- How to identify the safe components of tooling.
- The correct safe methods used to change tooling, such as blades.
- How to identify potential hazards in using equipment.
- The importance of teamwork for using equipment.
- The importance of reading and evaluating the manufacturer's operating instructions.

This section will look at each of the major power tools in turn, and identify these elements.

Safety tip

Use non-slip matting around machinery but make sure the edges do not present a trip hazard.

Find out

Take a look around your work site and measure it against the safety standards laid down in Table 30.01 – Does it meet all the criteria? Are there any areas where there are potential hazards?

Table saws

Table saws – also known as rip saws, circular table saws and bench saws – are available in a range of shapes and sizes.

The main functions of a table saw are flatting (cutting the timber to the required width) and deeping (cutting the timber to the required thickness).

Use the Internet and the manufacturer's instructions to discover the safest process for changing the blade on a circular table saw, and then carry out the change.

With advances in technology and a new variety of saw blades available, the table saw is capable of other tasks, including creating housings.

Table saw parts

- **Saw bench or table** – the table or bench on which the saw is fitted. This should be extendable via rollers or extra out-feed tables to allow larger materials such as sheet timber to be machined.
- **Blade** – the cutting tool.
- **Crown guard** – a guard suspended over the top of the blade. It must be adjustable to ensure that as much of the blade is guarded as possible.
- **Riving knife** – thicker than the blade, this acts as a spreader to prevent the cut from closing, which could cause binding.
- **Fence** – a guide to give straight and accurate cutting, which should be accurately labelled and easily adjustable.
- **Finger plate** – a cover piece that sits over the spindle of the saw.

Figure 30.01 Flatting (note that the guards have been adjusted for photographic purposes)

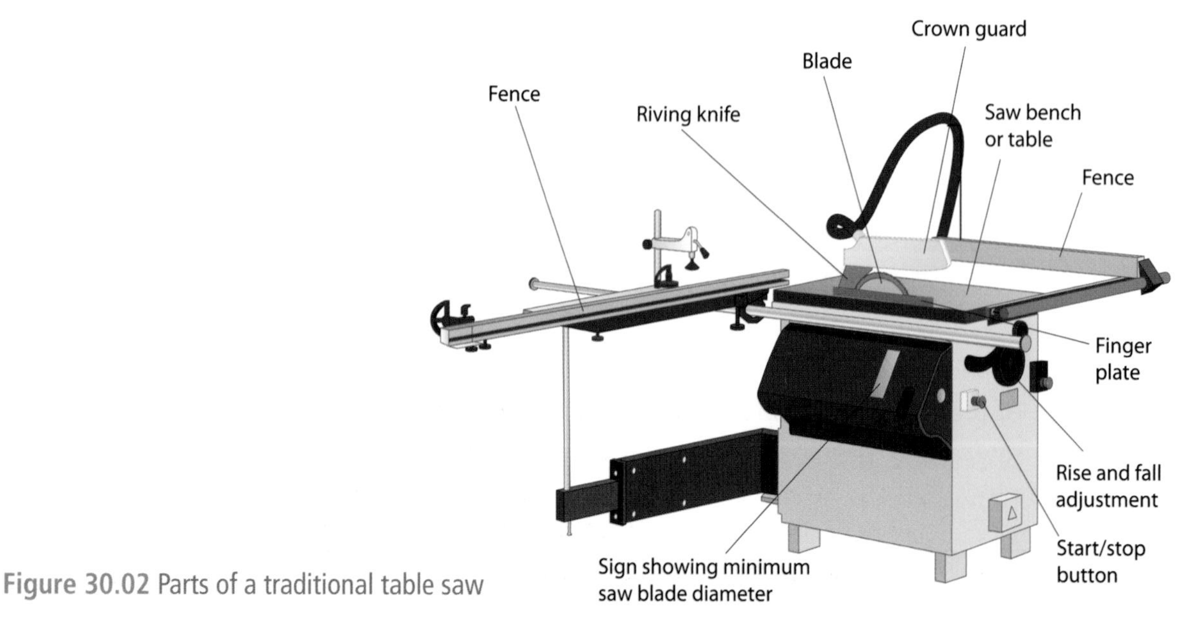

Figure 30.02 Parts of a traditional table saw

- **Rise and fall adjustment** – a wheel used to adjust the height of the blade.
- **Start/stop button** – this should be clearly labelled and within easy reach of the operator.
- **Sign showing minimum saw blade diameter** – a legal requirement.

Saw blades

Several different types of saw blade are available. The standard components of a saw blade are as follows (see also Figure 30.03).

- **Pitch** – the distance between two teeth.
- **Hook** – the angle at the front of the tooth. A positive hook is required for ripping, with an angle of 20–25° for softwoods and 10–15° for hardwoods.
- **Clearance angle** – ensures that the heel clears the timber when cutting.
- **Top** – the angle across the top of the tooth.
- **Set** – the amount each tooth is sprung out to give clearance and prevent the saw from binding.
- **Gullet** – the space between two teeth, which carries away the sawdust.
- **Kerf** – the width of the saw cut (not the width of the blade).

Tungsten carbide-tipped (TCT) blades are now preferred, as they stay sharper for longer and do not have a set.

Find out

Use the manufacturer's instructions and the Internet to sketch a circular saw blade, showing the pitch, clearance angle, gullet, hook, etc.

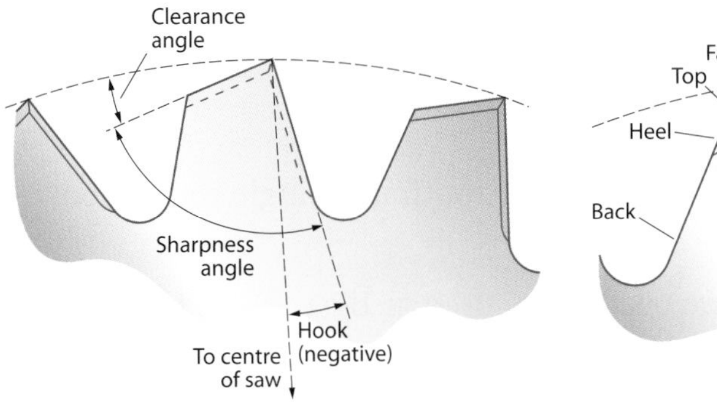

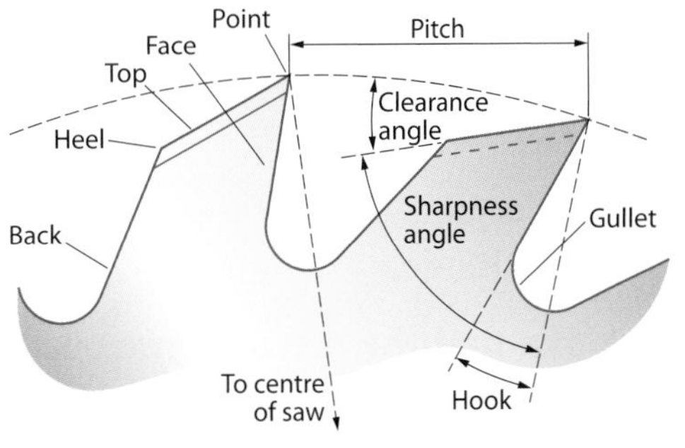

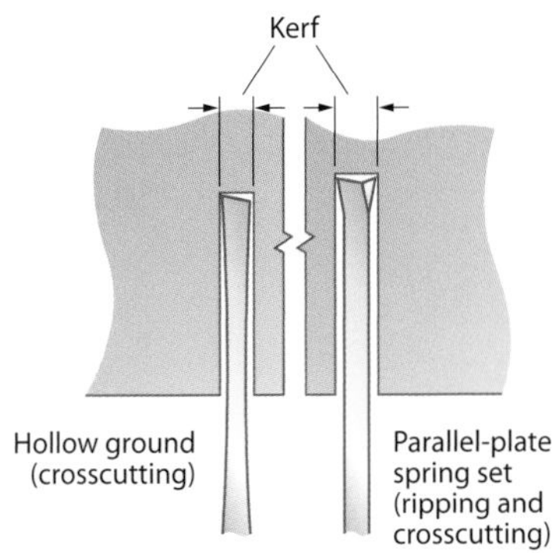

Figure 30.03 Saw blade terminology

Changing the saw blade

Saw blades are fitted on a spindle between two flanges or collars. When a saw blade needs to be fitted or removed, the front collar can be taken off to allow access. The rear cradle is fixed to the spindle. In some cases, circular saws will have a locating pin in the rear collar, which assists in the saw blade taking up the drive of the motor.

Safety tip

Care must be taken when replacing any saw blades and it is important to check the manufacturer's instructions.

Did you know?

Man-made boards, such as chipboard or plywood, contain glues which will also blunt the blade.

Remember

If the run speed is too fast, the blade will become stressed. A slow rim speed, on the other hand, means that the saw will struggle to cut the material.

Remember

When changing the blade for one of a different size or style, it is vital to check that the replacement blade is suitable.

As Figure 30.02 shows, a saw should have a label on it stating the minimum/maximum size of a blade to be fitted to the machine. This is important as the spindle on the saw runs at a certain speed to drive the blade. The blade runs at what is called peripheral speed.

The life of the saw blade will vary greatly and depends upon the materials that it is being used to cut. Some timbers are very strong, such as oak, and these may dull the blade more quickly than timbers such as Douglas fir.

Calculating the rim or peripheral speed of a blade

The relation of the saw blade size to the speed of the spindle is crucial for the safe running of the saw (this is why the minimum and maximum blade size must be shown clearly on the machine). A blade with a rim/peripheral speed of 100 metres per second can become stressed, causing a dangerous situation.

To calculate the rim/peripheral speed, you must first know the diameter of the saw blade and the spindle speed. The diameter of the saw blade can be measured or found from the blade manufacturer; the spindle speed can be found from the saw manufacturer's handbook.

For this example we will use a saw blade with a diameter of 600 mm and a spindle speed of 1600 revolutions per minute.

First, work out the distance around the rim (the circumference of the blade), using the following calculation:

Circumference = $\pi \times$ diameter (where π is 3.142)
Circumference = 3.142×0.600
Circumference = 1.8852

This equals the distance travelled by 1 tooth in a single revolution of the blade.

Next, find the distance travelled every minute by 1 tooth, by multiplying the circumference by the spindle speed:

$1.8852 \times 1600 = 3016.32$ metres per minute

Finally, to find the answer in metres per second, divide this answer by 60:

$3016.32 \div 60 = 50.272$ metres per second (mps)

A rim/peripheral speed of 50 metres per second is considered suitable.

If you are unsure about which blade size to fit, contact the manufacturer.

Machine set-up

The set-up of any table saw depends on the manufacturer and the type and model of saw, so be sure to check the manufacturer's handbook prior to using any machine. Whatever the make or model of your saw, there are certain safety measures you must take before using it. You must ensure that:

- the machine is isolated from the power source prior to setting up
- all blades are in good condition, securely fixed and running freely
- all guards, guides, **jigs** and fences are set up correctly and securely
- suitable push sticks are available and at hand, ready to use
- suitable PPE is available.

Safe operation

The operation of each table saw varies slightly depending on the task you are doing, but the basic operating principles remain the same.

First, ensure that the above checks have been done and the machine is set up safely. Wear the appropriate PPE and have a push stick at hand, ready for use.

The machine can then be started and allowed to get up to speed before you offer the timber to the blade. Stand at the side of the piece being sawn and not directly behind it, as **kickback** may occur and can cause injury.

Using the fence/jig as a guide, slowly ease the timber into the blade using a steady amount of pressure: applying too much pressure will cause the blade to overheat, leaving burn marks on the face of the timber and even causing the blade to wobble, resulting in kickback. The right amount of pressure varies from material to material and between sizes of material but, generally, if you hear the saw straining or if there is a burning smell from scorching timber, you are applying too much pressure.

Keep the timber hard against the fence to ensure accuracy, and continue to feed it into the saw until there is no less than 300 mm remaining to be cut. The last 300 mm or more must be fed through using a push stick.

Teamwork when using table saws

For some functions, when working with table saws, you will find it very useful to work as a team to complete the jobs safely.

Key terms

Jig – a guide or template attached to the machine, allowing the machining of timber to a variety of shapes.

Kickback – where the timber being machined is thrown back towards the operator at speed, usually because the timber is binding against the blade.

Did you know?

Binding is where the timber being sawn closes or 'binds' over the blade, so that contact with the fence of the blade is made. It can be caused by overexerting the blade by applying too much pressure and using a blunt or damaged blade. The most common cause is high moisture content in the timber. This has a tendency to move while being machined, which can cause the kerf to close at the rear of the riving knife, applying pressure to the blade and causing binding.

Safety tip

The push stick should also always be used to remove the cut piece between the saw blade and the fence and any other off-cuts.

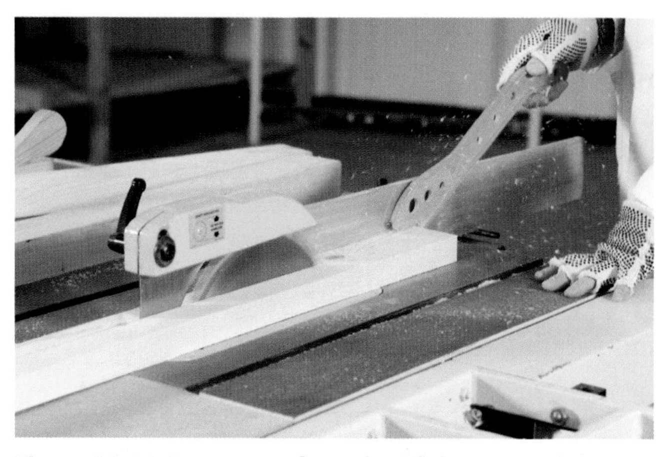

Figure 30.04 Correct use of a push stick (Note: guards have been adjusted for this photograph in order to reveal the blade)

Binding can be prevented by having a second person standing at the rear of the saw blade, making sure this person is kept well away from the blade. The second person can then drive the timber wedges into the kerf to keep it from closing.

You may also need to ask another worker to help you rip long lengths of timber. They can help you guide the timber through the saw and support the weight of the timber as it passes through.

Working life

You have been asked to change the blade on a circular table saw. You work out the peripheral speed, and it comes out at almost 100 mps.

What are the hazards of using this blade? What causes these hazards? What should you do?

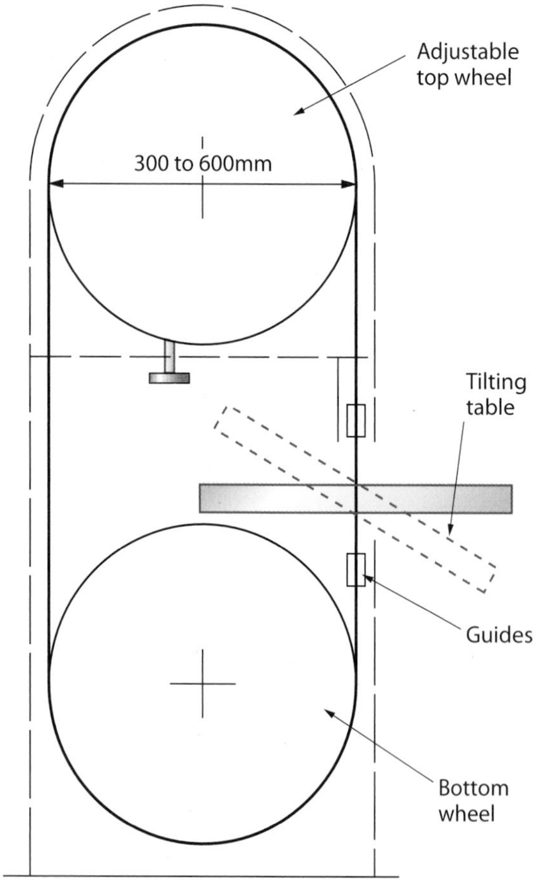

Band saws

The band saw is mainly used for curved or shaped work, but with the addition of fences a band saw can be used for ripping and crosscutting too. A band saw

Figure 30.05 Wheels, table and blade for saw blade

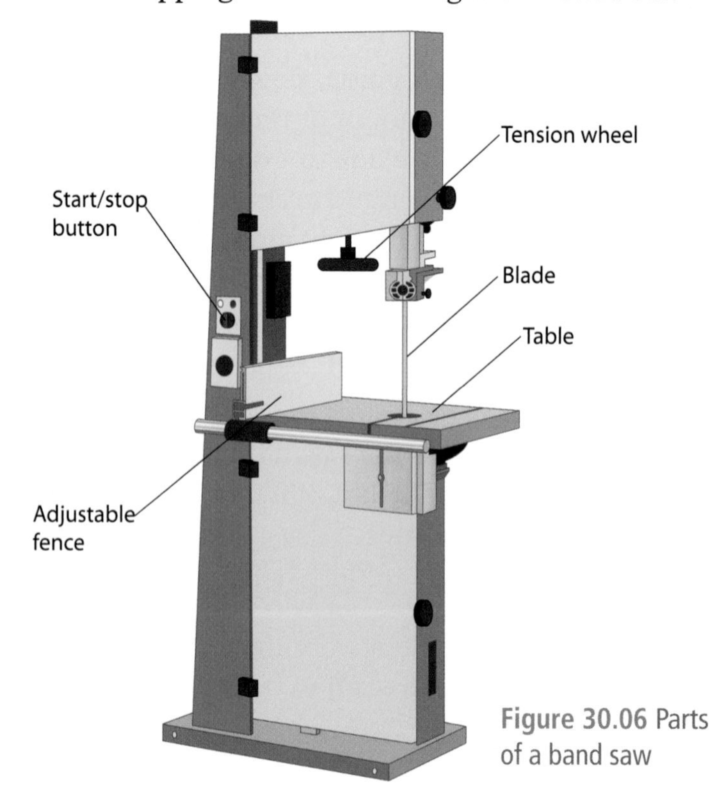

Figure 30.06 Parts of a band saw

consists of an endless blade that runs around two wheels, with one wheel mounted above the other. The wheels are encased in the machine to protect the user, and an adjustable table sits between the wheels, which is where the cutting action occurs.

The bottom wheel of the band saw is driven directly by the motor, while the top wheel is driven by the blade. The top wheel is also adjustable to allow the blade tension to be set. Rubber tyres attached to the rims of the wheels help to stop the blade slipping, as well as preventing damage to the saw blade.

As well as running on the wheels, the blade is supported by saw guides situated above and below the saw table. The top guide is fitted with a guard to protect the user. The guides are fitted with thrust wheels, which prevent the saw blade being pushed back when the work is pressed against it. Blades come in a variety of sizes from 3 mm upwards: the smallest blades are for more intricate, curved work; the larger blades are for ripping large, sectioned timber.

Find out

Use the manufacturer's specifications to create a sketch showing how the guides at the top and bottom of a band saw are set up.

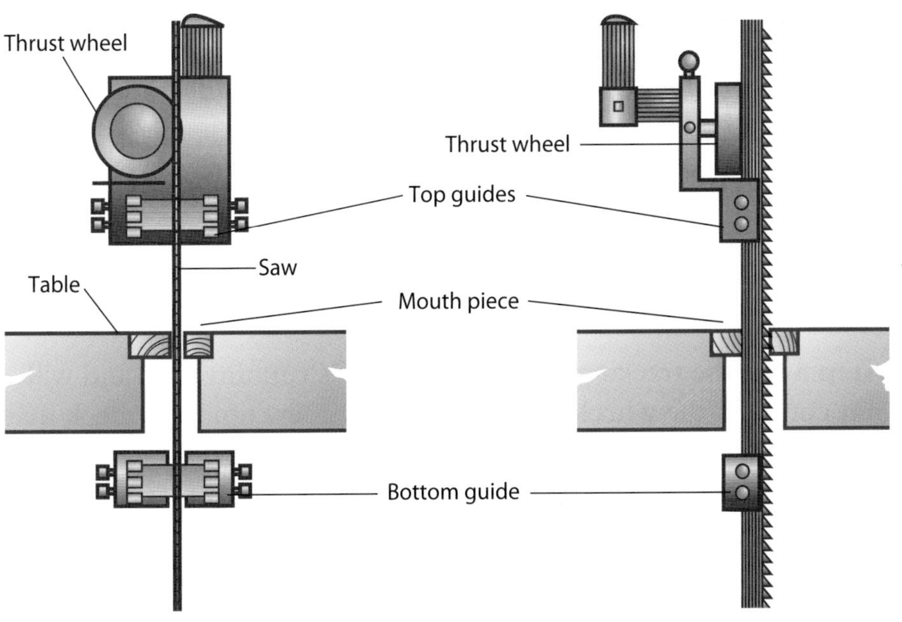

Figure 30.07 Guides

Remember

The exposed part of a band saw blade is guarded with an adjustable guard. It is vital that the guard is set prior to every use to leave as little of the blade showing as possible. The machine should already be set to the correct tension, but it is always better to check before use. A blade that is loose will come off the wheels and may snap; a blade that is too tight can snap too.

Setting up the band saw

As with all machines, the set-up depends on the manufacturer. Not every machine can do all tasks, so it is always best to refer to the manufacturer's handbook when setting up the machine.

Before use, it is important to check the tracking. If the wheels are not set correctly, the blade could come off the wheels or tilt backwards, running against the guide and damaging or even breaking the blade.

Safety tip

Always check the manufacturer's instructions before making any adjustments to tension or tracking and, if you are unsure, don't use the machine.

Remember

Once a blade has been broken or blunted, it cannot be repaired. It is best to replace it and throw out the old blade, making sure you dispose of it safely.

Remember

A sign or panel on the machine should tell you what tension you need to set the blade to. If there is no sign, contact the manufacturer for more information.

Did you know?

For ripping large timber you will need to change a thin blade to a thicker one.

Using the machine

The machine's operations will be listed in the handbook. Always ensure the machine is set up correctly for the operation you need, prior to use. When working intricate curves into timber with a band saw your hands can get close to the blade, so it is vital to use push sticks and the correct fence attachments. If a band saw blade does snap, there can be a loud bang (and it can be quite scary), but all band saws must be fitted with a brake designed to stop the machine as quickly as possible, preventing the wheels from continuing to drive the broken blade.

Changing the blade

If a blade does break, or the machine needs to be set up for a different use, the blade will need to be replaced. Again, the manufacturer's handbook should be followed, but here is a basic guide to changing a blade:

Step 1 Ensure that the machine is switched off and isolated from the power supply before making any adjustments.

Step 2 Open the top and bottom guard doors and move aside any other obstructions such as the blade guards and table mouthpiece.

Step 3 Lower the top wheel to remove the tension, then remove the old/broken blade.

Step 4 Fit the new blade, ensuring that the teeth are pointing in the right direction (downward in most cases), then re-adjust the tension so that it is set correctly.

Step 5 Set the tracking so that the blade is running true and will not come off the wheels. Spin the wheels a few times to check that the tracking is correct.

Step 6 Re-attach the guards, etc., re-set the thrust wheels and guides, and close the top and bottom guard doors.

Step 7 Switch the machine on and let it run for a little while to ensure that it is tracked and tensioned properly before attempting to cut any materials.

Teamwork for band saws

Teamwork is rarely needed for band saws, as in a standard workshop situation they are mainly used for intricate or curved work and the table saw or crosscut is used to rip or cut larger sections of timber. In saw mills, however, larger band saws are used to rip logs into sectional timber and, although these are mainly machine operated, teamwork is required during this process.

Figure 30.08 Large band saw

Working life

You have been tasked with machining circular pieces on a band saw. You have changed the blade to a smaller blade to allow for the cutting of curves. You start the machine up and there is a smell of burning rubber coming from the machine. What has caused this? What can be done to rectify it? What will happen if you do not rectify it?

Functional skills

This exercise will allow you to practise **FM 2.3.1** Interpret and communicate solutions to multistage practical problems and **FM 2.3.2** Communicate solutions to answer practical problems.

Planers

A hand-fed power planer is another essential tool in a woodwork shop, as it can do in seconds what a craftsperson would take hours to do by hand. This sort of planer is in essence the same as a portable power planer, but on a much larger scale and more accurate. All planers comprise an in-feed table, an out-feed table and a cutter block, into which either two or three cutting blades are housed.

The three main types of planer available are:

- surface planers
- thickness planers
- combination planers.

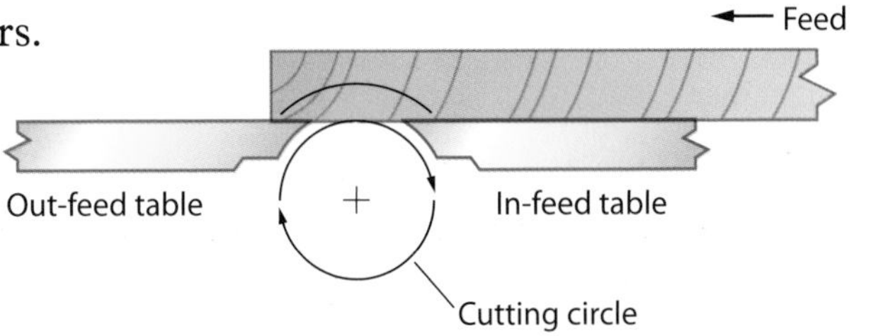

Figure 30.09 Cutter block, with out-feed and in-feed plates

Safety tip

When operating planing machines, you should wear boots, gloves, goggles, ear defenders and, depending on the timber being machined, a dust mask. You must also be sure that you have no loose clothing that could get caught in the cutter block.

Remember

Planers can be used to face, edge and finish materials to a desired thickness.

The combination planer is by far the best machine to have, as it can carry out all the operations of both surface and thickness planers. Given that the combination planer is essentially a combination of the other two, we will only look at the surface and thickness planers here.

Surface planer

Did you know?

With the use of a fence, a surface planer can also be set up to create rebates and splayed or angled pieces.

As its name suggests, the surface planer creates a smooth and even surface on the piece of timber. It is most commonly used for two main operations:

- **Facing** – when the timber is planed flat and even on the widest side of the timber.
- **Edging** – when the timber is placed flat and even on the narrowest side of the timber.

The surface planer works by passing the timber smoothly over an in-feed and out-feed table, in between which the cutter block is situated. The height of the in-feed table is adjustable to regulate the amount of timber removed in a single pass.

Remember

With a surface planer, it is not recommended to remove more than 3 mm in a single pass.

The in-feed and out-feed tables are machined to be perfectly flat. When facing, provided the timber is held firmly to the surface of the out-feed table, a perfectly flat surface is produced. Edging

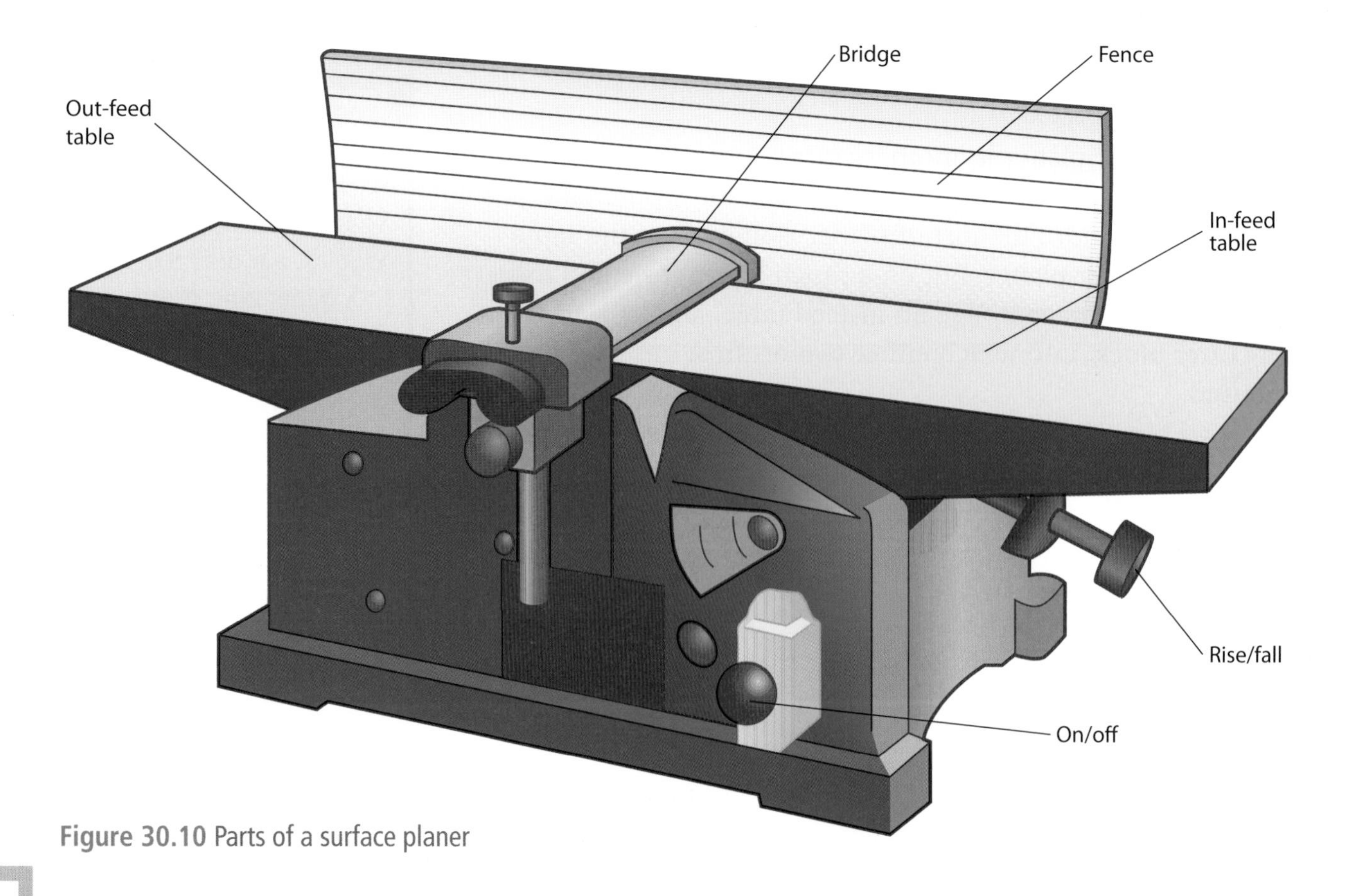

Figure 30.10 Parts of a surface planer

is done by placing the faced side of the timber against the fence and running it over the machine.

The surface planer is fed mostly by hand, so great care must be taken to avoid your hands coming into contact with the cutter block.

The main way of protecting the user is through the use of suitable guards. The main guard used on a surface planer is the bridge guard. This is strong and rigid, and is usually made from aluminium so that if it comes into contact with the cutter block neither the guard nor block will disintegrate. Telescopic bridge guards are advisable as they can cover the full length of the cutter block; they must be wider than the cutter block.

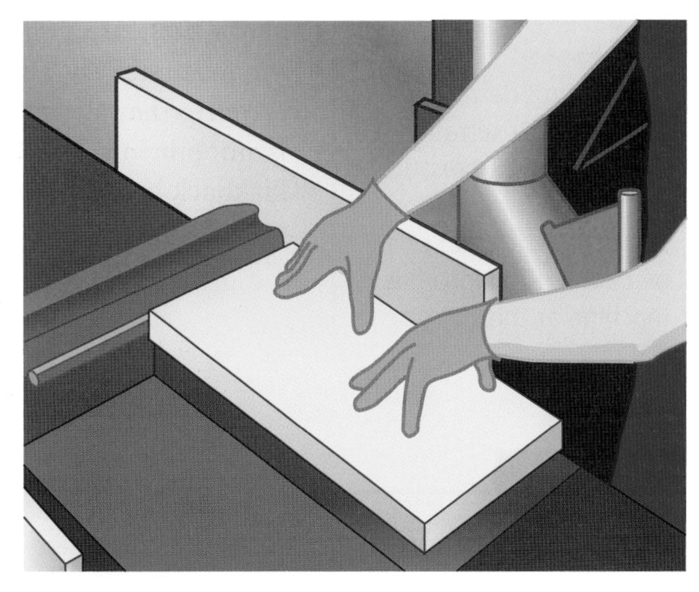

Figure 30.11 Telescopic bridge guard (Note: for visual purposes, this diagram shows the operator standing behind the surface planer; however, when operating a surface planer yourself, you should stand in front of it)

Tunnel guards can also be used in conjunction with a push stick; push blocks should also be used when appropriate (see Figure 30.12).

There can be a tendency to feed timber over the cutter block too fast, which will leave a bad finish; to produce a good finish, it is best to go slowly.

Find out

Using the manufacturer's instructions, draw a sketch showing how a tunnel guard is used when surface planing.

Thickness planer

The thickness planer planes the timber to the required width or thickness, usually after the surface planer has faced and edged the timber. The thickness planer can also be used with a variety of jigs to create simple mouldings such as window beads.

The thickness planer has a power feed system, usually in the form of four rollers. Two idle rollers are fitted at the bottom on the rise and fall table, and the two rollers

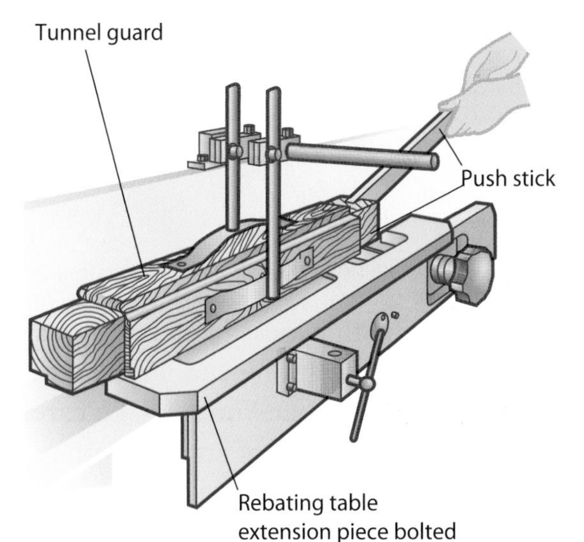

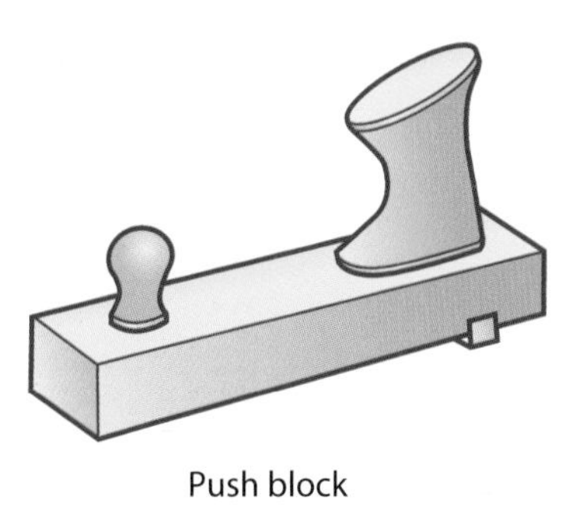

Figure 30.12 Tunnel guard and push block

Did you know?

On most modern machines the serrated roller and front pressure bar are made in sections to allow more than one piece to be fed into the machine at the same time.

Safety tip

The thickness planer should carry a label stating whether more than one piece can be fed into the machine at one time. If you feed more than one piece into a thickness planer that doesn't have a sectional serrated roller, it may result in one of the pieces being thrown back at you at high speed.

above these are driven rollers, which propel the timber through the machine. The first upper roller is usually serrated so that it can grip the sawn timber better, while the second power roller is smooth so as to not damage the finished surface. To prevent kickback both upper rollers are spring-loaded, as are the two pressure bars.

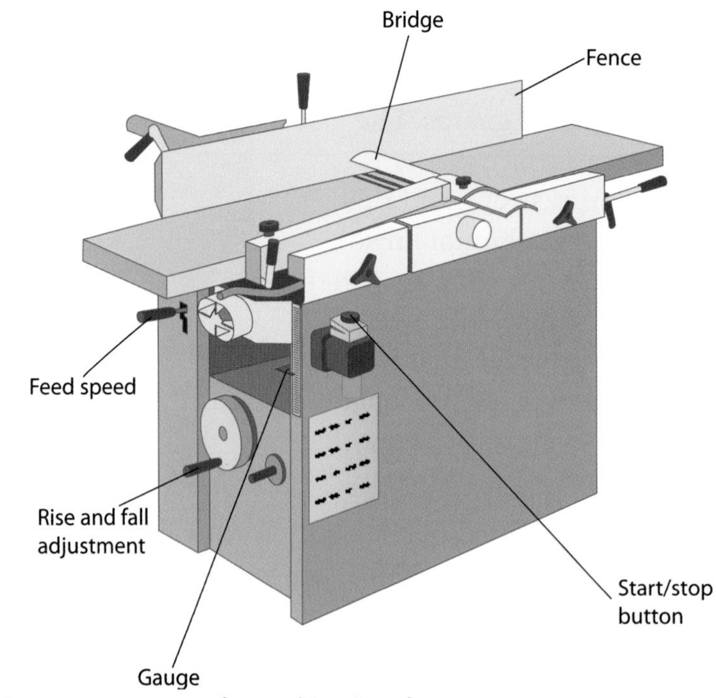

Figure 30.13 Parts of a combination planer

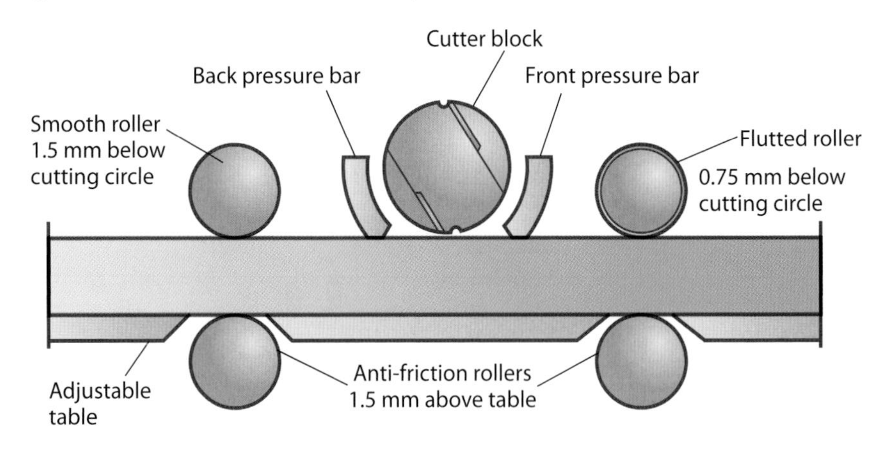

Figure 30.14 Section through a thickness planer

Operating the thickness planer is very simple: the timber is fed into the machine at one end and comes out planed to the required thickness at the other. You must be careful to ensure that the timber being fed in is of an appropriate thickness: too thick and planing the timber becomes very difficult; too thin and the timber will just pass through without being planed at all. The rise and fall table can be adjusted through a wheel or, on more modern machines, electronically via a button. A depth gauge should be in easy view so that the required depth can be set easily.

The thickness planer should have an adjustable roller-feed speed setting to help create a better finish. One of the most common problems with planing machines is pitch marks, where the surface of the timber is rippled. Pitch marks can be caused by having a fast in-feed speed or a slow cutter block speed.

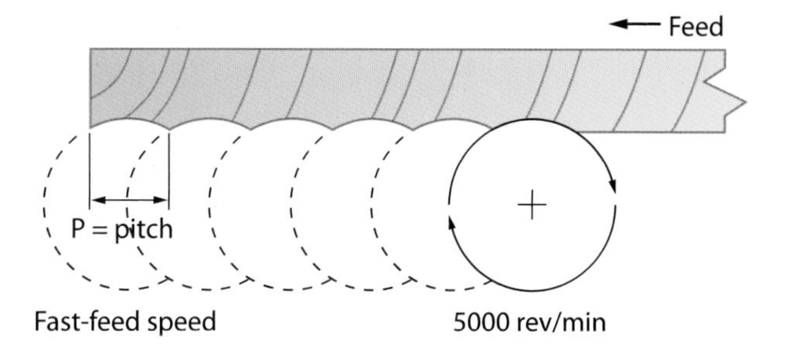

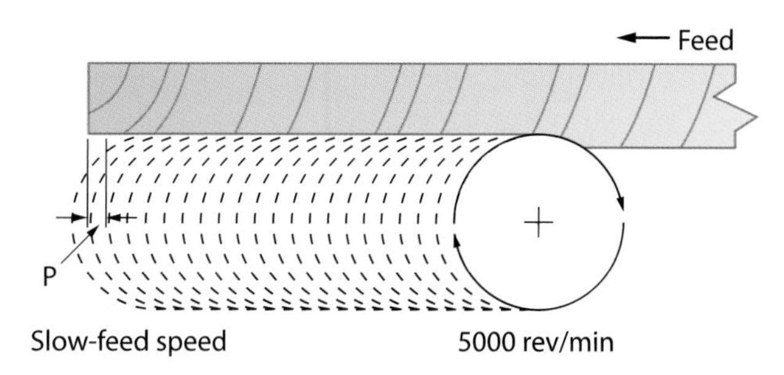

Figure 30.15 Pitch marks for slow- and fast-speed feeds

Changing planer blades

Blades should be changed in line with the manufacturer's guidance. Special care must be taken to ensure that new blades are fitted correctly and in line with each other. Wrongly aligned blades may result in a poor finish: with one blade set above the other, only one blade is doing all the cutting, resulting in pitch marks. Poorly set blades may also pose a safety risk, as they may shatter or chip.

Teamwork when using planers

Working with planers is largely a one-man job unless large or awkward-sized components are to be cut, at which point help may be needed to support the piece during cutting.

To get a good finish when surface planing long lengths of timber it is best to have two people – one at the front and the other at the back, both supporting the weight.

Remember

If the thickness planer is creating pitch marks, check the machine's feed speed and adjust it so that the machine feeds more slowly.

Did you know?

Planer blades are also known as knives.

Working life

You have been tasked with running 200 pieces of timber through a thickness planer to machine them to a desired finish size. The machine is a single-roller feed, meaning that you should only feed one piece of timber in at a time. The materials need 7 mm removed, which means that each piece should be fed into the machine three times. Your boss needs the job done quickly, and asks you to feed four pieces in at the same time, taking the full 7 mm off. What hazards can this create? What effect will this have on the finish? What should you do?

Mortise machines

The hollow-chisel mortise machine is designed predominantly to chisel out mortises for mortise and tenon joints, but it can also be used to chisel out mortises for any purpose.

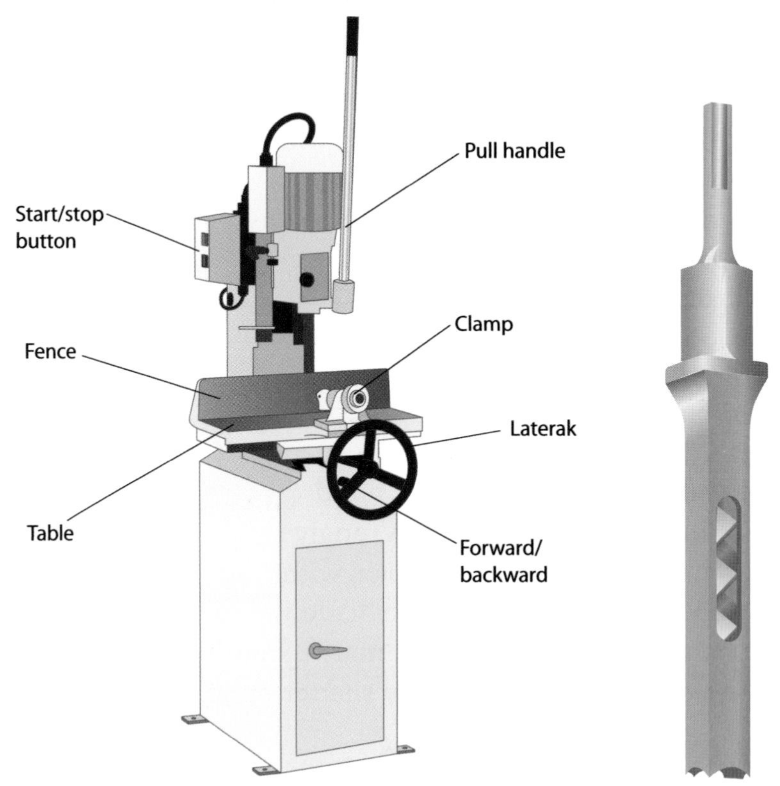

Figure 30.16 Parts of a mortise machine

Figure 30.17 Hollow square chisel

Remember

Most machines come with a depth stop that can be set to any depth. This will let you know when you have mortised half-way through.

Did you know?

If you want to mortise a 21 mm mortise but only have a 18 mm and 22 mm chisel available, then you must use the 18 mm chisel and adjust the table laterally to complete the mortise.

Safety tip

When machining timber of long lengths, make sure that they are adequately supported on any overhanging ends to prevent the work piece from lifting.

A mortise machine consists of a revolving auger housed inside a hollow square chisel (this comes in various sizes, producing a range of mortises). The auger cuts most of the timber, while the chisel squares the hole up. The machine is operated much like a pillar drill, with a lever forcing the cutting action into the work piece. When setting the machine up it is important to ensure that there is a clearance of 2–3 mm between the tip of the auger and the chisel: this prevents the tips of the auger and chisel overheating.

The work piece sits on a table and is clamped in place to prevent it from moving. The table can be moved forwards and backwards as well as from side to side; this means more than one hole can be drilled without removing the timber from the clamp.

Setting up and using the mortise machine

The mortise machine should be set up, and any tooling changed, as per the manufacturer's handbook. If the tooling has been changed, it is a good idea to run a test piece to ensure that the mortise is set up square; this prevents stepped mortising (see Figure 30.18).

Once the machine has been set up correctly, the work piece can be clamped in place and any lateral or sideways adjustments can be made so that the chisel cuts in the correct place. When mortising full mortises, it is good practice to go only a little over half-way down the depth, then turn the timber over and complete the mortise from the other side; this avoids splitting out the bottom of the timber.

Did you know?

'Bits' is a name also given to the chisels and auger bits in the mortiser. General tooling, except blades, are also referred to as bits.

Teamwork when using mortise machines

As with other machines, when cutting large or awkward-sized materials it is best to have an additional person helping to support the weight, particularly during transit and machining. Care must be taken, particularly when inserting or removing any large or awkward-shaped components into the machining area, as a slip or confusion with instructions could see the work piece damaged, or even the operator injured.

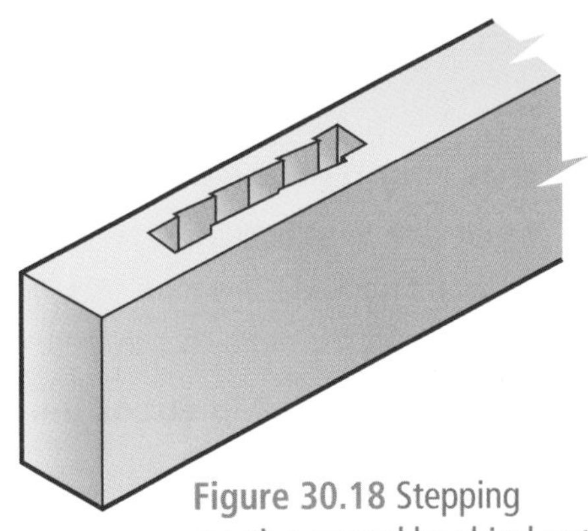

Figure 30.18 Stepping mortise caused by chisel not being set correctly

Working life

You mortise the first piece of a series of through mortises for a frame and, when you remove it from the machine, the back of the mortise is split.

What could have caused this? What can be done to prevent it from happening again?

FAQ

The timber that I am ripping down comes out a different measurement from the one I set the saw to. Why is this?

The gauge on the fence is out and must be recalibrated.

The saw is not cutting straight. Why is this?

There are two main causes. Either the fence is not secured properly and is moving when you are feeding the timber through, or the timber is not being held tight against the fence throughout the operation.

What length does a push stick need to be and why?

A push stick must be between 300 mm and 450 mm. Any less than 300 mm and your hand will be too close to the operating blade; any more than 450 mm and you are too far away to have proper control. Also, the longer the push stick the more chance there is of springing it out of place.

There is a small chip out of the blade on the surface planer I am using? Do I need to change the blades?

No. The fence adjustment should be moved so that you are planing in the area without the chip in it. Replacing blades every time there is a small chip can be an expensive practice.

How do I know if I can feed more than one piece of timber into a thickness planer at the same time?

There should be a label on the machine stating that only one piece can be machined at any time. If there is not a label, check the manufacturer's instructions. If you are in doubt, feed just one piece at a time.

There are sparks appearing when I use a band saw. What causes this?

Sparks can occur when the thrust wheel is not set correctly or when the tracking or tension is not correct. Switch off the machine, isolate it and check the thrust wheel, tracking and tension.

Check it out

1. Describe the information contained in manufacturers' instructions, collecting examples from different types of machinery. Use these to draw up your own key guidelines for working with machinery.
2. What regulations are used to govern the use of woodworking machines? Describe the conditions that these regulations impose.
3. What are the main functions of a table saw? Describe the purposes of the main components of a table saw, with reference to the crown guard, saw blade and finger plate.
4. Explain the importance of peripheral speed.
5. Prepare method statements explaining the two main operations carried out by a surface planer.
6. Describe why the tension of a band saw blade should be checked, and explain the purpose of tracking when using a band saw.
7. Explain the purpose of chisels on mortise machines and explain what causes stepped mortises.

Getting ready for assessment

The information contained in this unit, as well as continued practical assignments that you will carry out in your college or training centre, will help you with preparing for both your end of unit test and the diploma multiple-choice test. It will also aid you in preparing for the work that is required for the synoptic practical assignments.

The information contained within this unit will aid you in learning how to identify the components of fixed and transportable machinery, as well as learning about their safe set-up and operation.

You will need to be familiar with:

- inspecting and maintaining fixed and transportable power machinery and equipment
- using machinery efficiently and safely
- how to use fixed and transportable machinery safely.

Using fixed and transportable machinery can be very dangerous if you do not use them correctly. You will need to remember to follow health and safety guidance at all times to make sure you, and everyone around you, is safe while you work.

For learning outcome one, you have seen the potential hazards that exist when working with various types of machinery, such as circular saws, as well the current legislation that governs the use of machinery. You will need to be able to follow these guidelines to check for potential hazards, such as faulty or missing guards, faulty or wrongly fitted tooling or damage to equipment. You will also need to be sure that you have set up the machinery in line with the manufacturer's recommendations and current legislation.

For learning outcome two, you will need to be able to set up machinery, using information taken from drawings, rods and cutting lists. You will then need to use this machinery to produce work, cut to size in a safe and efficient manner, using all necessary safety aids. When working with equipment, you will also need to be able to co-operate effectively with others in a team to lift or cut large or heavy sections of material using the machinery.

Before you start work on the synoptic practical test, it is important that you have had sufficient practice and that you feel that you are capable of passing. It is best to have a plan of action and a work method that will help you. You will also need a copy of the required standards, any associated drawings and sufficient tools and materials. It is also wise to check your work at regular intervals. This will help you to be sure that you are working correctly, and help you to avoid problems developing as you work.

Your speed at carrying out these tasks will also help you to prepare for the time limit of the synoptic practical task. But remember that you should not try to rush the job as speed will come with practice and it is important that you get the quality of workmanship right.

Make sure you are working to all the safety requirements given throughout the test and wear all appropriate personal protective equipment (PPE). When using tools, make sure you are using them correctly and safely.

Good luck!

CHECK YOUR KNOWLEDGE

1. Why does there need to be sufficient lighting in a workshop?

- **a** So there is a good glare for working with
- **b** So the dials and gauges are visible
- **c** Because tinted safety goggles are used
- **d** All of the above

2. No person should use woodworking machinery unless:

- **a** they are over 18
- **b** they are over 21
- **c** they have been trained
- **d** their supervisor says so

3. What is the cover piece that sits over the spindle of the saw on a table saw called?

- **a** Crown guard
- **b** Riving knife
- **c** Finger plate
- **d** Fence

4. What is the name for the distance between two teeth on a saw blade?

- **a** Hook
- **b** Pitch
- **c** Gullet
- **d** Kerf

5. What is a suitable rim/peripheral speed?

- **a** 25 mps
- **b** 50 mps
- **c** 75 mps
- **d** 100 mps

6. Which planer is the most useful?

- **a** Surface planer
- **b** Combination planer
- **c** Thickness planer
- **d** None of the above

7. What does a sectional serrated roller on a thickness/combination planer mean that it can do?

- **a** Plane faster
- **b** Plane smoother
- **c** Handle more than one piece being fed into the machine at any one time
- **d** Only handle one piece being fed into the machine at any one time

8. What should be checked prior to using a band saw?

- **a** Blade tension
- **b** Wheel tracking
- **c** Fitting of the guards
- **d** All of the above

9. The mortise machine is operated in a similar way to a:

- **a** crosscut saw
- **b** pillar drill
- **c** spindle moulder
- **d** table saw

10. The clearance between the tip of the chisel and the auger should be:

- **a** 0–1 mm
- **b** 1–2 mm
- **c** 2–3 mm
- **d** 3–4 mm

Functional skills

In answering the Check it out and Check your knowledge questions, you will be practising **FE 2.2.1** Select and use different types of texts to obtain relevant information and **FE 2.2.2** Read and summarise succinctly information/ideas from different sources.

You will also cover **FM 2.3.1** Interpret and communicate solutions to multistage practical problems.

Unit 3031

Know how to manufacture complex shaped joinery products

While the carpenter is generally employed on site carrying out first and second fixing and carcassing, the joiner is most likely to be found in a workshop manufacturing the components that the carpenter will fit on site. Whichever qualification you follow, you will need a good understanding of how basic components such as stairs are made.

This unit contains material that supports the following NVQ unit:

- VR 29 Manufacture complex shaped products

This unit will cover the following learning outcomes:

- Know how to select correct materials
- Know how to manufacture complex shaped bench joinery

K1. Know how to select correct materials

You may be familiar with the expression 'a good carpenter always has the right tools for the job'. What you might not know is that tools mean more than just those of the hand and power varieties. Tools used in that context means equipment, but a carpenter and joiner also needs to have the right timber, sheet material, or man-made board at his or her disposal. After all, the best equipped toolbox in the world is useless when there is nothing around to work with.

Identification of timber and materials

The identification of timber has been covered in Unit 2008, pages 153–56. As you become more skilled, you will find that you are working with a variety of materials, not all of them timber. Sheet materials are often used in joinery work.

The standard size for sheet materials is 2400 × 1200 mm, although it is also available in 1800 × 1200 mm, 1800 × 900 mm and 1200 mm × 600 mm, in increments of 300 mm. The thickness of the sheet varies in multiples of 3 mm, and so can be 3 mm thick, 6 mm thick, 9 mm thick, and so on.

Timber sizes vary greatly with all dimensions and can be made available or machined to any size. Lengths of timber usually start at 1200 mm and rise in increments of 300 mm. The width and thickness of timber are still usually referred to in imperial sizes such as 3″ × 2″ as opposed to 50 mm × 75 mm. These sizes are numerous by usually going up by 1″ or 25 mm increments, such as 2″ × 1″, 2″ × 2″, 3″ × 2″.

Wood-based manufactured boards can also be used:

- Plywood is made from thin layers of timber called veneers glued together to form boards. The grain alternates across and along the sheet, giving strength and stability. The different grades of plywood are shown in Table 31.01.
- Laminated board is made from strips of wood laminated together and sandwiched between two veneers.
- Chipboard (particle board) is made from compressed wood chips and wood flakes bonded with a synthetic resin glue.
- Fibre board is made from pulped wood mixed with an adhesive and pressed into sheets. Available as hardboard, insulation board (softboard) and MDF.

Table 31.01 Plywood grades

Stamp	Grade	Use
INT	Interior	Internal use only – has low resistance to humidity or dampness
MR	Moisture resistant	Has a fair resistance to humidity and dampness
BR	Boil resistant	Has a fairly high resistance in exposed conditions
WBP	Weather and boil proof	Can be used in extreme conditions under continuous exposure (boats, buildings, etc.)

Working life

Wayne goes to the store to get some timber for a job. When he lifts the timber from the shelf, he notices that the wood is in wind (i.e. twisted).

How could this have happened? What could have been done to prevent it from happening? Think about what causes twisting in timber and the measures you could put in place to stop it from happening. Think about the storage of the materials.

Do you think Wayne can still use the timber? Give reasons for your answer. If he does use it what might the impact be? What is different about the wood that might affect its use?

Functional skills

When answering questions you will be practising **FE 2.2.1** Select and use different types of texts to obtain relevant information, **FE 2.2.2** Read and summarise succinctly information/ ideas from different sources and **FE 2.2.3** Identify the purposes of texts and comment on how effectively meaning is conveyed. Answering questions clearly and carefully will allow you to practise **FE 2.3.1–2.3.5** Write documents, including extended writing pieces and communicate information, ideas and opinions effectively and persuasively.

Plastics

The construction sector is the second highest user of plastics after packaging. Plastics in construction are mainly used for seals, profiles (windows and doors), pipes, cables, floor coverings and insulation. There are potentially further uses for plastics as they do not rot, rust or need regular re-painting. They also have strength with a lack of weight and are easily formable. Their light weight enables them to be easily transported and moved on site.

Plastic is used in the following areas:

- **Piping and conduit** – this accounts for 35 per cent of the production and use of plastics in the industry, producing cabling, rainwater goods and large diameter pipes for sewage, drainage and potable water; made from PVC and polyethylene.
- **Cladding and profiles** – these are used for windows, doors, coving and skirting made from uPVC. Exterior cladding using phenolic is replacing timber and traditional resins because it has a minimum fire risk.
- **Insulation** – this is generally produced from polystyrene rigid foam which is incorporated into panels or in the construction of walls and roofs. Plastic installation is simple to install, as well

as being very light and strong. This allows contractors to meet energy conservation regulations.

- **Seals and gaskets** – these are made from polymers and are able to resist the effects of weather and being deformed by external factors. PVC is used in windows and doors as a seal for roofs and linings.

Glass

Remember

The wrong type of glass used in the wrong place will not only be unsatisfactory to the client, it could also present a serious hazard to the people using the building.

Glass can serve two purposes in a building. The most obvious use is its function to let light into a room. It can also be used as a decorative feature. Whichever function you require, you will need to make sure that you have chosen the correct kind of glass. This will ensure that the final job is effective, attractive and, above all, safe.

Some of the most common types of glass are covered below. There are a large number of types of glass, in a range of patterns and tints.

'Ordinary' sheet glass

This glass is made by passing molten glass through rollers, giving it a flat finish. It can be used in domestic windows, but is more often used in greenhouses and garden sheds. This is because the rollers can introduce distortions into the glass, which can affect the view through the glass and the light coming in. This is less important for greenhouses and sheds than it is for domestic windows.

Float glass (plate)

Unlike ordinary glass, this glass is flat and distortion free. Rather than being pressed between rollers, the glass is 'floated' onto a bed of molten tin. This type of glass is often used in domestic windows as it has no distortion effect.

Energy-efficient glass

This is a type of float glass, but with a special thin coating added on one side. This coating prevents the thermal heat energy in the room from escaping, while still allowing heat and light from the sun to enter the room.

This special coating can give a slight tint to the glass – either brown or grey. The coating is not very robust, so cannot be cleaned normally or be exposed to weather conditions. As such, it is normally used in sealed double-glazed units, with the coating placed on the inside.

Self-cleaning glass

Again, this is a type of float glass with a special coating on one side. This photocatalytic coating uses the ultraviolet rays from the sun to steadily break down any organic dirt on the surface of the glass. It also has hydrophilic properties, which means that it uses rainwater to wash away the dirt that has been loosened. Combining these effects allows the glass to remain clean.

Patterned (obscured) glass

Made from flat glass, this glass has a pattern rolled onto one side during its manufacture. It can be used for both decorative effects and to provide privacy – for example, in a bathroom. It can be purchased in a range of coloured tints as well as plain.

Each pattern design has a rating that reflects the distortion level offered by the glass. This varies from 1 to 5, with 5 being the highest level of distortion.

Did you know?

On external glass, the distortion is usually on the inside, so that any dirt from the atmosphere and environment can be easily removed.

Toughened (safety) glass

This type of glass is made from applying a special treatment to ordinary float glass, after it has been cut to size and finished. The glass is heated so it begins to soften, and is then rapidly cooled. This produces a glass which, when broken, breaks into small pieces without any sharp edges. The treatment can increase the surface tension of the glass, meaning that it 'explodes' rather than shatters if broken, however, this effect is not a great hazard. It is ideal for glazed doors, low level windows and table tops (as it can resist the effects of high temperatures).

The treatment should only be applied after all the cutting and processing of the glass has been completed. If you attempt to cut toughened glass, it will shatter.

Wired glass

This glass incorporates a wire mesh (usually spaced at about 10 mm) in the middle of the glass. If the glass cracks or breaks, the wire can hold the glass together. It is ideal to use for roofing in areas such as garages or conservatories, where the unattractiveness of its 'industrial' look is less important. Wired glass will still break with sharp edges, so it is not considered a safety glass. It is clear and unobscured.

Laminated glass

This is made from two or more sheets of glass (or plastic) bonded together by a flexible, normally transparent, material.

Did you know?

Some glass is laminated for decorative internal finishes or to act as fire breaks.

If this type of glass is cracked or broken, the flexible material is used to hold any fragments of the glass in place. Any type of glass can be laminated, although it will still retain its original breaking properties.

Fabrics

Fabrics are textile materials made up from fibres to produce a cloth-type material. There is limited use within the construction industry for fabrics, but they are mainly used as vapour barriers or within the fibreglass process of covering a flat roof.

Metals

Key term

Anodising – a method of protecting metals by coating them with an oxide to prevent corrosion.

Ferrous metals contain iron. Non-ferrous metals do not contain iron, are not magnetic and are usually more resistant to corrosion than ferrous metals.

Non-ferrous metals are used widely within the construction industry as they do not rust, and are used for fixings among other things.

Table 31.02 Metals commonly used in the construction industry

Name	Identification	Properties/description	Uses
Copper		Extremely ductile and malleable non-ferrous metal when either hot or cold. A good conductor of electricity and heat and a warm reddish colour, which tarnishes and oxidises quickly. Can be easily damaged during use and so must be stored carefully to protect it.	Available in tube form as well as sheet, wire, rod and flat bar. Mostly used for water pipes and electrical wiring.
Aluminium		Extremely malleable, ductile and lightweight non-ferrous metal and a good conductor. Its natural colour is a dull greyish-white but it is often polished to a silver-white colour. A non-toxic metal that has a number of uses depending on its purity, but often alloyed with other elements to make it useful for construction purposes. Highly resistant to most forms of corrosion and if left untreated the metal's coating of aluminium oxide provides very effective protection.	Some types of window frames are made from aluminium instead of timber, due to its strength and conductivity. Excellent for stamping and forming, and can be dyed with different colours through a process called **anodising**.

Name	Identification	Properties/description	Uses
Lead		Very soft malleable non-ferrous metal, heavy in weight and highly resistant to corrosion. A bluish-white metal that tarnishes to a dull grey colour when exposed to air.	Has very poor electrical conductivity and is poisonous, so great care must be taken when melting and shaping this metal into flashings, etc. when used during roofing construction.
Galvanised steel		Highly resistant to corrosion due to being alloyed with a metal called zinc. Zinc helps to protect the iron from corrosion and will deteriorate before the iron in the steel. The combination of the two metals means the metal can withstand any corrosion-inducing circumstances such as salt water and moisture, rain and snow. Lightweight, fire resistant, basically maintenance free and extremely durable and resistant to scratches and abrasion.	An essential fabrication component used in construction, as well as other industrial industries such as the marine industry. Used for girders, frames, roofing, support beams, piping and handrails, etc. Matt-grey in colour and comes in tube form, sheets, ropes and flat bar.

K2. Know how to manufacture complex shaped bench joinery

To manufacture complex shaped bench joinery items you will need to know how to work with work programmes. These were covered in some detail in Unit 3002, pages 43–47.

A work programme is a method of showing very easily what work is being carried out on a building and when. The most common form of work programme is a bar chart. Used by many site agents, or supervisors, a bar chart lists the tasks that need to be done down the left side and shows a timeline across the top. A work programme is used to make sure that the relevant trade is on site at the correct time and that materials are delivered when needed. A site agent, or supervisor, can quickly tell from looking at the chart if work is keeping to schedule or falling behind.

Safety tip

When heated, galvanised steel is very toxic. You must remember this if you work in a situation where heated galvanised steel is likely to be present.

Methods of work

The work programme and methods of work you will need to use are different in a woodworking machine shop. Because you are working in a controlled environment, the planning of the work can be more regimented and timely, particularly as the delays that may affect site work, such as weather delays, will not affect the work. The planning involved with woodworking shops can be more specific, and the sequence of operations can also list the machinery used in task.

The way that the machines are set up in a workshop will help the flow of the job and aid in the planning process. A good workshop will have the machines set up in a specific way. For example, the table saw will be the first machine next to the store, as this machine is used first in the machining process. Next will be the planer, then the crosscut saw and any shaping machines such as the spindle moulder or mortise machine, and finally an assembly area.

Woodworking joints

At the end of this section you should be able to:

- understand simple jointing methods used on doors and windows
- state the correct jointing methods used on a common staircase
- identify the main joints used in the assembly of units and fitments.

Halving joints

Halving joints are used where two pieces of timber overlap, and can be used for lengthening or on corners. They are relatively simple joints to construct, but variations can make them stronger and more difficult, such as dovetail or mitred halving.

Edge joints

Edge joints are where the timber is joined along its edge and is used in lengthening, usually in the form of a finger-type joint.

Lengthening

Other edge-type joints are simple butt joints where a decorative piece of timber is placed along the edge of a material such as plywood to disguise the rougher edge. In some cases the edge joint can take the form of a tongued and grooved joint.

Joints used on doors and windows

During the manufacture of doors and windows, the mortise and tenon joint is extensively used. The type of mortise and tenon will depend on its location. Examples of this joint are described over the next few pages.

Scribes are often combined with other joints, particularly in doors and window construction where a straight-edged tenon will not fit over the shaped stiles or jambs.

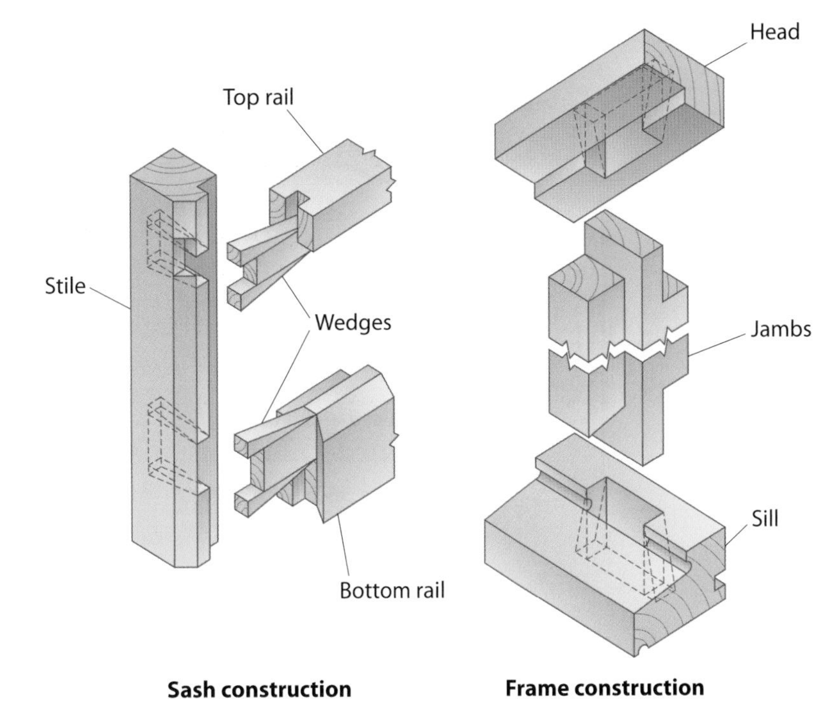

Figure 31.01 Joints in frames and sashes

Through mortise and tenon

In a through mortise and tenon joint, a single rectangular tenon is slotted into a mortise.

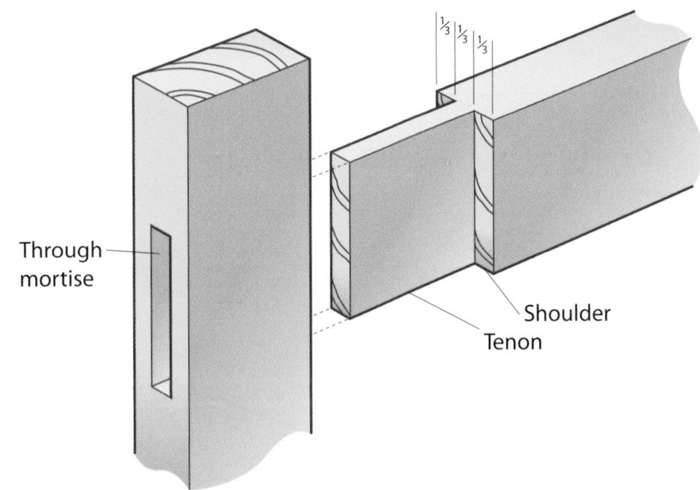

Figure 31.02 Through mortise and tenon

Stub mortise and tenon

In a stub mortise and tenon joint, the tenon is stopped short to prevent it protruding through the member.

Haunched mortise and tenon

In a haunched mortise and tenon joint, the tenon is reduced in width, leaving a shortened portion of the tenon protruding, which is referred to as a haunch (see Figure 31.04 on the next page). The purpose of the haunch is to keep the tenon the full width of the timber at the top third of the joint. This will prevent twisting. A haunch at the end of the member will aid the wedging-up process and prevent the tenon becoming **bridled**.

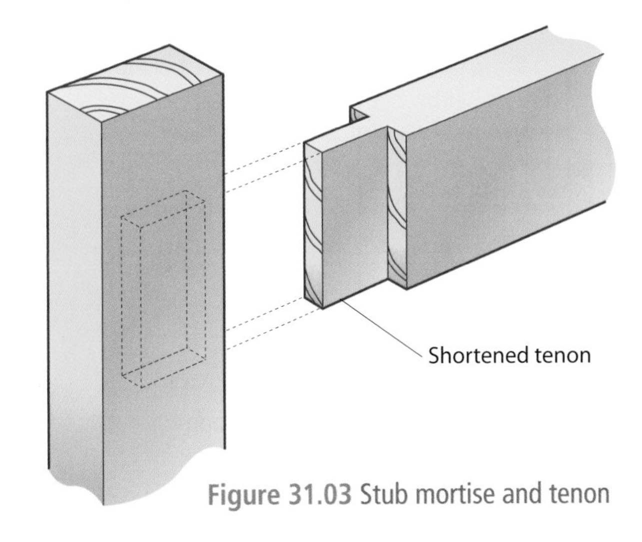

Figure 31.03 Stub mortise and tenon

Key term

Bridled – an open mortise and tenon joint. A tenon that has bridled is one that has no resistance and so is not secure.

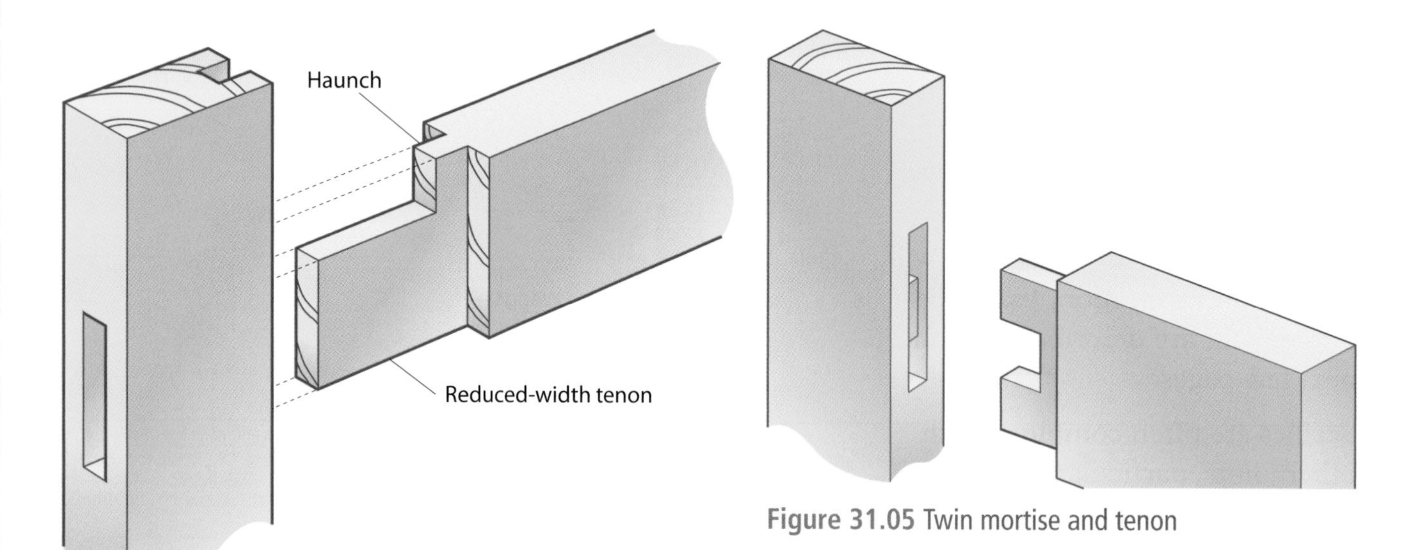

Figure 31.04 Haunched mortise and tenon

Figure 31.05 Twin mortise and tenon

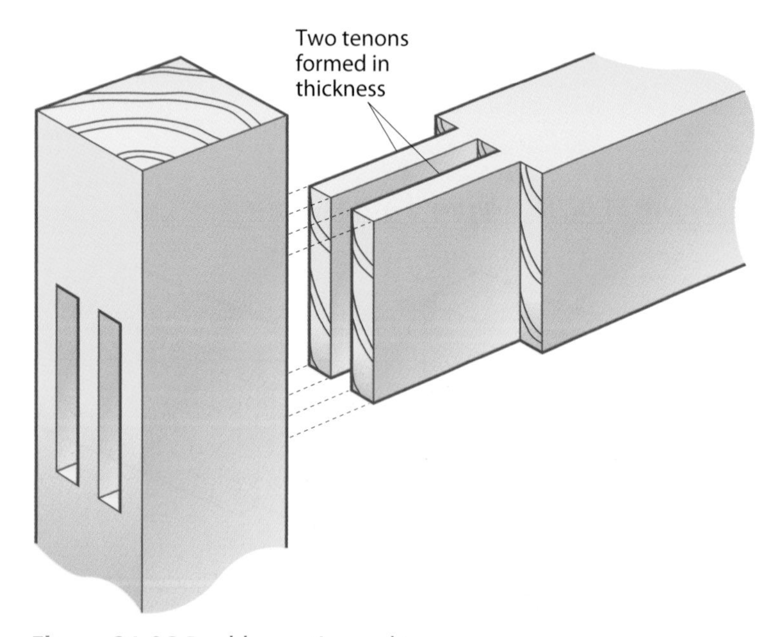

Figure 31.06 Double mortise and tenon

Twin mortise and tenon

In a twin mortise and tenon joint, the haunch is formed in the centre of a wide tenon, creating two tenons, one above the other.

Double mortise and tenon

For a double mortise and tenon joint, two tenons are formed within the thickness of the timber.

Stepped shoulder joint

Used on frames with rebates, a stepped shoulder joint has a shoulder stepped the depth of the rebate. This joint can also be combined with haunched, twin or stub tenons.

Twin tenon with twin haunch

A twin tenon with twin haunch joint is used on the deep bottom rails of doors.

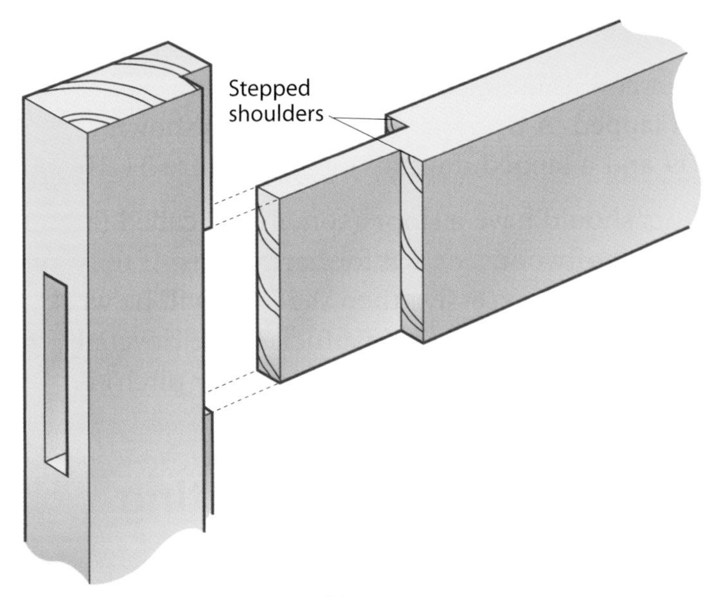

Figure 31.07 Stepped shoulder joint

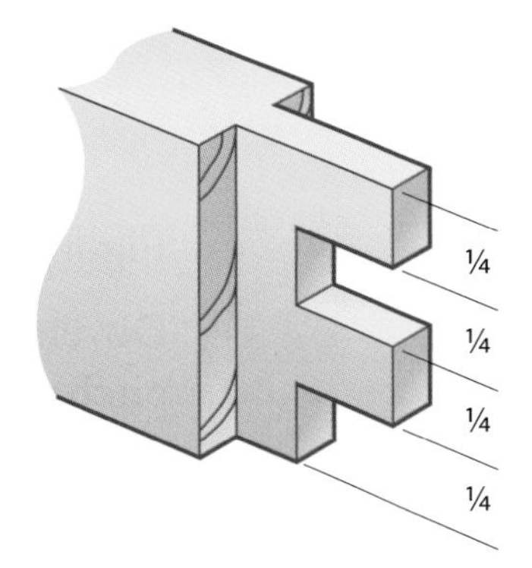

Figure 31.08 Twin tenon with twin haunch

Basic rules on mortise and tenon joints

The proportions of mortise and tenon joints are very important to their strength. Some basic rules are as follows:

- Tenon width should be no more than five times its thickness. This prevents shrinkage and movement in the joint. If more than five times, then a haunch should be introduced.
- The tenon should be one-third of the thickness of the timber. If a chisel is not available to cut a mortise at one-third, the tenon should be adjusted to the nearest chisel size.
- When a haunch is being used to reduce the width of a tenon, then about one-third of the overall width should be removed. The depth of a haunch should be the same as its thickness.
- Although a tenon should be located in the middle third of a member, it can be moved either way slightly to stay in line with a rebate or groove.

Remember

Components that are formed with mortise and tenon joints include:

- jambs
- sills
- transoms
- mullions
- stiles
- rails
- bars.

Joints used in units and fitments

During the design and setting out of units and fitments the most common joint used is the mortise and tenon, but this type of joint is not the best if there are forces likely to try to pull the joint apart. These are called **tensile forces**.

Key term

Tensile forces – forces that are trying to pull something apart.

Parts of a unit or fitment subject to such forces must incorporate a joint design that will allow for this. A drawer on a unit is often subject to tensile forces, so a dovetail joint is used.

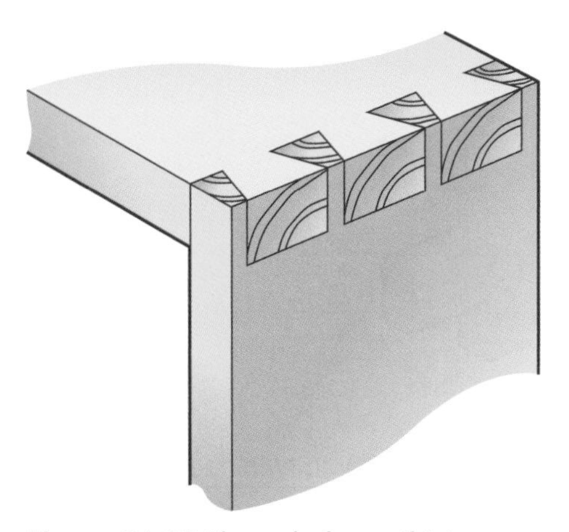

Figure 31.09 Through dovetail joint

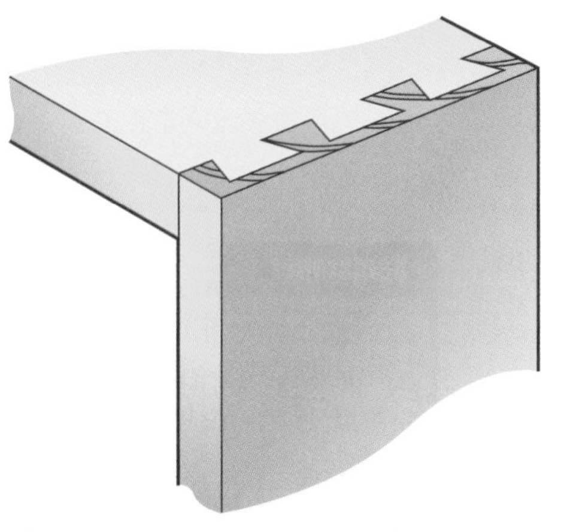

Figure 31.10 Lapped dovetail joint

Dovetail joints

The two most common types of dovetail joint are through and lapped. A through dovetail joint is shown in Figure 31.09 and a lapped dovetail joint in Figure 31.10.

Dovetail joints should have a slope (sometimes called the pitch) of 1:6 for softwoods, or 1:8 for hardwoods. If the slope of the dovetail is excessive then the joint will be weak due to short grain. If the slope is insufficient the dovetail will have a tendency to pull apart. The slope (or pitch) is shown in Figure 31.11.

Measure dimensions on existing work on site and compare to specifications

One of the most important jobs you will need to carry out when manufacturing bespoke and made-to-measure components is to ensure that these are made to the exact size required. This allows the component to be fitted in place first time, without any need for timely and costly alterations.

The best way to do this is to go on site and double check any measurements, so you can be certain that no slight errors have been made in the measurements supplied to you.

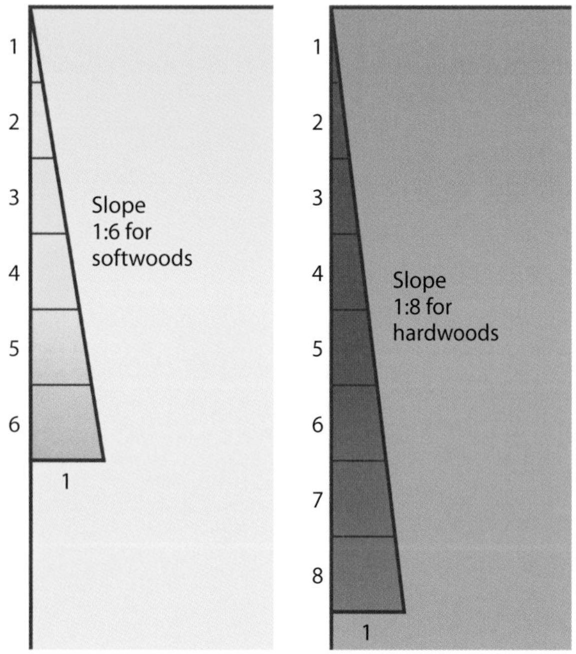

Slope 1:6 for softwoods

Slope 1:8 for hardwoods

Figure 31.11 Slope of a dovetail joint

Apply curvature techniques and skills to produce components

There are occasions when curved sections are required on a job. This could be to match existing work, or for aesthetic reasons. This can range from the structural work, such as the heads of doors and windows, or to more decorative work, such as tables and units. As timber is machined straight, you need to work with it to make the curves required. This can be done through lamination with thin strips or by gluing together larger sections.

Laminated components

Laminated components are widespread throughout the construction industry, and are used more and more nowadays, the most basic form of laminate being plywood. Laminates are simply thin strips of timber that are glued together. In carcassing, particularly for roofing, laminates have an advantage over solid timber as laminated beams can be made to any length. Most laminated beams are now mass produced and machined with specialist machines, but for the purpose of this book we will look at traditional methods for laminating a curved piece and a long beam.

Laminating a curved piece

Laminated curved pieces can be used for almost any purpose, and can be made to any radius, or even a variety of radii on a single piece to create a serpentine effect. Laminated timber can even be used for shaped door heads.

The amount of laminate needed depends on the thickness required, which in turn depends on the curve. For our example, the finished timber will be 60 × 60 mm with a curve that is not too severe, so we will be using 3 mm thick laminate.

Common sense might say that 20 × 3 mm laminates will give you a 60 mm finish, but this will probably not be the case, as there will be a thin layer of adhesive between each laminate, increasing the finished size to more than 60 mm. Nineteen laminates may be enough, but this may leave you slightly short of the 60 mm. One option, depending on the precision of the finish required, might be to use 20 laminates, then plane the laminated component to the exact size using a spoke shave plane.

It is good practice to machine more laminates than you will use, as this allows for any strips with shakes or dead knots, and any damage that might occur in the manufacturing process. The 3 mm laminates should be sawn and ideally planed to the exact thickness. Ensure that all the laminates are uniform and clean to create a good bond.

Once the timber is prepared to the correct size, it is ready for shaping. Shaping can be achieved by the following three methods:

- **Dry clamping** – the timber is placed into the jig and forced into shape by the clamp. This method yields poor results as the timber needs to be kept in the clamps for a long time, with no guarantee that it will keep its shape.
- **Wet clamping** – the timber is soaked, then clamped in place. This produces better results than dry clamping, but again the timber must be kept clamped until it has dried out, which can be time-consuming. Also, with this method there is a good

Did you know?

A steam box must be manufactured from a suitable material such as WBP plywood – otherwise the timber making up the box will break down from the effects of the steam.

Safety tip

When using a steam box, always wear the correct personal protective equipment (PPE) as the timber will be hot and placed immediately into the jig and clamped in place.

chance that the timber will spring back and not meet the required radius.

- **Steam clamping** – the timber is placed into a steam box for a set time, then clamped into place. This is by far the best method: it does not take as long and gives the best results.

  Timber is placed into the steam box and steam is pumped in. The heat allows the timber to be bent, while the moisture stops the timber becoming brittle and snapping. The length of time that the timber should be in the box depends on the type and thickness of the timber and the severity of the bend required, but 1 hour per 6 mm thickness is a good guide time (it is better to oversteam the timber rather than understeam). Timbers should be placed into the box with piling sticks between them to allow the steam to circulate. After the required amount of time the timber should be removed.

The clamping of the timber can be done in several ways, but the best way by far is to make up a jig. The jig in Figure 31.13 is an example only: any type of jig can be used.

With modern advances in adhesives it is now possible to glue up the timber straight out of the box and clamp it into the jig, but traditionally the timber is placed into the jig dry (i.e. not glued) and left overnight to dry out, then glued and clamped the next day.

Use the adhesive manufacturer's guidelines to decide the length of stay in the jig. Once dry, the laminated component can be cleaned up ready to be fitted. There may be a slight spring back once the clamps are released. This can be overcome by making the radius on the jig slightly more than required: when the spring back occurs it will leave the laminate at the correct radius.

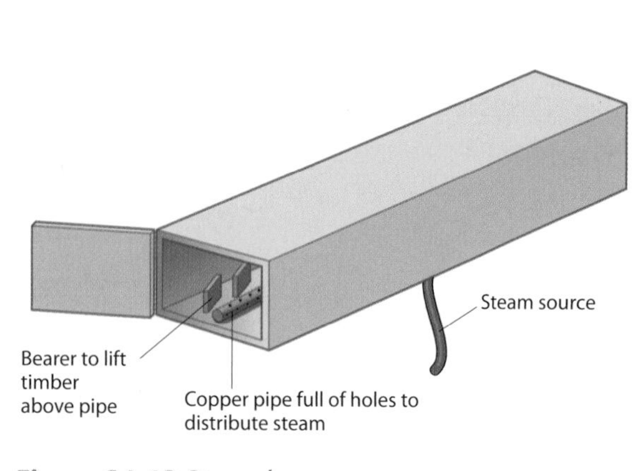

Figure 31.12 Steam box

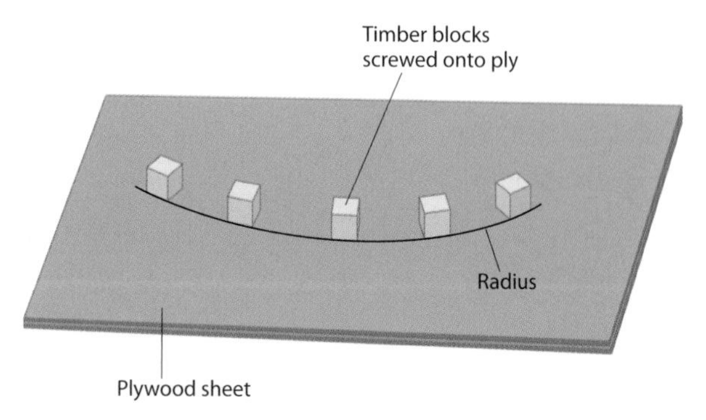

Figure 31.13 Typical jig

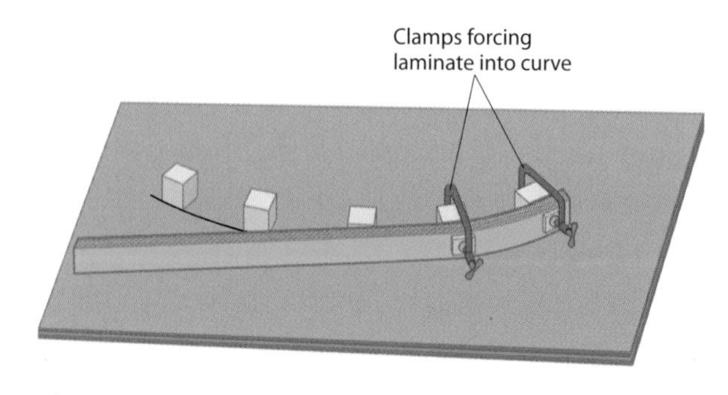

Figure 31.14 Timber clamped into a jig

Laminating a long beam

Laminating a long beam is a simple process with no need for steaming (unless the beam has a bend in it). A laminated beam can be made to any length because the laminates can be staggered. Staggered joints give strength to the beam.

The procedure starts with the required number of laminates planed to the required thickness, which for this example is 50 laminates at 5 mm. The laminates are then glued together and clamped.

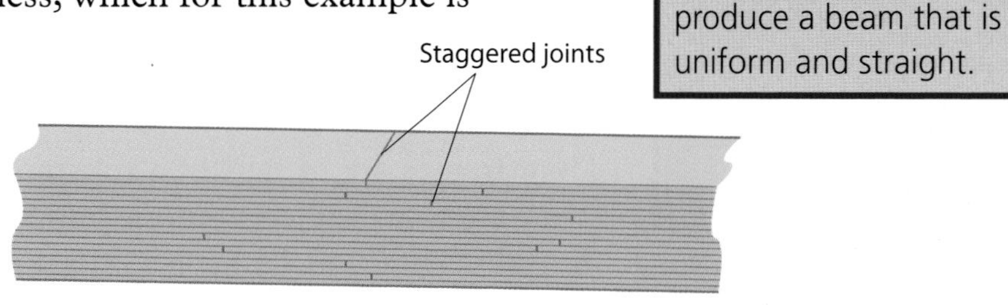

Figure 31.15 Staggered laminates

Remember

When gluing and clamping a long beam, take care to ensure that the laminates are tightly clamped together and laid flat on bearers. This will produce a beam that is uniform and straight.

Working life

You have been asked to create a curved laminated component that will be used in the head of a door frame. Your employer tells you to cut up some laminates and put them in the steam box for about an hour.

Do you have all the information you need? What other information do you require? What could the outcomes be if you don't find out the extra information?

Solid and built up

Creating curved work in solid timber is an easier process than laminating. However, this can be more costly as there is often a lot of wastage. A thin plywood template is accurately created and then secured onto the solid work piece.

The work piece is roughly cut to the shape of the template using a band saw. It is then cleaned to the exact shape of the template, using either a router or spindle moulder.

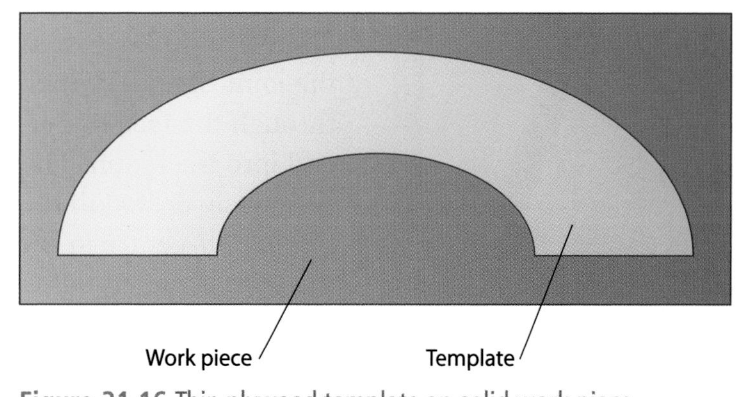

Figure 31.16 Thin plywood template on solid work piece

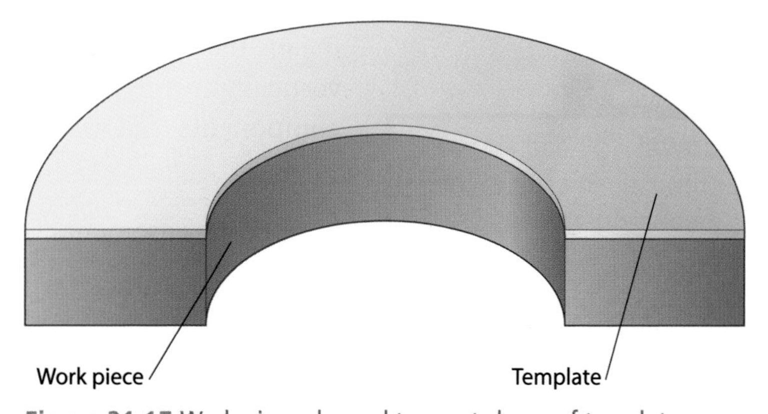

Figure 31.17 Work piece cleaned to exact shape of template

Fit, assemble and finish bench joinery

A joiner's main task is fitting and assembling bench joinery in place. This section will look at some of the basic joinery you will need to be able to fit in place.

Remember

It is very important to make sure the frame or lining fits the opening well and that it is level. If this is not done, you will end up with doors and windows that don't hang properly and don't open and close properly.

Frames and linings

Window and door frames/linings are fitted into openings left in masonry and hold the window or door in place.

Door frames and linings

Door frames are usually of a substantial and solid construction. They are mortised and tenoned and comprise heads, jambs and sills. They usually have a rebate cut from the solid timber (see Figure 31.18).

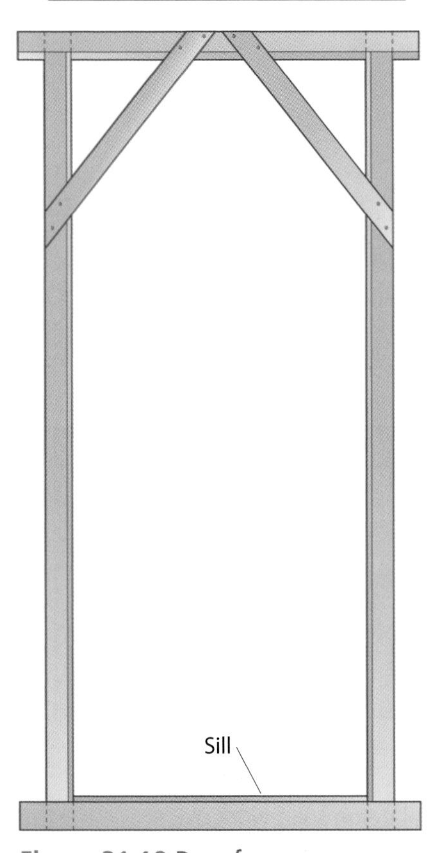

Figure 31.18 Door frame

Door linings are of much lighter construction than frames and are used exclusively for internal doors. They don't usually have sills and are normally jointed between head and jambs with a form of housing joint, such as tongued housing (see Figure 31.19). Door linings usually have planted (nailed on) stops.

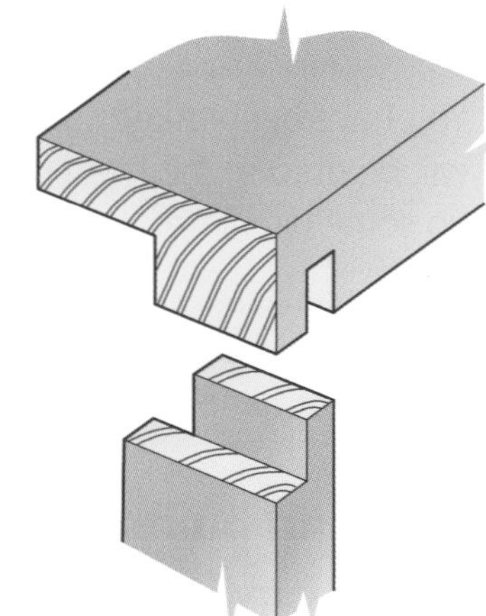
Figure 31.19 Tongued housing joint

To give additional strength and to pull the joint tight, a timber **dowel** is fixed through the face of the frame's head and into the tenon. The hole should be previously drilled in the face and then slightly off-centre in the tenon. When the dowel is knocked home the joint is pulled tight. This method is known as draw boring. The hole needs to be offset as shown in Figure 31.20.

Window frames and linings

The majority of modern windows are casement windows. This means they are hinged on the side allowing them to swing open vertically (similar to the way a door opens). The two main components that make up a casement window are the:

- frame
- opening casement.

Like a door frame, a window frame consists of a head, sill and jambs. When the frame is to be divided, members called **mullions** (vertical dividers) and **transoms** (horizontal dividers) are included (see Figure 31.21).

Key terms

Dowel – a headless wood or metal pin used to join timber together.

Mullions – vertical dividers in a window frame.

Transoms – horizontal dividers in a window frame.

The opening part of the window, known as the sash, consists of a top rail, bottom rail and two **stiles**. When the opening of the casement sash is to be divided up further, glazing bars are used (see Figure 31.22). The procedure for hanging a casement sash is exactly the same as for a door, only usually on a smaller scale.

When the dowel is knocked in, the joint is pulled tight

Figure 31.20 Draw boring and pin

Hanging casement window sashes

1. Mark the hanging side on both the frame and the sash.
2. Cut off any horns (the waste stock left overhanging on the stiles to aid cleaning up and prevent damage to sash corners prior to installation).
3. Plane to fit the hanging stile and plane the sash to the required width, running parallel with the side of the frame.
4. Plane to fit the top and bottom of the frame.
5. Mark out and cut the hinges and screw one leaf of each hinge to the sash.
6. Offer up the sash to the opening and screw the other leaves of the hinges to the frame.
7. Make fine adjustments if needed and fit specified ironmongery.

Key term

Stiles – the vertical members of the sash. The hinges are fitted on to one of them.

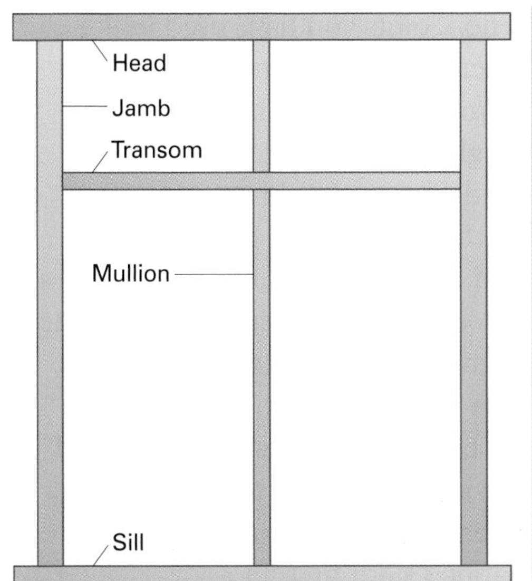

Figure 31.21 Window frame

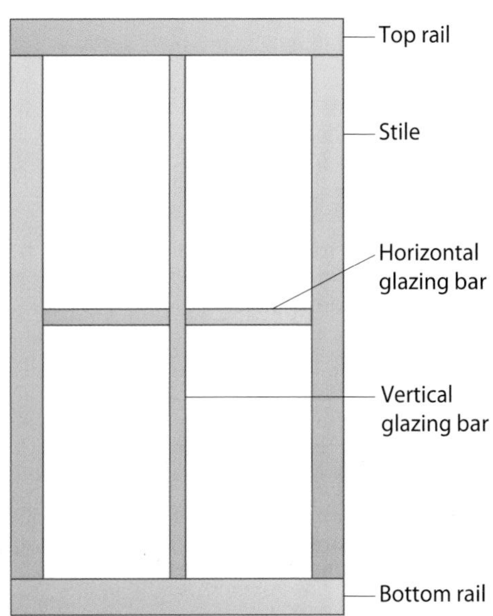

Figure 31.22 An opening casement

Remember

Frames should be assembled 'dry' (without glue) first to check that all the joints are a good fit and the frame is the correct size, square and not winding (see page 308).

Remember

Once the frame is assembled dry, check the sizes one more time. As soon as the frame is glued together it will be too late to adjust anything.

Squaring up

When the frame has been assembled, glued and cramped up (held together while drying with clamps), it should be tested to make sure that it is square, meaning that the corners are at right angles. The most accurate way of doing this is to compare the diagonals using a squaring rod, which is a piece of rectangular timber with a small nail or panel pin knocked into the end (see Figure 31.23). The protruding nail is placed in one corner of the frame and the corner of the opposite diagonal is marked on the rod with a pencil. This procedure is repeated for the two other corners of the frame. If both pencil lines match up, then the frame is square.

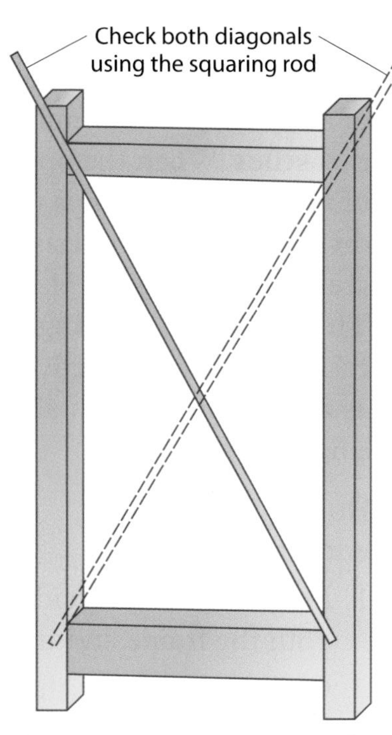

Figure 31.23 Squaring up with a squaring rod

Remember

If the pencil lines do not match up, the frame needs to be adjusted by angling the clamps and pushing the frame. This should be done until the two pencil marks match up, meaning that the diagonals are the same length and the frame is square.

Remember

When you have made any adjustments to remove winding, go back and make sure the frame is still square.

Checking for winding

You will need to check that the frame you have assembled is not twisted, by using winding rods. These are simply two pieces of timber which are laid parallel across the frame when it is lying flat on the workbench. Close one eye and look across the winding rods. They should be parallel. If they are, the frame has no twist.

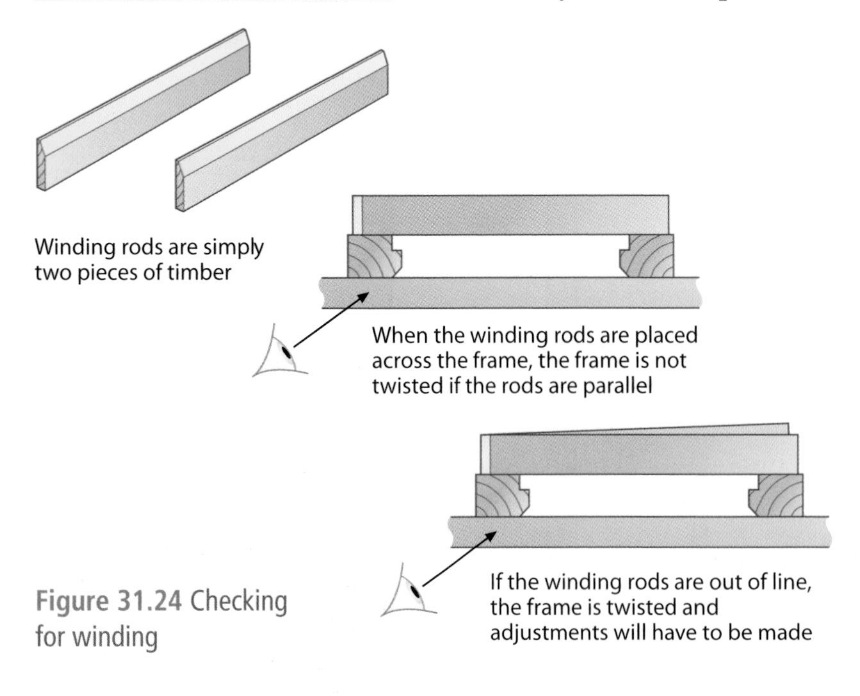

Figure 31.24 Checking for winding

Wedging up

The haunched mortise and tenon joints you have used in the assembly of your door or window frame are not only held in place by glue, but they are also wedged. Wedging up should be done after the frame has been glued, squared and checked for winding. It involves placing a small wedge on either side of the tenon and carefully driving it in to ensure a good tight joint. To help keep the frame square and the joints pulled in tight, it is best to drive in the external wedge (the haunch side) first.

Doors

There are a number of different types of door assembled by carpenters and joiners. In this section we will only look at panelled and glazed doors, since they incorporate all the principles that you need to know about to be able to assemble any door.

Panelled doors have a frame made from solid timber rails and stiles. When made by the bench joiner in the workshop, they will almost certainly incorporate a mortise and tenon type of joint. The frame will either be grooved or rebated to receive a number of either plywood or timber panels. Glazed doors are made in a similar fashion to panelled doors; however, one or more of the panels is replaced by glass.

The following assembly procedure can be carried out for all types of framed door:

1. Assemble the frame dry to ensure that all joints are tight, the right size, square and not winding.
2. Before final assembly, clean up the inside edges of all components as this will be extremely difficult once the frame has been glued.
3. Glue, assemble, cramp up and check again for square and winding.
4. When you are satisfied that everything is correct, the door can be wedged up.
5. Clean up the rest of the frame and prepare for finishing.

Find out

How many different types of door are available? Why do you think there are so many?

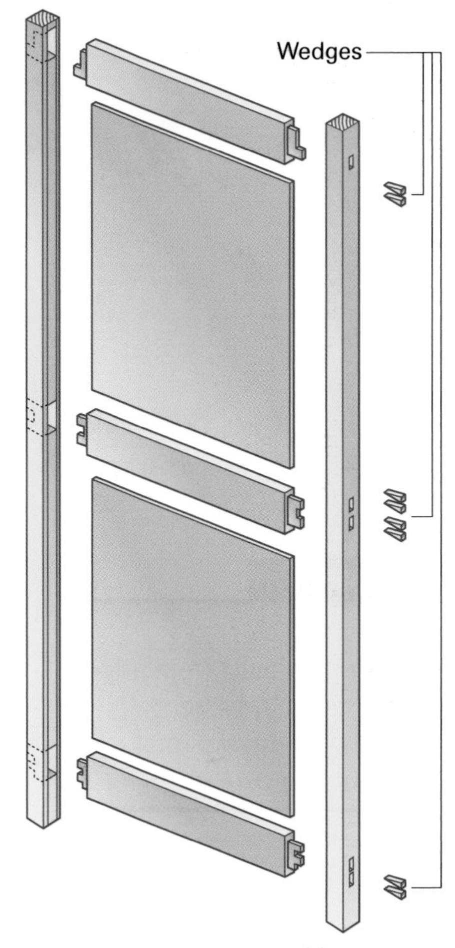

Figure 31.25 Door assembly

Stairs

Stair assembly was looked at in Unit 2008, pages 166–69. You may want to look back at this section to refresh your understanding before reading on.

This section will look at assembling stairs in more detail. Geometrical staircases are covered in Unit 3032, pages 334–36.

Start by laying one string on the bench with the housings facing upwards. Put in place all the treads and risers, ensuring that the tread/riser ends and the housings are glued. Place the other string on top of the treads and risers, glue, then apply clamps.

Drive glued wedges into the housings to force the treads and risers to the front of the housing, making sure there are no gaps. This helps to strengthen and secure the treads and risers.

Remember

Strings are the main boards which the treads and risers are attached to. Treads are the flat, horizontal parts of the step; risers are the vertical parts of the step. Newel posts are the uprights at the bottom of the stairs and at each turn in the staircase.

Figure 31.26 Tread and riser assembly

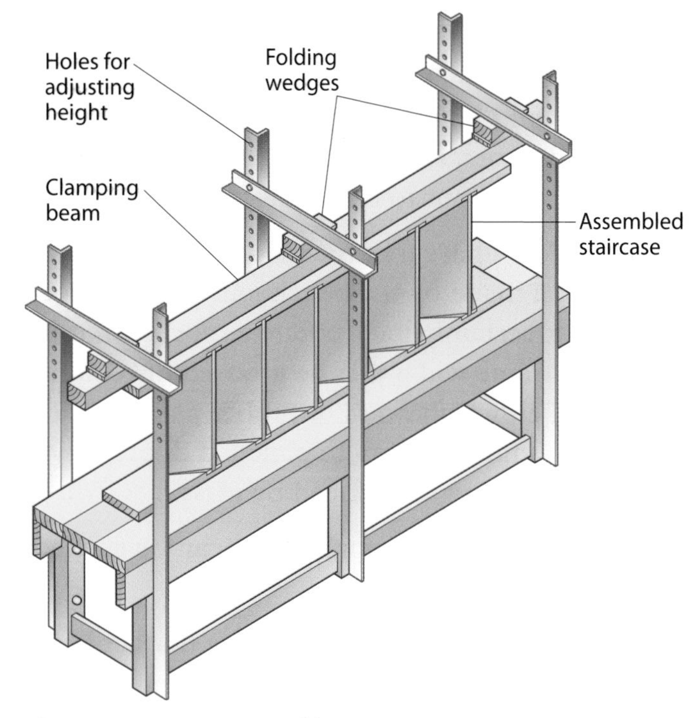

Figure 31.27 Stair assembly

Remember

The bench you are working on must be sturdy and level as it is essential that the strings are totally straight and parallel.

Once all the wedges are in place, check the front of the staircase to ensure that the treads and risers are situated perfectly. Now screw the bottom of the risers to the back edge of the treads. Finally, fit glue blocks where the tread meets the riser, at the internal angle.

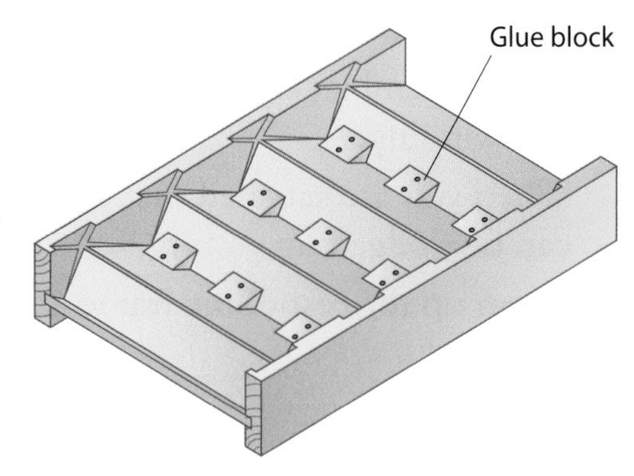

Figure 31.28 Glue blocks fitted

Once the adhesive has dried, remove the clamps, then clean up the staircase, protect it with hardboard or bubble wrap, and send it on site to be fitted.

Remember

If the string is not going to be seen or painted, it is good practice to screw through the strings into the treads, making the stairs stronger.

Working life

You have been tasked with manufacturing a straight flight staircase. You mark out the strings individually and then machine the treads, risers, etc. You then router out the strings and prepare the staircase for assembly. Once the frame is assembled, you see that the stair seems to be out of square: one string projects past the other by at least 2", so when the staircase is laid flat, the treads are well out of level. What could have caused this? What can be done to rectify it? What could have been done to prevent it?

Spandrel panelling

Spandrel panelling is used on a stairway that is open on one side (not in between two walls). Spandrel panelling is a way of finishing the underside of the stairs between the bottom of the string and the floor.

The method used for constructing spandrel framing is similar to the way that wall panelling is constructed. The basic framing is constructed using timber studwork with the panelling being fitted to it.

Figure 31.29 Spandrel panelling on stairs

Wall panelling is designed to provide a decorative finish to a room and can be found in places such as courts of law and executive offices. It is usually set at one of three heights:

- **Dado panelling** – where the panelling runs to the height of the dado rail.
- **Three-quarter panelling** – where the panelling runs to the top of the door.
- **Full-height panelling** – where the panelling runs from floor to ceiling.

The panelling can be made up in a variety of ways, depending on the type and style of the house. The first thing to do is to make a frame or fix battens to the wall, onto which you can then fix your panelling. Once the panelling is in place, fix the capping pieces and skirting to finish the panelling off.

On the next page, three different types of panelling are illustrated:

- matchboard panelling
- flush dado panelling
- framed dado panelling.

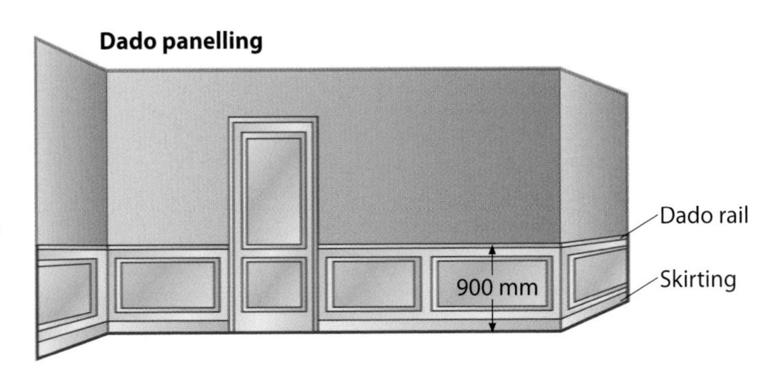

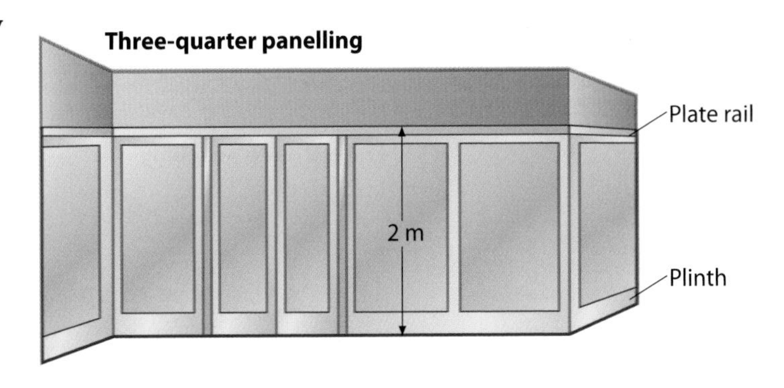

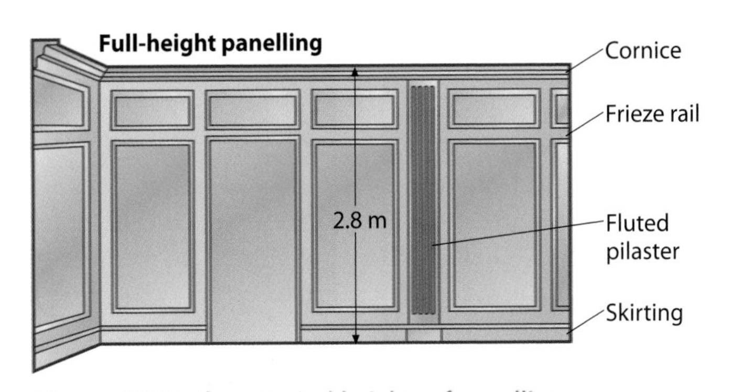

Figure 31.30 Three typical heights of panelling

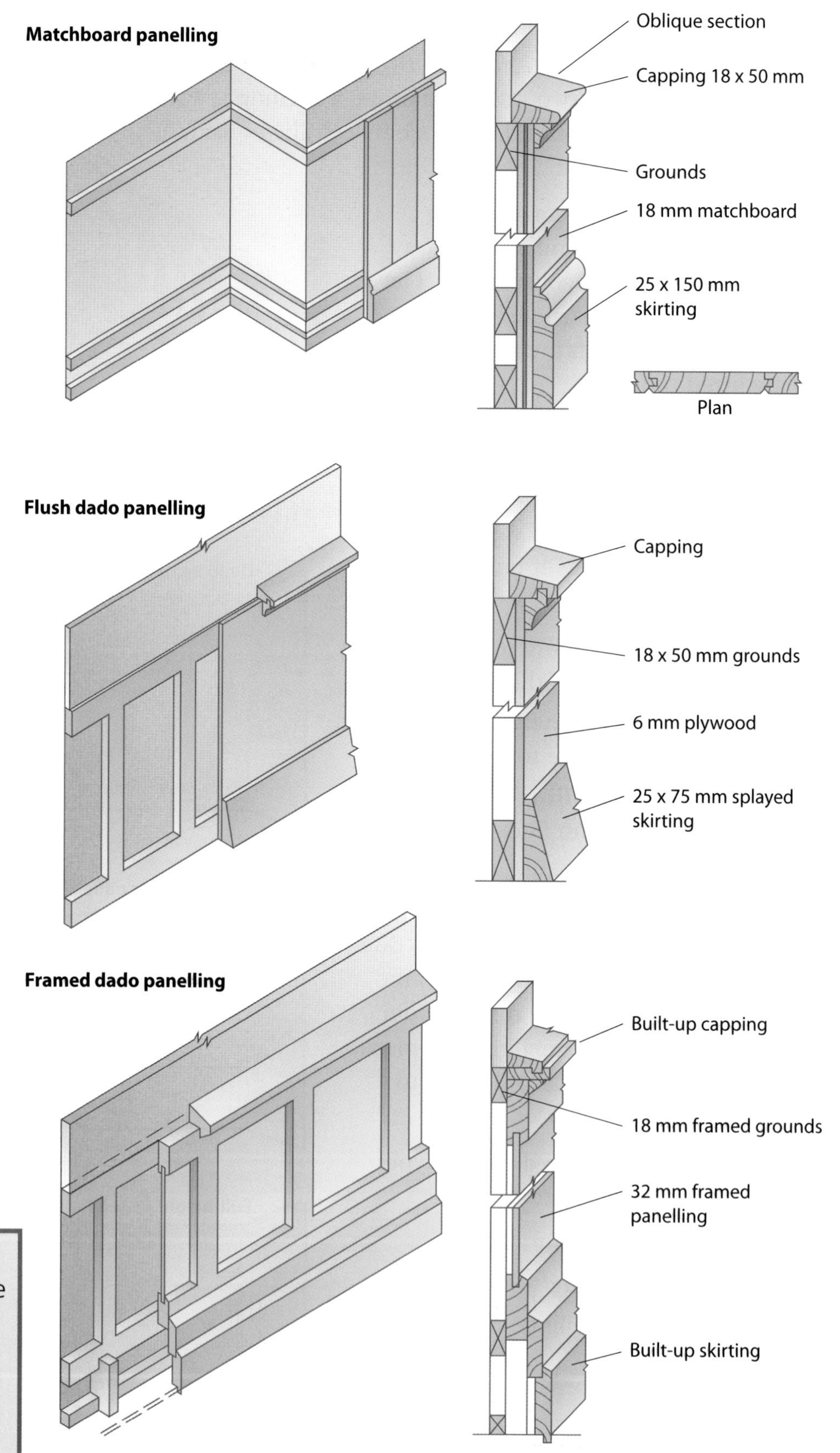

Figure 31.31 Three types of panelling

Find out

Using the Internet and other sources, locate some sections through framed dado panelling and sketch how this panelling would be completed for stairways.

Units

The most common type of unit you will probably work with is kitchen units, although if you understand the basic principles of assembling these, you will find that the assembly of other types of unit (for example, bathroom, bedroom) can easily follow a similar procedure. Most types of unit are made from melamine-faced chipboard, MDF, block board, plywood and, sometimes, solid timber.

There are a number of different methods of unit construction, but we will only look at probably the two most common: box and framed.

Box construction

This is also known as 'slab construction'. The unit is composed of vertical standards, rails and shelves, and the plinth and bottom shelf are often an integral part of the unit. The back panel holds the frame square.

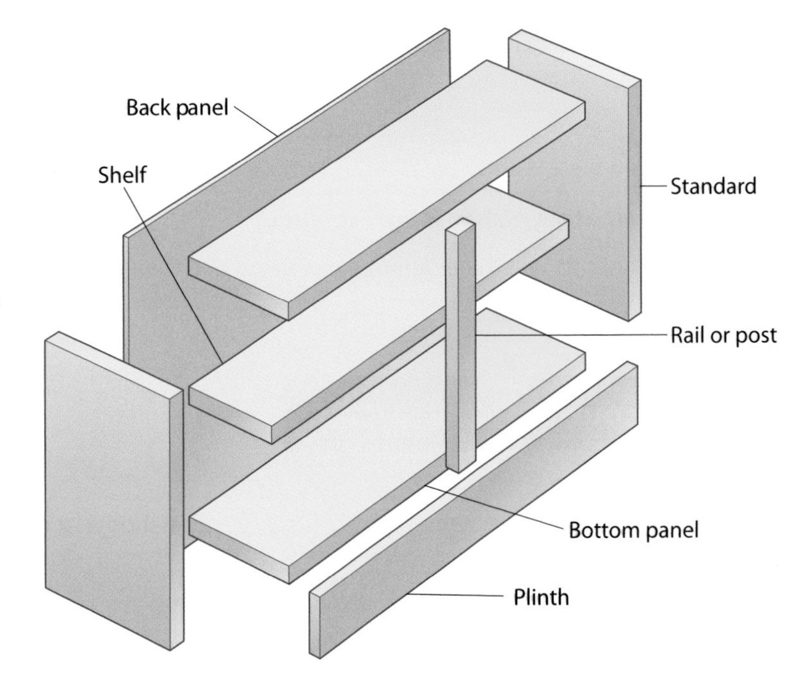

Figure 31.32 Box construction

Framed construction

This is also known as 'skeleton construction'. The units are composed of either a pair of frames (a front and a back frame) joined together with rails, or cross frames joined together by rails at the front and back (see Figure 31.33). The plinths and drawers are usually built separately in this type of unit. The frames are mortise and tenoned and assembly follows the same procedure as doors or casement window sashes.

As with the other items we have covered, units must be dry assembled to ensure that the frame is the right size, square, and not winding. You must remember to square up, check for winding and wedge up, using the procedures outlined on page 308.

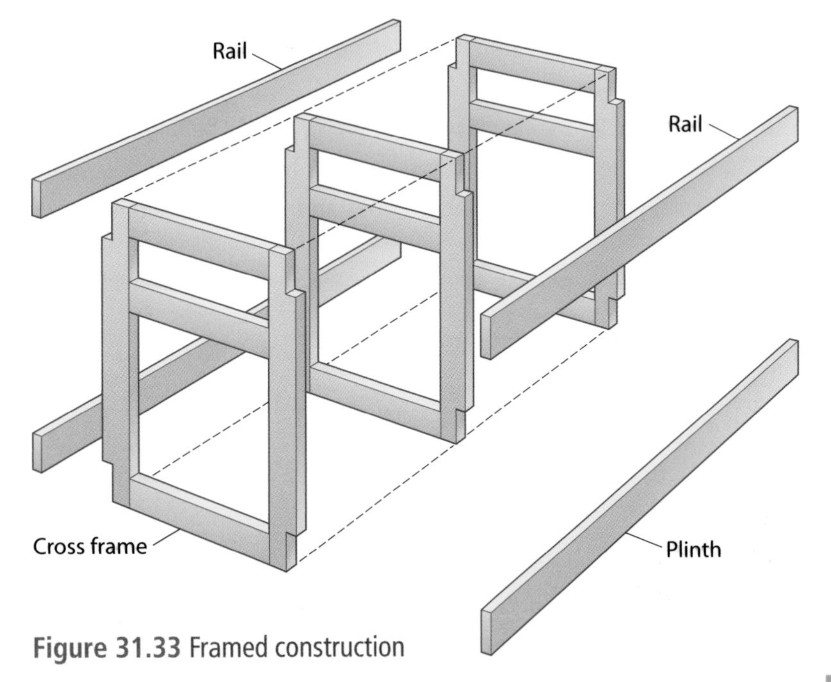

Figure 31.33 Framed construction

FAQ

Do face and edge marks have to be used?

No, they don't, but without them there is a good chance that mistakes will be made. Using face and edge marks is considered to be good practice.

Which type of joint is best for a door: the mortise and tenon or dowel?

The mortise and tenon is the more traditional way, but the dowel is far easier. Both joints work well and the choice of which to use is down to the joiner and the client.

Why are glue blocks fitted in stairs instead of just screwing down through the face of the tread?

Glue blocks prevent any movement once the stair is in use. They are preferred to screwing, as screwing is unsightly and there is a risk of splitting the nosing.

When laminating, what is spring back?

Spring back occurs when the clamps are released and the timber springs back into its natural state.

Check it out

1. Prepare a method statement explaining the information that should be used to choose the correct material for a job.
2. Using diagrams and evidence, demonstrate and explain some of the different uses for joints.
3. Explain why laminated beams are preferred to solid timber and suggest a suitable time for steaming laminates.
4. Explain when a tenon with a stepped shoulder should be used.
5. Prepare a method statement describing the general assembly procedure used for both doors and windows.
6. Explain what the first stage in assembly should be and why.
7. Explain the best way of cutting out the housings on a stair string.
8. Explain the difference between box and framed unit construction.

Getting ready for assessment

The information contained in this unit, as well as continued practical assignments that you will carry out in your college or training centre, will help you with preparing for both your end of unit test and the diploma multiple-choice test. It will also aid you in preparing for the work that is required for the synoptic practical assignments.

The information contained within this unit will aid you in learning how to identify and calculate the materials and equipment required for manufacturing complex shaped joinery products.

You will need to be familiar with:

- selecting the correct materials
- manufacturing complex shaped joinery products.

This unit will have made you familiar with the manufacture of complex joinery products. This builds on the setting out and marking out of joinery products covered at Level 2. You will need to remember the information you learned in those units in order to meet the requirements of this unit. For learning outcome two, this unit has introduced the correct techniques to produce complex components. You will need to use this knowledge during your practical assignments to be sure that you are creating the correct components with suitable methods. As with marking out and setting out, you will need to complete work within the time allowed in the work programme and to deadlines. You will also need to be sure that you are maintaining tools correctly as the work progresses.

Before you start work on the synoptic practical test, it is important that you have had sufficient practice and that you feel that you are capable of passing. It is best to have a plan of action and a work method that will help you. You will also need a copy of the required standards, any associated drawings, and sufficient tools and materials. It is also wise to check your work at regular intervals. This will help you to be sure that you are working correctly, and help you to avoid problems developing as you work.

Your speed at carrying out these tasks will also help you to prepare for the time limit of the synoptic practical task. But remember that you should not try to rush the job as speed will come with practice and it is important that you get the quality of workmanship right.

Make sure you are working to all the safety requirements given throughout the test and wear all appropriate personal protective equipment (PPE). When using tools, make sure you are using them correctly and safely.

Good luck!

CHECK YOUR KNOWLEDGE

1. Which of the following sheet materials is made from pulped sugar cane?
- **a** Hardboard
- **b** Chipboard
- **c** Block board
- **d** Laminated board

2. Which type of glass has a photocatalytic coating?
- **a** Wired glass
- **b** Laminated glass
- **c** Energy-efficient glass
- **d** Self-cleaning glass

3. Tenon width should be no more than:
- **a** 3 × thickness
- **b** 4 × thickness
- **c** 5 × thickness
- **d** 6 × thickness

4. How many methods are there for shaping laminated timber?
- **a** 1
- **b** 2
- **c** 3
- **d** 4

5. Which is the best method for shaping laminated timber?
- **a** Dry clamping
- **b** Wet clamping
- **c** Steam clamping
- **d** Fire clamping

6. What is a good guide time for steaming timber?
- **a** 1 hour for every 6 mm thickness
- **b** 2 hours for every 6 mm thickness
- **c** Half an hour for every 6 mm thickness
- **d** 3 hours for every 6 mm thickness

7. What tools are best to use when steaming timber?
- **a** Screws
- **b** Jig and clamps
- **c** Nails
- **d** All of the above

8. Laminated beams are preferred because:
- **a** they can be made to any length
- **b** they are more economical
- **c** they warp and twist
- **d** they cannot be cut cleanly

9. What are most units made from?
- **a** Plastic
- **b** Melamine-faced chipboard
- **c** Aluminium
- **d** Hardboard

10. How many main methods of construction are used in basic units?
- **a** 2
- **b** 3
- **c** 4
- **d** 5

Functional skills

In answering the Check it out and Check your knowledge questions, you will be practising **FE 2.2.1** Select and use different types of texts to obtain relevant information and **FE 2.2.2** Read and summarise succinctly information/ideas from different sources.

You will also cover **FM 2.3.1** Interpret and communicate solutions to multistage practical problems.

Unit 3032

Know how to produce complex shaped joinery product details

Having learnt the basics of component manufacture in Unit 3031, you will now be able to look at the construction of more complex components. These joinery items are the more skilled jobs that clients will look for from a professional carpenter. As with more routine items, selecting the wrong timber, or using the wrong tools for the job, will result in poor quality work and will lead to schedule delays and added costs if work needs to be re-done.

This unit contains material that supports the following NVQ unit:

- VR 27 Produce complex shaped product details

This unit will cover the following learning outcomes:

- Know how to interpret information for setting out
- Know how to select resources for setting out
- Know how to set out for complex shaped bench joinery

K1. Know how to interpret information for setting out

Setting out requires a knowledge and understanding of a wide range of information. You will need to be able to use both site documentation and nationally recognised sources of information to make sure that you are building something as it is required by the client and which meets legal requirements.

Information used for setting out

Remember

The specification for joinery components will state things that are not available on any drawings, ranging from materials used to the type of finish.

We have covered some of the information you will need to use to decide on setting-out details earlier in this book. For more information on setting out, look back at the sections on plans, specifications and schedules in Unit 3002, pages 20–34.

The setting-out process will rely heavily on certain contract documents, such as plans. When producing joinery components these will be the assembly or detail drawings. The drawings are usually at a scale of no more than 1:10 and will show how components are assembled, giving cross-sectional details. Schedules are only used where there is a range of different components to be made.

The other main items you will use to make decisions about setting out are covered below.

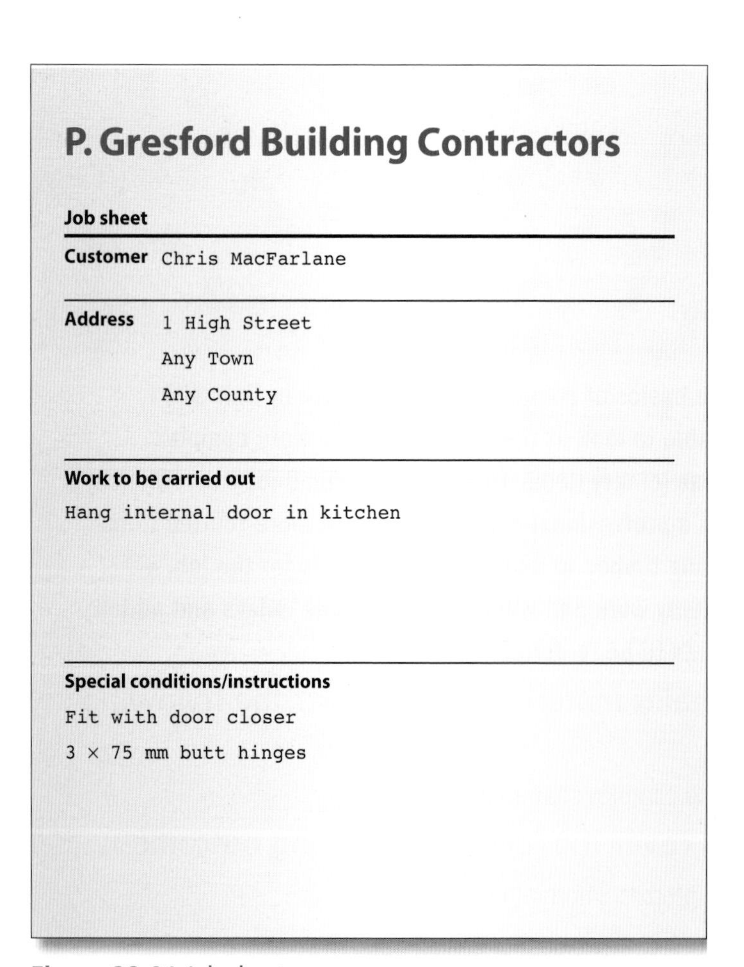

P. Gresford Building Contractors

Job sheet

Customer Chris MacFarlane

Address 1 High Street
Any Town
Any County

Work to be carried out
Hang internal door in kitchen

Special conditions/instructions
Fit with door closer
3 × 75 mm butt hinges

Figure 32.01 Job sheet

Job sheets

Job worksheets are used to record work that has been done. They are used when the work has already been priced. Job sheets enable the worker to see what needs to be done and the site agent or working foreman to see what has been completed.

Building Regulations

The Building Regulations cover the planning and construction of buildings. The current law is the Building Regulations 2000, covering England and Wales. It was amended in April 2006 to take into account wheelchair access and more environmentally friendly practices.

Scotland is governed slightly differently and is covered by the Building (Scotland) Act 2003. Northern Ireland is covered by the Building (Amendment) Regulations (Northern Ireland) 2006, which came into effect in November 2006.

The main purpose of the Building Regulations is to ensure the health, safety and welfare of all people in and around buildings, as well as to further energy conservation and to protect the environment. The regulations apply to most new buildings as well as any alterations to existing buildings.

The regulations are broken down into categories which contain an 'approved document', detailing what is covered:

Figure 32.02 The Building regulations help to protect the environment

Approved document A (Structural safety)

A1 – Loading

A2 – Ground movement

A3 – Disproportionate collapse

Approved document B (Fire safety)

B1 – Means of warning and escape

B2 – Internal fire spread (linings)

B3 – Internal fire spread (structure)

B4 – External fire spread

B5 – Access and facilities for the fire service

Approved document C (Resistance to moisture/weather)

C1 – Site preparation and resistance to contaminates

C2 – Resistance to moisture

Approved document D (Toxic substances)

D1 – Cavity insulation

Find out

For more about the Building Regulations, visit www.ukbuildingstandards.org.uk

Remember

The Building regulations can change over time. Always be sure you are using the most up-to-date version.

Approved document E (Resistance to sound)

E1 – Protection against sound from other parts of the building and adjoining buildings

E2 – Protection against sound within a dwelling-house, etc.

E3 – Reverberation in the common internal parts of buildings containing flats or rooms for residential purposes

E4 – Acoustic conditions in schools

Approved document F deals only with ventilation

Approved document G (Hygiene)

G1 – Sanitary conveniences and washing facilities

G2 – Bathrooms

G3 – Hot water storage

Approved document H (Drainage/waste disposal)

H1 – Foul water drainage

H2 – Wastewater treatment systems and cesspools

H3 – Rainwater drainage

H4 – Building over sewers

H5 – Separate systems of drainage

H6 – Solid waste storage

Approved document J (Heat producing appliances)

J1 – Air supply

J2 – Discharge of products of combustion

J3 – Protection of building

J4 – Provision of information

J5 – Protection of liquid fuel storage systems

J6 – Protection against pollution

Approved document K (Protection from falling)

K1 – Stairs, ladders and ramps

K2 – Protection from falling

K3 – Vehicle barriers and loading bays

K4 – Protection from collision with open windows, skylights and ventilators

K5 – Protection against impact from and trapping by doors

Approved document L (Conservation of fuel and power)

L1 A – Conservation of fuel and power in new dwellings

L1 B – Conservation of fuel and power in existing dwellings

L2 A – Conservation of fuel and power in new buildings other than dwellings

L2 B – Conservation of fuel and power in existing buildings other than dwellings

Approved document M (Access to and use of buildings)

M1 – Access and use

M2 – Access to extensions to buildings other than dwellings

M3 – Sanitary conveniences in extensions to buildings other than dwellings

M4 – Sanitary conveniences in dwellings

Approved document N (Glazing safety)

N1 – Protection against impact

N2 – Manifestation of glazing

N3 – Safe opening and closing of windows, skylights and ventilators

N4 – Safe access for cleaning windows, etc.

Approved document P (Electrical safety)

P1 – Design and installation of electrical installations

When are the Building Regulations needed?

The types of work classified as needing the Building Regulations approval include:

- the erection of an extension or building
- the installation or extension of a service or fitting which is controlled under the regulations
- an alteration project involving work which will temporarily or permanently affect the ongoing compliance of the building, service, or fitting with the requirements relating to structure, fire, or access to and the use of the building
- the insertion of insulation into a cavity wall
- the underpinning of the foundations of a building
- work affecting the thermal elements, energy status or energy performance of the building.

If you are unsure whether the work you are going to carry out needs Building Regulations approval, you should contact the local authority.

Did you know?

The building inspector's role is vital in enforcing the Building Regulations. Think of a medium-sized job that you are familiar with. What do you think a building inspector would need to check on that job?

Enforcement

The Building Regulations are enforced by two types of building control bodies: Local Authority building control and Approved Inspector building control. If you wish to apply for approval, you must contact one of these bodies. If you use an approved inspector, you must contact the local authority to tell them what is being done where, stating that the inspector will be responsible for the control of the work.

If you choose to go to the local authority, there are three ways of applying for consent.

- **Full plans** – plans are submitted to the local authority along with any specifications and other contract documents. The local authority scrutinises these and makes a decision.
- **Building notice** – a less detailed amount of information is submitted (but more can be requested) and no decision is made. The approval process is determined by the stage the work is at.
- **Regularisation** – this is a means of applying for approval for work that has already been completed without approval.

The inspector will review your plans and make regular visits to ensure the work is carried out to the standards set down in the application.

Working life

You are part way through building an extension when the client asks for an alteration to the original plans. You think that this alteration may need Building Regulations approval, but applying now would put the job back a few weeks, and you are already under time pressure. The client says they do not care about the Building Regulations; they want the work done now, or they will stop paying you.

What should you do? What could the repercussions of your actions be? You will need to think about the repercussions not only to yourself, but also to the project and the client.

Cutting lists

The cutting list is an accurate, itemised list of all the timber required to complete a job. The cutting list will need to be referred to throughout the manufacturing process. It is, therefore, good practice to include the cutting list on the setting-out rod wherever possible.

Although there is no set layout for a cutting list, certain information should be clearly given in all lists. It should include:

- a reference for the setting-out rod, i.e. rod number
- the date the list was compiled
- a brief job description
- the quantity of items required
- the component description (e.g. head, sill, stile, etc.)
- the component size, both sawn and finished (3 mm per face should be allowed for machining purposes)
- any general remarks.

An example of a cutting list is shown in Figure 32.03.

Timber cutting list						
Job description: Two panel doors			Date: 8 Sept 2011			
Quantity	Description	Material	Length	Width	Thickness	Remarks
2	Stiles	S wood	198 1	95	45	Mortise/groove for panel
1	Mid rail	"	760	195	45	Tenon/groove for panel
1	Btm rail	"	760	195	45	Tenon/groove for panel
1	Top rail	"	760	95	45	Tenon/groove for panel
1	Panel	Plywood	760	590	12	—
1	Panel	"	60 0	590	12	—

Figure 32.03 Cutting list

Did you know?

One of the main points of reference used during the planning and setting-out process are manufacturers' catalogues, particularly where there are components from the manufacturer that you will be placing into the work, such as locks, etc.

Checking information and reporting discrepancies

One of the last things you need to do before the manufacturing process can begin is to ensure that all the information that you have been given is completely accurate. This includes checking that the work will meet the Building Regulations, double checking the measurements to ensure that the manufactured component will fit, and checking with the client that all the information you have is what they want.

Any discrepancy in the information must be reported to your supervisor as soon as possible to prevent any timely and costly delays.

Remember

You will need to make sure that you are using scales correctly when working with plans and drawings. Look back at pages 35–37 for more information on scales.

K2. Know how to select resources for setting out

We covered the materials you will need to use for setting out in Unit 2008, pages 153–56. Refer back to this section for more information about the properties, qualities and defects of timber.

Information on the standard available sizes of materials was covered in Unit 3031, page 292.

When working on setting out, you will also need to work with a range of marking tools, making sure that these are properly maintained.

Measuring and marking out tools

Folding rule

Folding rules are used in the joiner's shop or on site. They are normally one metre long when unfolded and made of wood or plastic. They can show both metric and imperial units.

Remember

Imperial measurements (such as yards, feet and inches) have been replaced by metric units (such as metres and millimetres), but most older items will have been constructed using imperial measurements.

Figure 32.04 Folding rule

Retractable steel tape measure

Retractable steel tape measures, often referred to as spring tapes, are available in a variety of lengths. They are useful for setting out large areas or marking long lengths of timber and other materials.

Remember

Check the sliding hook regularly for signs of wear.

Figure 32.05 Retractable steel tape measure

Metal steel rule

Metal steel rules, often referred to as bar rules, are used for fine, accurate measurement work. They are generally 300 mm or 600 mm long and can also serve as a short straight edge for marking out. The rule can be used on its edge for greater accuracy.

Did you know?

Metal steel rules may become discoloured over time. If so, give them a gentle rub with very fine emery paper and a light oil. If they become too rusty, replace them.

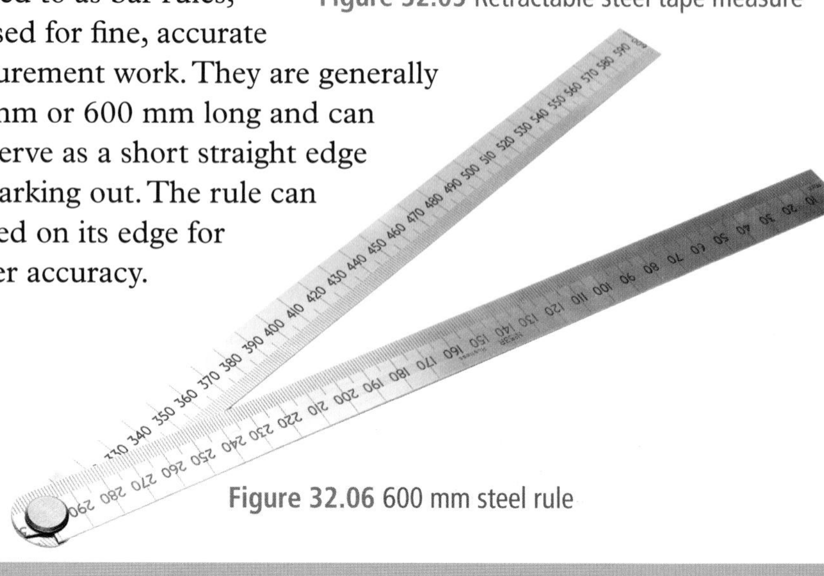

Figure 32.06 600 mm steel rule

Marking knife

Marking knives are used for marking across the grain and can be much more accurate than a pencil. They also provide a slight indentation for saw teeth to key into.

Figure 32.07 Marking knife

Try square

Try squares are used to mark and test angles at 90° and check that surfaces are at right angles to each other. They should be regularly checked for accuracy. To do this, place the square against any straight-edged spare timber and mark a line at right angles. Turn the square over and draw another line from the same point. If the tool is accurate the two lines should be on top of each other.

Figure 32.08 Try square

Sliding bevel

The sliding bevel is an adjustable tri-square, used for marking and testing angles other than 90°. When in use, the blade is set at the required angle, then locked by either a thumbscrew or set screw in the stock.

Mitre square

The blade of a mitre square is set into the stock at an angle of 45° and is used for marking out a mitre cut.

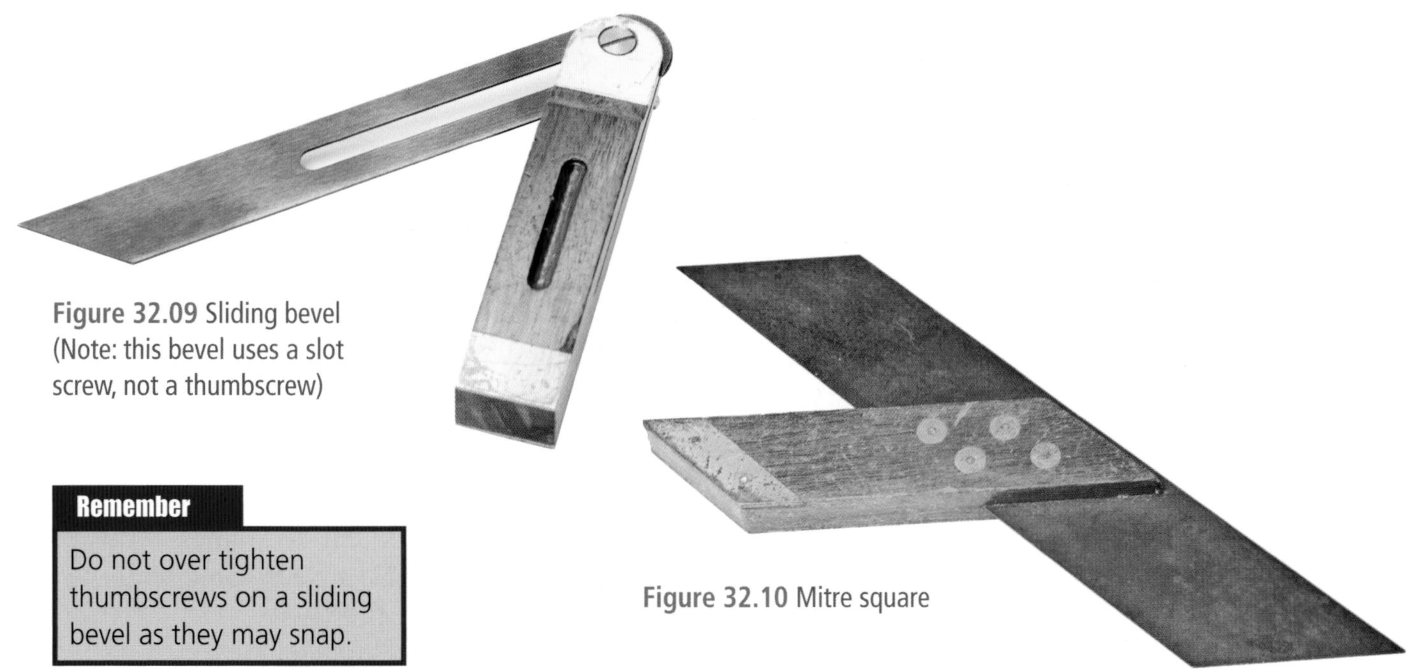

Figure 32.09 Sliding bevel (Note: this bevel uses a slot screw, not a thumbscrew)

Figure 32.10 Mitre square

Remember

Do not over tighten thumbscrews on a sliding bevel as they may snap.

Remember

All squares should be checked for accuracy on a regular basis.

Combination square

A combination square does the job of a try square, mitre square and spirit level all in one. It is used for checking right angles, 45° angles and that items are level.

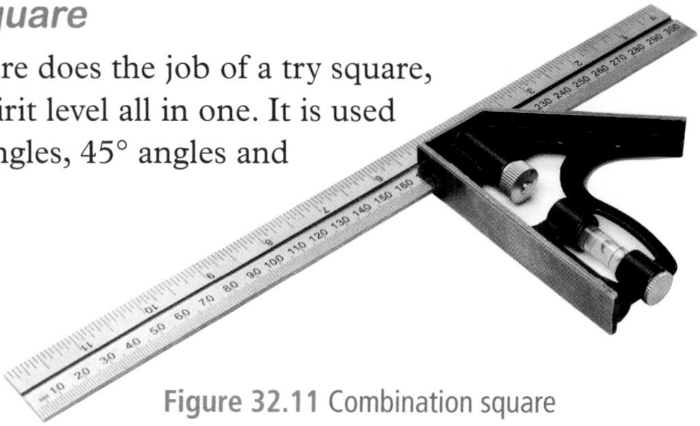

Figure 32.11 Combination square

Marking gauge

A marking gauge is used for marking lines parallel to the edge or end of the wood. The parts of a marking gauge include the stem, stock, spur (or point) and thumbscrew. A marking gauge has only one spur or point.

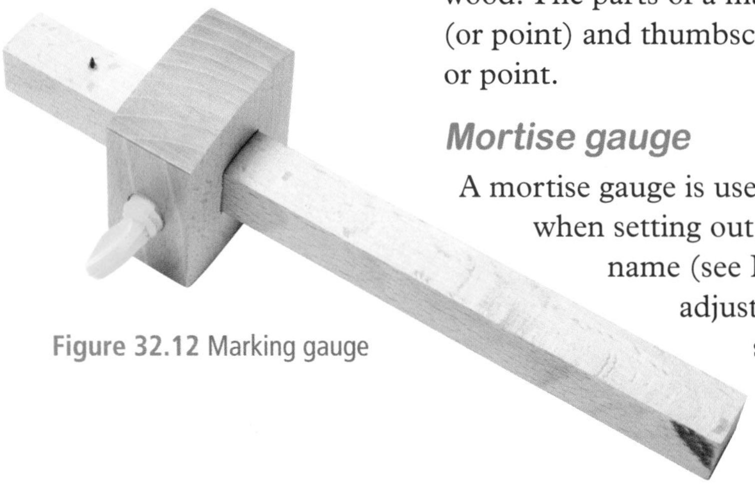

Figure 32.12 Marking gauge

Mortise gauge

A mortise gauge is used for marking the double lines required when setting out mortise and tenon joints, hence the name (see Figure 32.13). It has one fixed and one adjustable spur or point. Figure 32.14 shows setting of the adjustable point to match the width of a chisel.

Figure 32.13 Mortise gauge

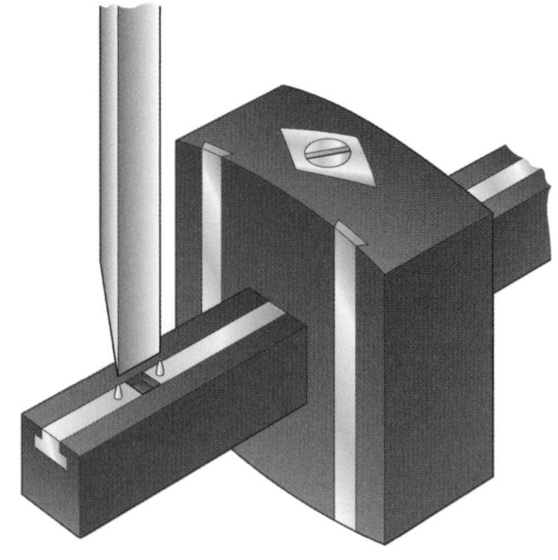

Figure 32.14 Setting mortise gauge to chisel blade width

Cutting gauge

The cutting gauge is very similar to the marking gauge, but has a blade in place of the spur. This is used to cut deep lines in the timber, particularly across the grain, to give a clean, precise cut (for example, for marking the shoulders of tenons).

Mitre box

Mitre boxes are usually made from wood or a hardwearing plastic and are used to cut materials with a hand saw without the need for marking out. They can be bought or, more often than not, they are made. The box is made up, then the angles to be cut are marked on and a solitary cut is made along the markings, leaving a guide for all future cuts. The initial cut has to be perfect or any future cuts done using the box will also be inaccurate.

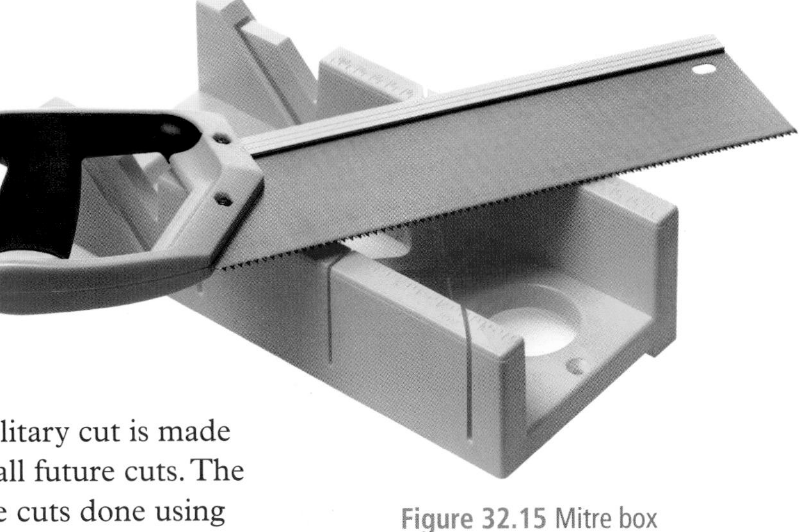

Figure 32.15 Mitre box

Trammel heads

Trammel heads are two small adjustable clamps with long pins that can be attached to a straight piece of timber or metal and then adjusted to any distance between them. They are ideal for marking out parallel lines and circles.

Figure 32.16 Trammel heads

K3. Know how to set out for complex shaped bench joinery

The joints used in the manufacture of joinery components are covered fully at Level 2. This section will give a brief re-cap of the marking and setting-out process.

Marking and setting out

Whatever you are setting out, it is best to start with a plan or a drawing. For joinery products, this is usually done to full scale, on a thin sheet of board (plywood, hardboard, etc.) better known as a setting-out rod. The setting-out rod is particularly crucial when manufacturing curved or complex work as it gives the joiner a true image of how the completed component will look.

Setting-out rods should include horizontal and vertical sections through the components, as well as elevations of any complex areas.

Rods can be re-used by painting over them, but it is a good idea to keep a record of the rods and store them for future reference.

Remember

Although rods are marked up full size and to scale, it is good practice to mark on the sizes of the sections/components. This will prevent the joiner having to measure the sizes, saving time and effort.

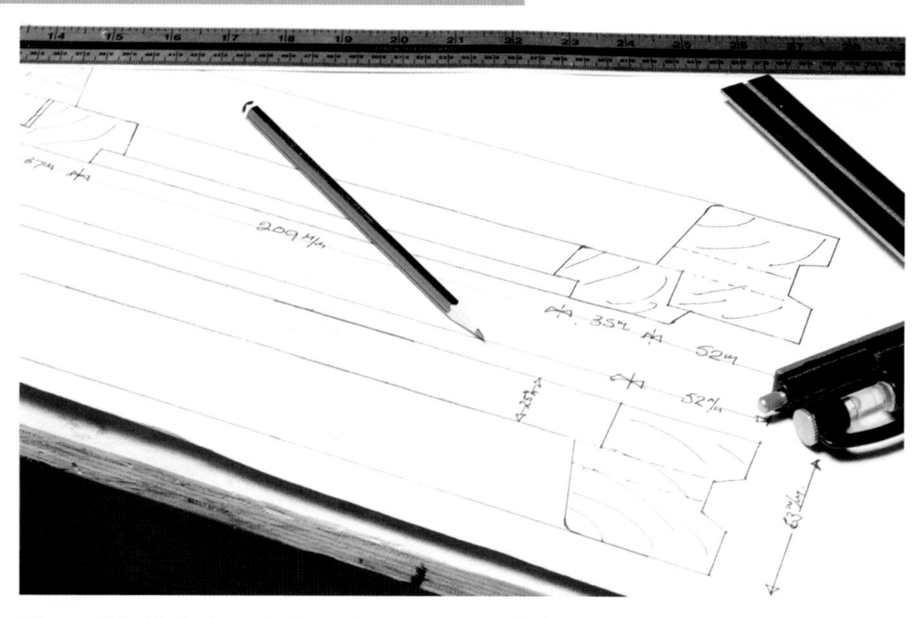

Figure 32.17 Rod marked out for casement window

Did you know?

The method of setting out and marking is virtually the same for every component. Remember this when you look at manufactured components.

Once the rods have been marked out, a cutting list can be made; the setting-out rods can be stored to protect them from damage and the materials can be machined.

Now select the timber and cut to size, bearing in mind which will be the face and the edge. The face and edge are usually the two best adjacent sides, as they are most likely to be seen. Note the position and severity of any defects, as it may be possible to remove these when machining any rebates or grooves.

Next, you must mark out the timber, which involves two main operations.

Remember

Marking out on the timber should be as clear and simple as possible with no unnecessary lines, as these cause confusion and possible errors. Whenever possible, marking out should be completed in one session, with any wrongly placed lines rectified immediately.

Figure 32.18 Transferring details from the setting-out rod

First, you must apply face and edge marks, which serve as a reference point for the rest of the marking out and machining. The marks are shown in Figure 32.19.

Now for the last stage in marking out: the position of the joints, etc. To do this, you transfer the information from the setting-out rods to the timber members. You should transfer the information as clearly as possible to avoid any confusion.

Now the joints can be cut and the component assembled.

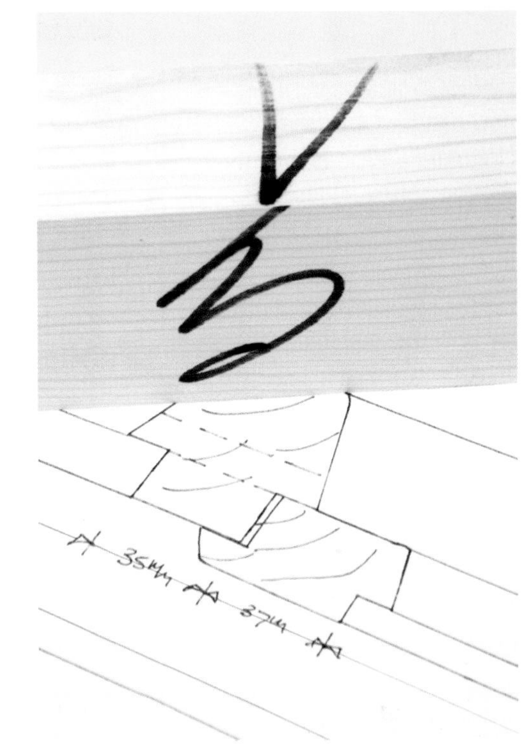

Figure 32.19 Face and edge of timber

Geometric and practical methods of producing complex curves and stairs

There are several different types of arch that can be constructed. The main types of arch are shown in Figure 32.20. The construction methods for the types of arches are all covered in the following section. There is also more information in the section covering geometric arches in Unit 3002, pages 30–32.

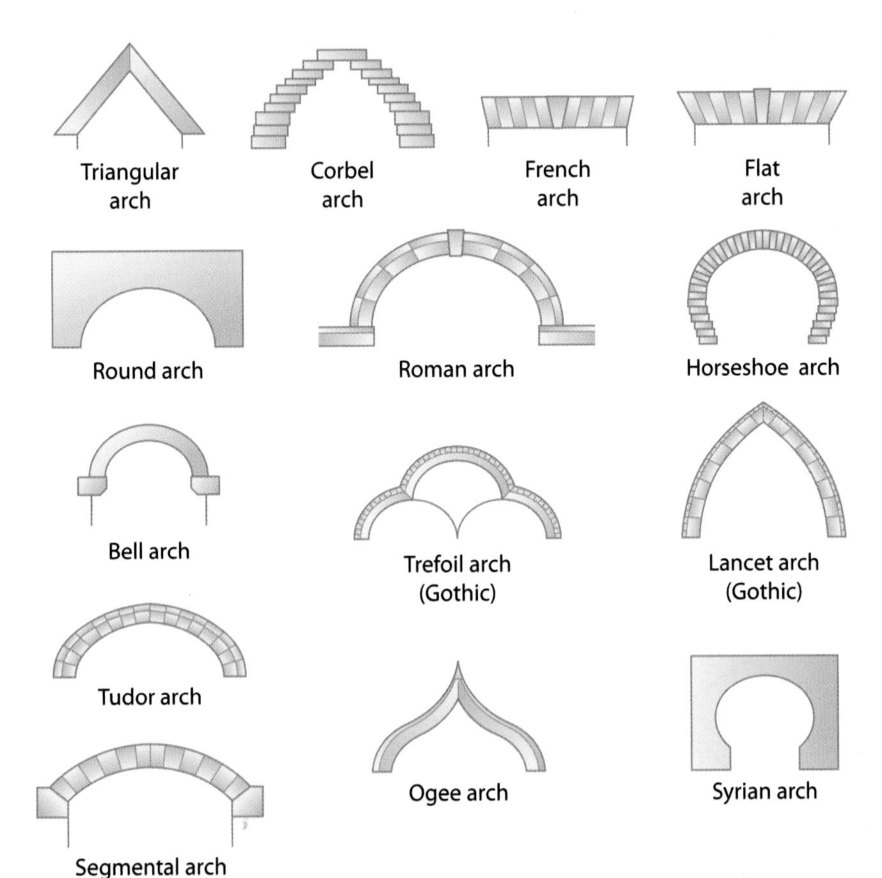

Figure 32.20 Types of arch

Did you know?

Ellipses are basically circular shapes that are not true circles. Any arch that has a true curve is classed as an ellipses arch. Examples of these include roman and segmental arches.

Did you know?

Most types of ventilator suit most situations, so usually a type of ventilator is chosen that best matches the architecture of the building and its surroundings.

Louvre ventilator frames

Louvre ventilator frames are commonly found in areas that require constant ventilation, such as boiler rooms. The ventilator is made up of a frame housing louvre boards, pitched at 30°, 45° or 60°. Louvre ventilator frames are usually rectangular or square, but for the purpose of this book we will look at shaped louvre ventilator frames. These are most commonly pitched, circular-headed or gothic. The first style we will look at is the pitched ventilator.

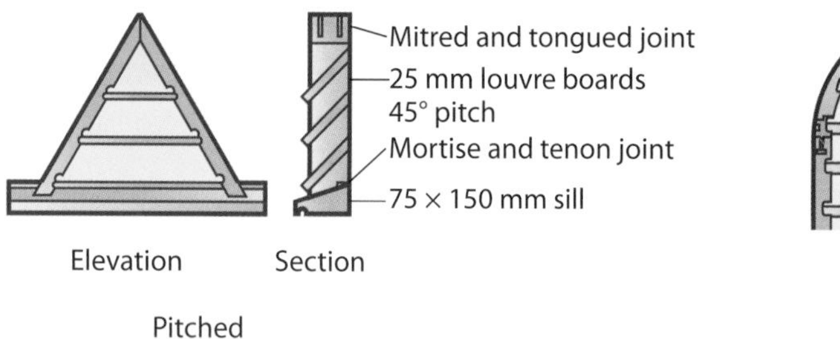

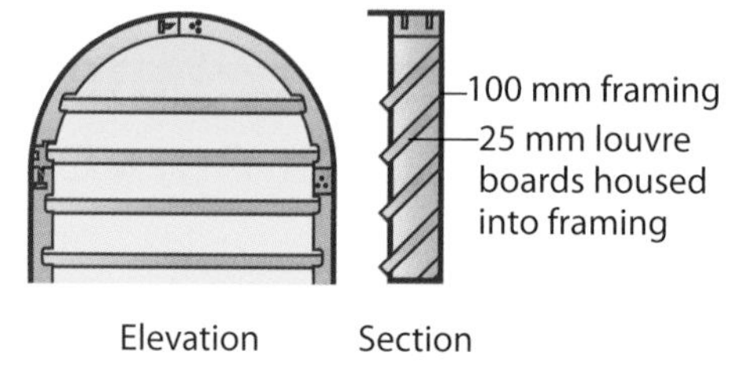

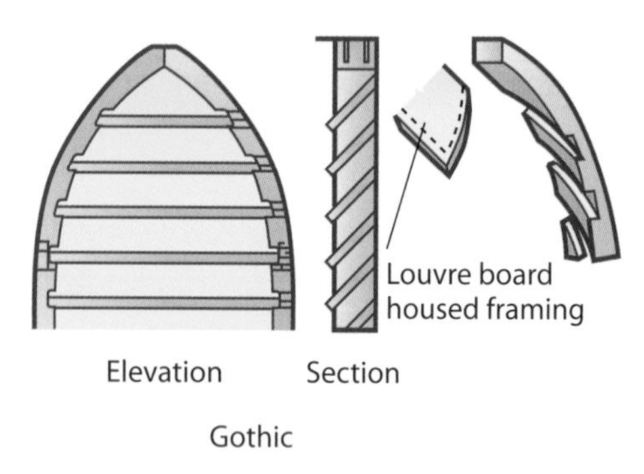

Figure 32.21 Pitched, circular-headed and gothic louvre ventilators

Pitched ventilator frame

The first step, as always, is to draw out the frame full size. From this drawing you can get the true width and shape of the louvre boards (see Figure 32.22).

The louvres can then be marked out and cut, and the stock material for the frame can be machined.

The next step is to mark out the frame joints and the housings, again using the full-scale drawing. Project the position of the housings from the section (A) onto the elevation (B) (see Figure 32.23). Machined stock (C) can be laid onto the drawing and the housings can be transferred from the elevation (B). The members should be marked out in pairs to ensure accuracy and that the finished frame is square (see Figure 32.24). The frame is now ready for assembly and finish.

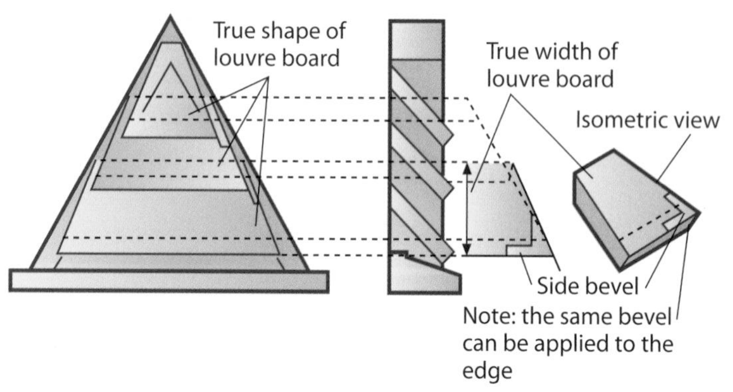

Figure 32.22 True shape of louvre boards

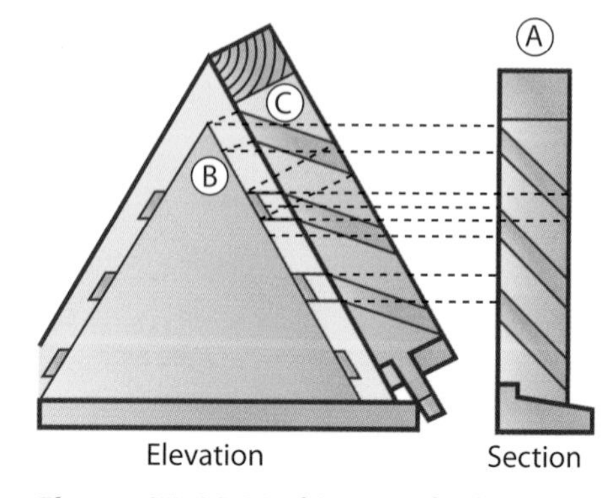

Figure 32.23 Marking out for housings

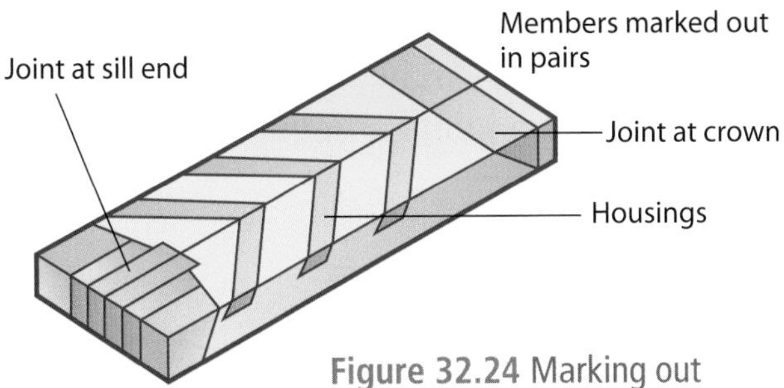

Figure 32.24 Marking out

Circular or curved ventilator frame

Circular or curved frames are more complex, and the method of producing curved members is explained more fully in the next section.

The setting-out process for circular-headed and gothic frames is broadly the same as for the pitch frame, but the marking of the louvre positions is more difficult. It is best to use a square frame and a straight edge to mark the positions of the louvres on the face, and then use a pitched block to mark out the inner side of the frame.

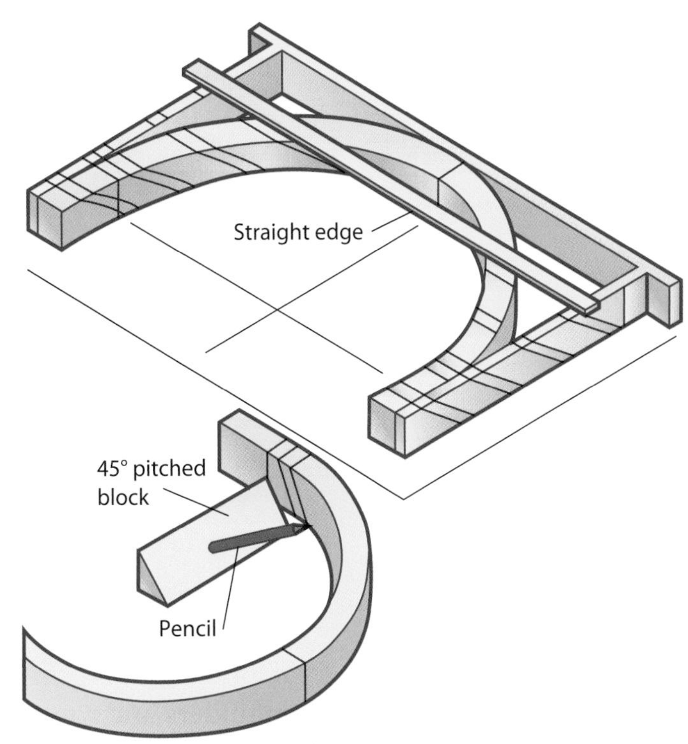

Figure 32.25 Marking for a circular head

Working life

You have been tasked with creating a pitched louvre frame. Your employer gives you the overall dimensions of the opening.

What other information do you need?

Complex curves

Doors with shaped heads can provide a sense of grandeur but have to be made to measure, which can prove costly in today's market where most doors are mass produced.

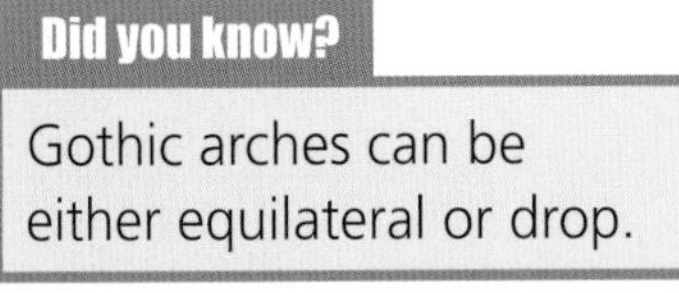
Did you know?

Gothic arches can be either equilateral or drop.

The four main types of shaped-head door are:

- segmental
- semicircular
- gothic
- parabolic.

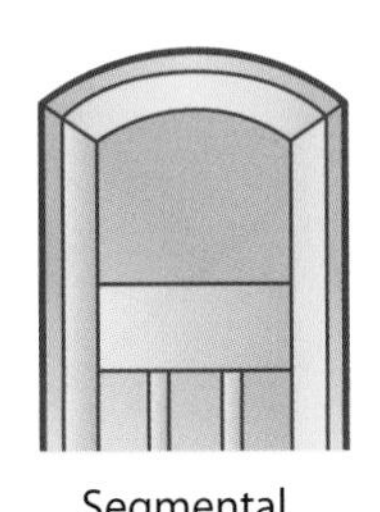

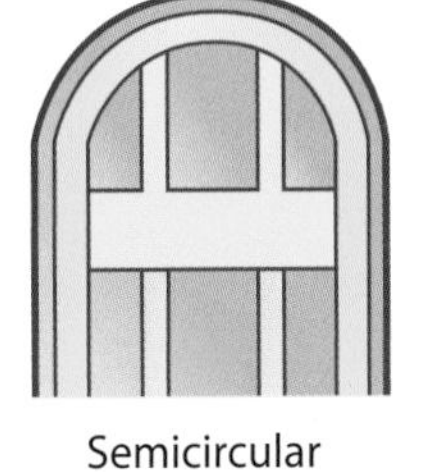

Figure 32.26 The four main types of shaped-head door

Did you know?

Doors with shaped heads are most commonly found in older buildings. Usually they are main entrance doors, but sometimes you may find a shaped head door at an entrance to a grand walled garden.

When working with shaped-head doors, many different types of joint can be used to create curved or arched members, for example hammer-headed keys, tenon joints or joints secured using handrail bolts.

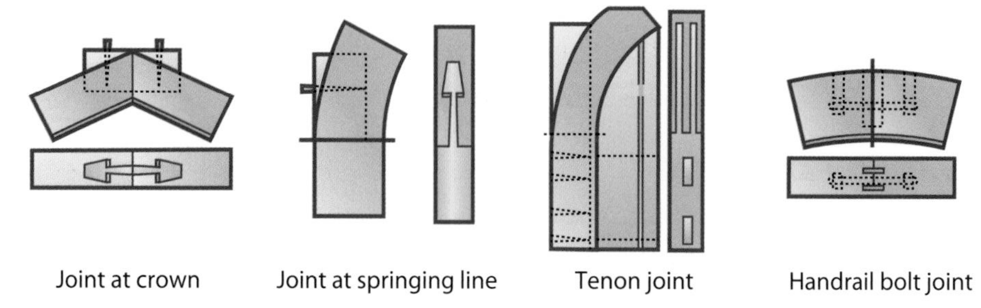

Figure 32.27 Hammer-headed keys, tenon joints and handrail bolt joints

Basic method for producing a shaped-head door

The example we will use here is a semicircular door but whichever type of door is to be made, and whichever joints are used, the method is the same.

Working shaped members involves similar operations to working straight members, though greater care and skill are needed to maintain a uniform standard of craftsmanship. Vertical spindle moulders and routers are the best tools to use for creating the curved sections, and accurately made face templates are vital for cleaning and shaping the curved members. The height and width rods should be made as normal, but the circular head section must be marked out full size.

To mark out the shaped head, begin with the centre line and the springing line, where the joints at the stiles will be. Next, mark on the stiles, and finally mark out the curved head with either a beam compass or radius rod.

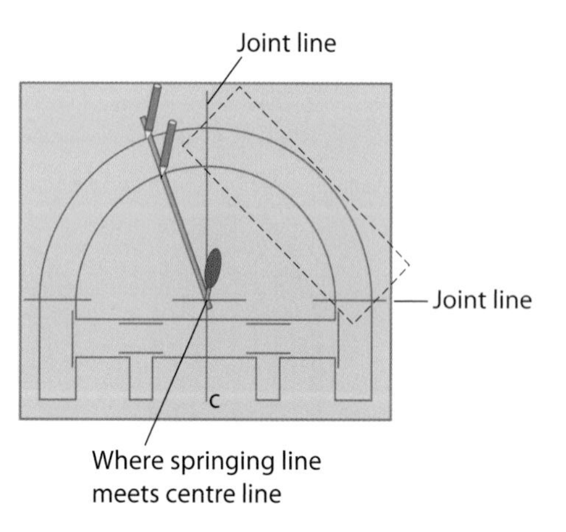

Figure 32.28 Setting out for a circular head

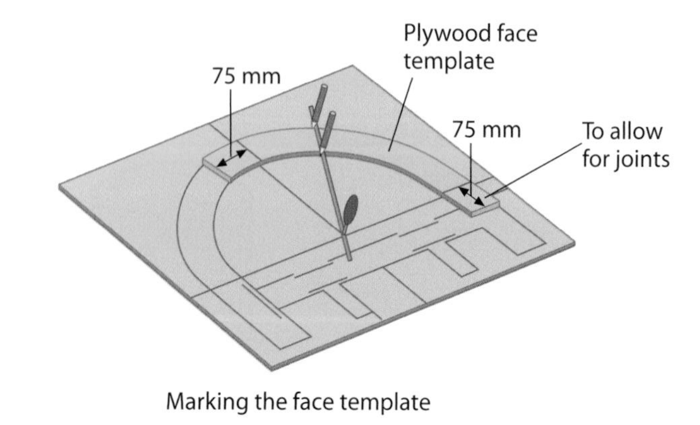

Figure 32.29 Face template

Next, mark the face template from the drawing, making the template a minimum of 75 mm longer at each end to allow for the joints to be constructed. The face template must be made from plywood at least 9 mm thick.

Now machine the materials. The curved sections can either be formed on the vertical spindle moulder, with the face template used as a jig, or cut roughly on the band saw and tidied up using the face template and a router. Once the material has been planed and cut to the correct dimension, mark it out, remembering to mark out the stiles in pairs and the bottom, middle and frieze rail together.

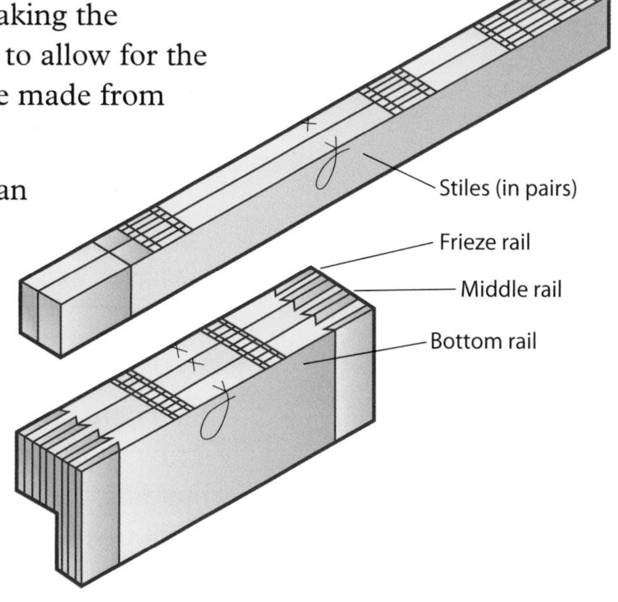

Figure 32.30 Rails marked out

The joints between the stiles and the rails are all mortise and tenon, so they can be manufactured and dry fitted to ensure a good tight fit. The joint between the stile and the circular head is a twin mortise and tenon – see Figure 32.31 for the detail.

Joint the crown using handrail bolts with cross tongues for additional strength, as in Figure 32.32.

Once all the joints have been machined and tested for dry fitting, the door can be glued, assembled and clamped. When clamping the door, take care to ensure that the joints remain tight and the door does not go out of shape. The ideal clamping method is shown in Figure 32.33.

Once the adhesive is set, remove the clamps, and glaze and finish the door.

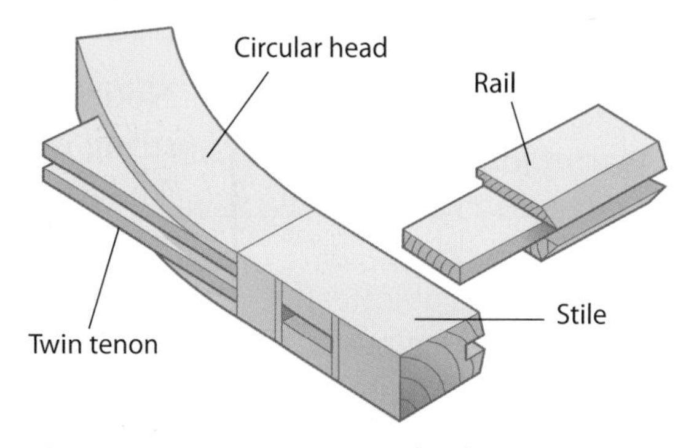

Figure 32.31 Twin tenon and rail

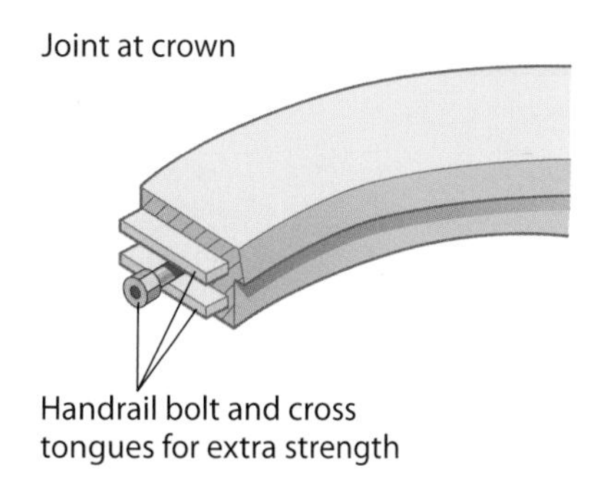

Figure 32.32 Crown joint

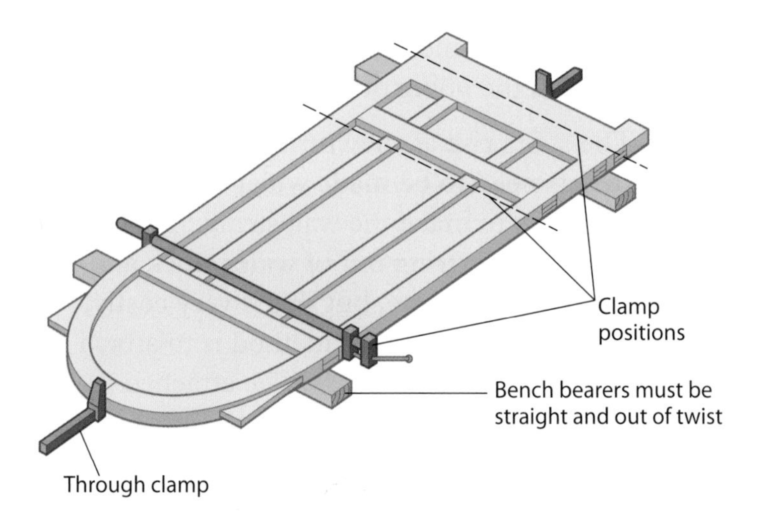

Figure 32.33 Door clamped in place

Working life

You have been tasked with creating a semicircular headed door. You mark out all the components and machine them. When you tidy up the curved pieces using a face template and router, there are several bumps and hollows in the finished piece.

What could have caused this? What needs to be done to remedy this? What should have been done to prevent it?

Producing a geometrical stair

Stairs are said to be geometrical when they have continuous strings and handrails. Geometric stairs are very complex, and building them requires a wide understanding of geometry. In this section you will find a brief overview of the geometry, and will see how to construct a basic geometrical stair.

Key terms

Wreathed string – a continuous string that rises while turning.

Biscuit joint – a type of tongued and grooved joint that works like a dowel joint, but instead of a dowel a flat oval plate (biscuit) fits into a slot. When PVA adhesive is introduced, the biscuit expands and makes a very strong bond.

Our example here is a geometrical stair over a **wreathed string** with a quarter-turn of winders (though this sounds complex, it is essentially a quarter-turn stair, as shown in Figure 32.34).

Notice that there is no newel post at the turn; instead the outer string is a continuous string, better known as a wreathed string. The wreathed string will be shown in more detail later, but first we will concentrate on the wall strings.

The first thing, as always, is to do a drawing so that you can work out the size of the winders, etc. You must remember to keep within these regulations:

- The rise of the tapered steps must be the same as the rise of the other steps.
- The tapered step must not be less than 50 mm at the narrowest point.
- The going of the tapered steps (measured at the centre of the steps) must be the same as the going of the other steps.

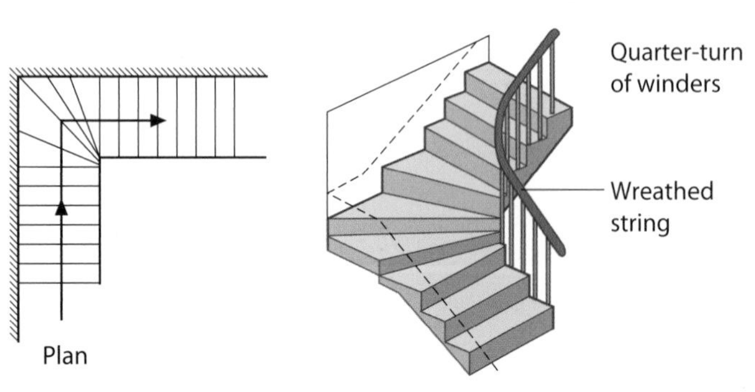

Figure 32.34 Quarter-turn stair

The stairs rise as they turn, so the wall strings need to be made wider. There are two ways to make the wall string: one is to make the string out of wider stock and cut away the waste, but this is very costly; the preferred, cheaper method is to attach pieces to the wall strings. The attached pieces should ideally be **biscuit-jointed** and glued for strength, although when the treads and risers are fixed to the string this will strengthen the joint.

Figure 32.35 shows why the attached pieces are needed, and how the wall string should be marked out for the treads and risers. To maintain accuracy, you should take the sizes for marking out the winders, etc. from the drawings.

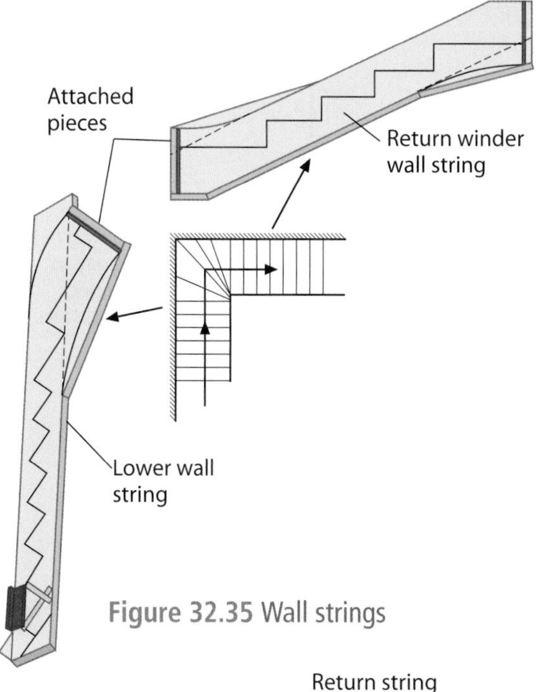

Figure 32.35 Wall strings

The housings for the risers and treads, and the tongued and grooved joint where the two wall strings meet, can all be machined.

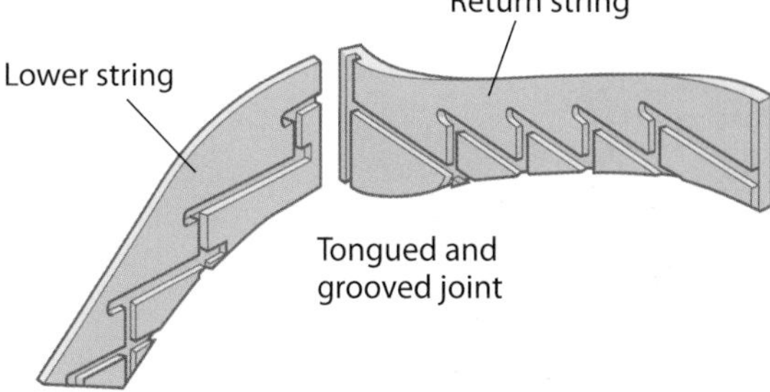

Figure 32.36 Joint between the wall strings

Now we will look at the wreathed string. Probably the most difficult aspect of making a wreathed string is finding the curved shape of the string – this is where a good grasp of geometry is required (see Figure 32.37).

1. Start with a full-scale plan drawing of the stair (for this example we will use a 40 mm string with a rise of 175 mm and a going of 250 mm).
2. Extend a horizontal straight line from point A the distance of the going (250 mm), creating line A–B.
3. From point B draw a line down at 60° to where it meets the outside of the string.
4. From point A draw a line down at 45° to meet the 60° line, forming point C.
5. Draw a horizontal line from point C and a vertical line down from point A: where these two lines meet is point D.
6. Distance A–D will give the radius for a going of 250 mm. From this the development of the string can be created.

The wreathed string is normally a cut string (as opposed to being routed) and consists of three parts (see Figure 32.38 on the next page). The wreathed part is the curved part, joined to a top and bottom cut string. This join can be done in three different ways, the most common of which is to use staves, which we will look at here.

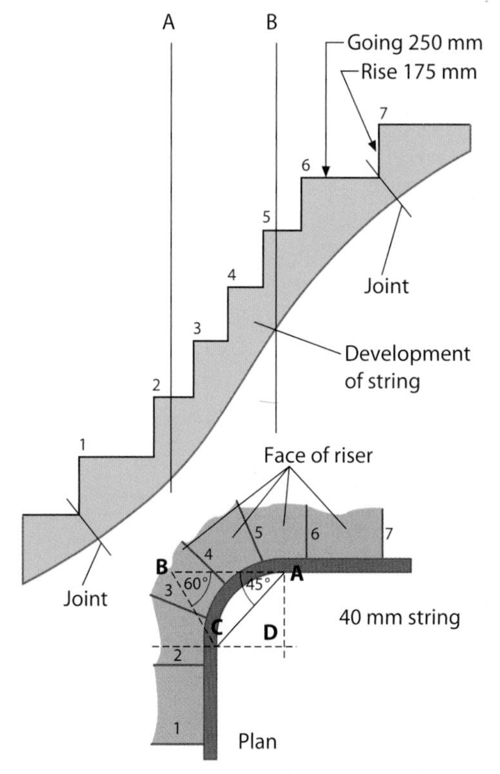

Figure 32.37 Geometry for wreathed string

Staves

The string starts out as a normal straight string piece. First, you must reduce the string's thickness at the curved area to a veneer thickness of around 3 mm. Once the section has been removed, place the string over a drum made to the shape of the curve and fix it in place,

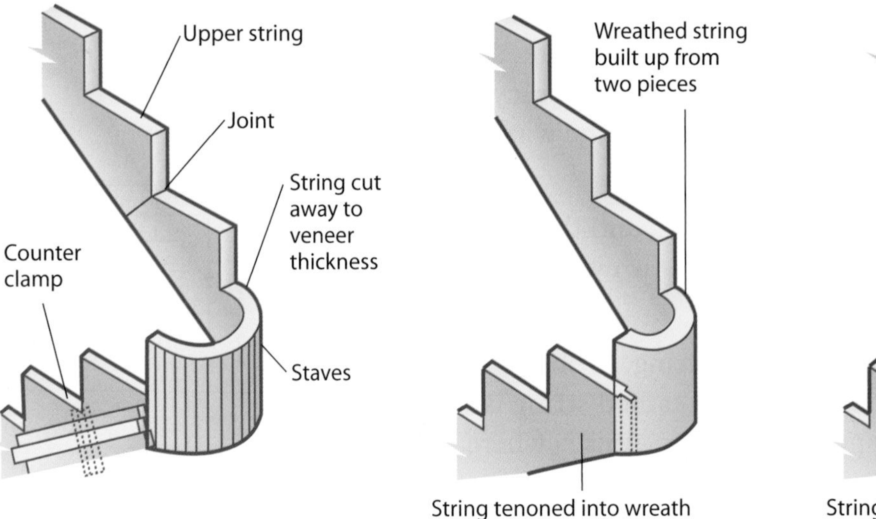

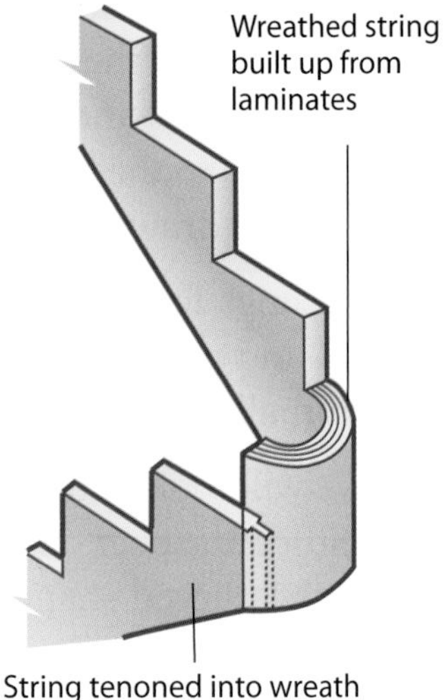

Figure 32.38 Wreathed string construction

Figure 32.39 String over drum

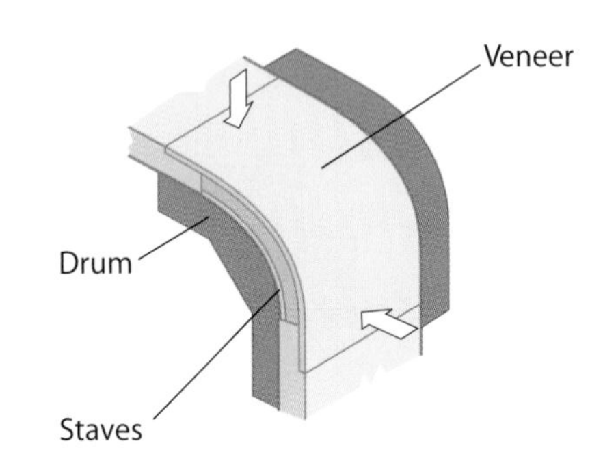

Figure 32.40 Veneer being fitted over staves

taking great care when bending the string. Glue a series of tapered staves into place and allow them to set.

If the string is going to be seen, and a quality finish is required, fit another veneer to the string to hide the staves (Figure 32.40).

Counter-clamp method

Once the adhesive has dried, remove the string from the drum and joint it to the top and bottom cut string. The joint used depends on the thickness of the string and how the curve is formed. Three of these are familiar to you already – bridle, halving, and mortise and tenon joints – so here we will look at the counter-clamp method.

The counter-clamp method is a simple but effective way of jointing the strings, working along similar lines to draw boring.

Once the wreathed strings are completed, fix the treads, winders and risers in place, and the staircase is ready to be fitted and finished.

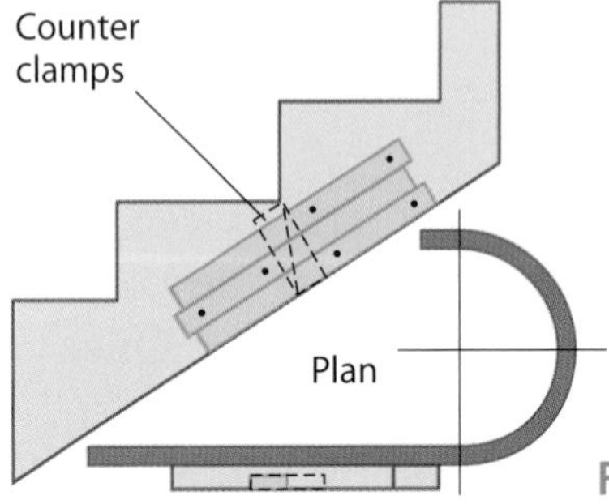

Figure 32.41 Counter clamps

Working life

You have been tasked with creating a wreathed string using a drum method. You reduce the string's thickness and then attempt to bend it over the drum. The string snaps, so you try again, and it snaps again.

What could be causing this? What needs to be done to rectify it?

Functional skills

This exercise will allow you to practise the interpreting elements of functional skills, such as **FM 2.3.1** Interpret and communicate solutions to multistage practical problems and **FM 2.3.2** Draw conclusions and provide mathematical justifications.

Organising work in sequence

One of the most important things to consider when organising work is the sequence of operations. If you get this wrong it can lead to work having to be re-done. When making windows, for example, you would not fully fit the glass and glazing bead before the window is installed, as you may need to fix the window through the rebate where the glass sits. To do this once the window was installed would mean having to remove the glass, fit the window and re-install the glass. This is extra work which is unnecessary, and will add time and money.

Experience helps when planning a sequence of operations. There are times when mistakes are made in the sequencing and things have to be re-done, but it is always a good idea to list the sequence of operations before starting the work.

FAQ

How many louvres should there be in a ventilation frame?

The number of louvres depends on the client's wishes, but the spacing is very important, as they should not allow birds or other animals into the building.

Which is the best way to create a shaped head: using a band saw and router, or using a spindle moulder?

Both ways are suitable, but using a spindle moulder creates the best finish and is quickest as it can be set to create moulded curved pieces in a single pass.

On a geometrical stair, does a wreathed string need to be a cut string?

No. The treads can be housed into the string, but with the curve this is very difficult so it is easier to have a cut string.

Do I need to cut veneers for both sides of the wreathed string?

This depends on the finish required. If the string is visible and has a varnish or stain varnish, it is good practice to have veneers on both sides, but a painted string does not need veneers.

Check it out

1. Explain the main purpose of the Building Regulations and state the main approved documents that deal with stairs, fuel and power.
2. Explain who to contact to check if the work you are doing requires Building Regulations approval.
3. What drawings should setting outs contain? Explain why the rods are stored after the component is made.
4. Explain the three angles that louvres are normally pitched at and why the louvre ventilator frames are necessary.
5. State three jointing methods used when creating shaped door heads and explain the benefits of using these.
6. Explain why a face template must be accurate.
7. Why are some stairs said to be geometrical?
8. Explain why tapered staves are fitted into the wreathed string.

Getting ready for assessment

The information contained in this unit, as well as continued practical assignments that you will carry out in your college or training centre, will help you with preparing for both your end of unit test and the diploma multiple-choice test. It will also aid you in preparing for the work that is required for the synoptic practical assignments.

The information contained within this unit will aid you in learning how to identify and calculate the materials and equipment required for marking out and setting-out details for complex joinery products.

You will need to be familiar with:

- interpreting information for setting out
- selecting resources for setting out
- setting out complex shaped bench joinery

Marking out involves transferring the information you have set out on drawings onto the timber itself. You will need to remember the skills you learnt at Level 2 in order to complete the work accurately. For learning outcome one, you have seen the documents that can be used to aid in the setting-out process. It is also important that you mark out with minimal errors. You will need to produce detailed cutting lists that take into account the information contained in the drawings.

For learning outcome three, you will need to be able to select and maintain all the marking out tools you require and then measure and mark out accurately with these materials. You will then need to set out bench joinery and work within an allotted time to complete work.

Before you start work on the synoptic practical test, it is important that you have had sufficient practice and that you feel that you are capable of passing. It is best to have a plan of action and a work method that will help you. You will also need a copy of the required standards, any associated drawings and sufficient tools and materials. It is also wise to check your work at regular intervals. This will help you to be sure that you are working correctly and help you to avoid problems developing as you work.

Your speed at carrying out these tasks will also help you to prepare for the time limit of the synoptic practical task. But remember that you should not try to rush the job as speed will come with practice and it is important that you get the quality of workmanship right.

Make sure you are working to all the safety requirements given throughout the test and wear all appropriate personal protective equipment (PPE). When using tools, make sure you are using them correctly and safely.

Good luck!

CHECK YOUR KNOWLEDGE

1. Which of the following documents forms part of the contract documents?
- **a** Specification
- **b** Schedule
- **c** Bill of quantities
- **d** All of the above

2. What does Part A of the Building Regulations deal with?
- **a** Fire safety
- **b** Toxic substances
- **c** Structural safety
- **d** Hygiene

3. What part of the Building Regulations deals with stairs?
- **a** Approved document J
- **b** Approved document S
- **c** Approved document K
- **d** Approved document D

4. What does Part K of the Building Regulations deal with?
- **a** Structural safety
- **b** Resistance to moisture and weather
- **c** Protection from falling
- **d** None of the above

5. What are dividers used for?
- **a** Accurate checking of widths and gaps
- **b** Marking out measurements
- **c** Measuring angles
- **d** Marking length and thickness

6. What angle are louvre boards usually pitched at?
- **a** 30°
- **b** 45°
- **c** 60°
- **d** Any of the above

7. How many main types of shaped-head doors are there?
- **a** 3
- **b** 4
- **c** 5
- **d** 6

8. Which type of joint can be used to create curved or arched members on a shaped-head door?
- **a** Hammer-headed key
- **b** Tenon joints
- **c** Handrail bolt joints
- **d** All of the above

9. When making a face template for a shaped-head door, the template must be longer to allow for the joints. How much longer should it be?
- **a** 25 mm
- **b** 50 mm
- **c** 75 mm
- **d** 100 mm

10. What is a continuous string on a staircase known as?
- **a** Warped string
- **b** Winding string
- **c** Wreathed string
- **d** Weathered string

Functional skills

In answering the Check it out and Check your knowledge questions, you will be practising **FE 2.2.1** Select and use different types of texts to obtain relevant information and **FE 2.2.2** Read and summarise succinctly information/ideas from different sources.

You will also cover **FM 2.3.1** Interpret and communicate solutions to multistage practical problems and **FM 2.3.2** Draw conclusions and provide mathematical justifications.

Index

Index

Index